D0044872

ALGEBRA

ABSTRACT AND CONCRETE

ALGEBRA

ABSTRACT AND CONCRETE

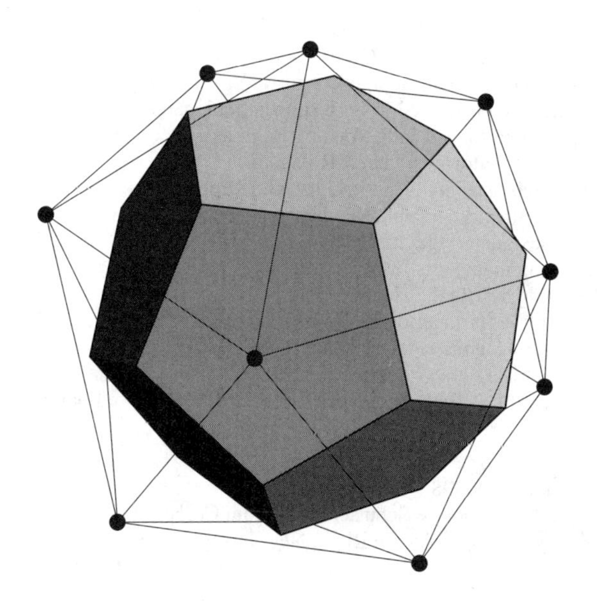

FREDERICK M. GOODMAN

UNIVERSITY OF IOWA

PRENTICE HALL, Upper Saddle River, New Jersey 07458

Library of Congress Cataloging-in-Publication Data

Goodman, Frederick M.
Algebra: abstract and concrete / Frederick M. Goodman
p. cm.
Includes index.
ISBN 0-13-283988-1
1.Algebra. I. Title
QA155.G64 1998
512'.02–dc21 97-19737
CIP

Editorial director: Tim Bozik
Editor-in-chief: Jerome Grant
Acquisition editor: George Lobell
Executive managing editor: Kathleen Schiaparelli
Managing editor: Linda Mihatov Behrens
Production editor: Nicholas Romanelli
Editorial assistants: Nancy Gross/Gale Epps
Creative director: Paula Maylahn
Cover designer: Jayne Conte
Marketing manager: Melody Marcus
Marketing assistant: Jennifer Pan
Assistant VP production and manufacturing: David Riccardi
Manufacturing manager: Trudy Pisciotti
Manufacturing buyer: Alan Fischer
Cover credit: Topkapi carpet (Turkish).©Namikawa Foundation

Simon & Schuster/ A Viacom Company
Upper Saddle River, New Jersey 07458

Printed in the United States of America

ISBN 0-13-283988-1

PRENTICE-HALL INTERNATIONAL (UK) LIMITED, LONDON
PRENTICE-HALL OF AUSTRALIA PTY. LIMITED, SYDNEY
PRENTICE-HALL CANADA INC., TORONTO
PRENTICE-HALL HISPANOAMERICANA, S.A., MEXICO
PRENTICE-HALL OF JAPAN, INC., TOKYO
SIMON & SCHUSTER ASIA PTE. LTD., SINGAPORE
EDITORA PRENTICE-HALL DO BRAZIL, LTDA., RIO DE JANEIRO

To Katie and the Zoo

Contents

Preface

This text is an introduction to "modern" or "abstract" algebra for undergraduate students. The book addresses the conventional topics: groups, rings, and fields, with symmetry as a unifying theme.

Presenting these topics is undoubtedly important, as the subject matter is central and ubiquitous in modern mathematics.

However, the more important goal of this book is to introduce students to the active practice of mathematics and to draw them away from the view of mathematics as a system of rules and procedures. Students are asked to participate and investigate, starting on the first page.

Exercises are plentiful, and working exercises should be the heart of the course.

This text is suitable for several different courses:

- An introduction to modern algebra for beginners. Part I of the text provides ample material for a two-semester course.
- A course in modern algebra for students who have already had a modest introduction or who, by prior experience or ability, are prepared for a more vigorous course. For this, Part I of the text taken at a faster pace followed by a selection of material from Part II would provide suitable subject matter.
- An undergraduate topics course, for example, on geometric aspects of group theory or on Galois theory.

The text is also adaptable to different teaching styles. My own preference is increasingly to lecture little, and to use class time for discussing problems. But those who wish to present the material in systematic lectures will find the subject matter cleanly organized and presented in the text.

The required background for using this text is a standard first course in linear algebra. Most students will need to review ideas from linear algebra, as they are needed in this course. Where I need results from linear algebra which clearly go beyond what can be expected from a first course, I develop what is needed in the text. Complex numbers seem to have no fixed abode in the curriculum, and may need to be introduced or reviewed for this course. They are used freely in the text; for example the cyclic groups are modeled by the complex roots of unity. I have included an appendix on the complex numbers.

Previous experience with reading and writing proofs is not presumed. More experienced students will be able to do more, beginners less. I have attempted to summarize the small amount of set theory and logic which is needed in two appendices and have included as well an appendix on mathematical induction. Equivalence relations are discussed in the text proper, in connection with cosets in groups. I believe that the way to gain a *usable* knowledge of this material is to *use* it in dealing with some genuine and interesting subject in mathematics. For students without previous experience with proofs, the instructor will need to pay explicit attention to points of logic or set manipulation as they arise naturally in the subject at hand. For my own classes for beginners, I like to recommend a short supplementary text on set theory and logic to be used as a reference; several such texts are currently available.

Texts in algebra for undergraduates are of two types: those about school mathematics (which by convention are called "elementary," "intermediate," or "college" algebra) and those about more advanced ideas (which by convention are called "modern" or "abstract" algebra). It is necessary for an author to stay close to these conventions in order that texts of the two sorts can be distinguished by title.

In common language, "abstract" means both "difficult" and "impractical," and it is a little unfortunate to start out by labeling the subject as hard but useless! It takes some effort to remember that even the counting numbers were once (and in principle still are) an enormous abstraction. But they are familiar, they no longer seem difficult, and no one would doubt their usefulness. Abstractions with which we have become familiar eventually lose their aura of abstractness, but those with which we are not yet familiar seem abstract indeed. So it is with the ideas of this course. They are abstract today, but tomorrow – or a little later – they will seem more concrete.

New abstractions are difficult, but it is impossible to understand the concrete world without abstraction. Abstractions are introduced to overcome the intractable variety of the concrete world; that is in the nature of mathematics. Pick up a brick. The brick is concrete, and its symmetry is manifest. But how can one understand, keep track of and eventually compute with its symmetries? Pick up a Rubik's cube. It is a concrete, totally impractical, and very difficult geometric puzzle. How is the puzzle to be understood? Put a cup of water in a microwave oven. The oven heats the water by exciting the rotational motion of its molecules. How does this work? Understanding these questions requires abstract initiative. Strangely, exactly the same abstract initiative – the theory of groups – is required to understand all of these questions. The theory of symmetry and groups is the first topic and central theme of this text.

Mathematics involves a continual interplay between the abstract and the concrete: abstraction is necessary in order to understand concrete phenomena, and concrete phenomena are necessary in order to understand the abstractions. Meanwhile, as one continues to study mathematics the boundary between the abstract and concrete inevitably shifts.

I am grateful to the following reviewers for their perceptive comments: Joseph W. Fisher, Daniel King, Michael G. Neuhauer, David A. Schmidt, Jo Ann Turisco, Nicholas Vaughn, and David Weinberg. I would like to thank George Lobell, Nicholas Romanelli, and the staff at Prentice-Hall for expert guidance and assistance.

I intend to maintain a world-wide-web site with electronic supplements to the text, at `http://www.math.uiowa.edu/~goodman`. Materials eventually to be available at this site include:

- Additional exercises
- Additional chapters or sections
- Color versions of graphics from the text
- Manipulable three-dimensional graphics of symmetric figures
- Programs for algebraic computations
- Errata

I would be grateful for any comments on the text, reports of errors, and suggestions for improvements.

Frederick M. Goodman

A Note to the Reader

I would like to show you a passage from one of my favorite books, *A River Runs Through It*, by Norman Maclean. The narrator Norman is fishing with his brother Paul on a mountain river near their home in Montana. The brothers have been fishing a "hole" blessed with sunlight and a hatch of yellow stone flies, on which the fish are vigorously feeding. They descend to the next hole downstream, where the fish will not bite. After a while Paul, who is fishing the opposite side of the river, makes some adjustment to his equipment and begins to haul in one fish after another. Norman watches in frustration and admiration, until Paul wades over to his side of the river to hand him a fly:

> He gave me a pat on the back and one of George's No. 2 Yellow Hackles with a feather wing. He said, "They are feeding on drowned yellow stone flies."
>
> I asked him, "How did you think that out?"
>
> He thought back on what had happened like a reporter. He started to answer, shook his head when he found he was wrong, and then started out again. "All there is to thinking," he said, "is seeing something noticeable which makes you see something you weren't noticing which makes you see something that isn't even visible."
>
> I said to my brother, "Give me a cigarette and say what you mean."
>
> "Well," he said, "the first thing I noticed about this hole was that my brother wasn't catching any. There's nothing more noticeable to a fisherman than that his partner isn't catching any.
>
> "This made me see that I hadn't seen any stone flies flying around this hole."
>
> Then he asked me, "What's more obvious on earth than sunshine and shadow, but until I really saw that there were no stone flies hatching here I didn't notice that the upper

> hole where they were hatching was mostly in sunshine and this hole was in shadow."
>
> I was thirsty to start with, and the cigarette made my mouth drier, so I flipped the cigarette into the water.
>
> "Then I knew," he said, "if there were flies in this hole they had to come from the hole above that's in the sunlight where there's enough heat to make them hatch.
>
> "After that, I should have seen them dead in the water. Since I couldn't see them dead in the water, I knew they had to be at least six or seven inches under the water where I couldn't see them. So that's where I fished."
>
> He leaned against the rock with his hands behind his head to make the rock soft. "Wade out there and try George's No. 2," he said, pointing at the fly he had given me. [1]

This passage has a great deal to say about the practice of mathematics. In mathematical practice – by those studying mathematics for its intrinsic interest and also by those seeking to apply mathematics – the typical experience is to be faced by a problem whose solution – if there is one – is an opaque mystery. Only by patient involvement with the problem does one gradually gain insight and eventual understanding. There is no other way but to think things through for oneself; for even if one has at hand a collection of methods and rules, the problem does not come labeled with the applicable method, and the fit between the problem and existing rules is likely to be imperfect.

The purpose of this course is to introduce you to the practice of mathematics; to help you learn to think things through for yourself; to teach you to see "something noticeable which makes you see something you weren't noticing which makes you see something that isn't even visible." And then to explain accurately what you have understood.

Not incidentally, the course aims to show you some algebraic and geometric ideas which are interesting and important and worth thinking about.

It's not at all easy to learn to work things out for yourself, and it's not at all easy to explain clearly what you have worked out. These are arts which have to be learned by thoughtful practice, and perhaps your previous mathematics education has not given you all that much practice in these arts.

You must have patience, or learn patience, and you must have time. You can't learn these things without getting frustrated, and you can't learn

[1]From Norman Maclean, *A River Runs Through It*, University of Chicago Press, 1976. Reprinted by permission.

them in a hurry. If you can get someone else to explain how to do the problems, you will learn something, but not patience, and not persistence, and not vision. So rely on yourself as far as possible.

But rely on your teacher as well. An extraordinary student may be able to do all this on his/her own, but most students will not. Your teacher will give you hints, suggestions, and insights which can help you see for yourself. A book alone cannot do this, because it cannot listen to you and respond.

I wish you success, and I hope you will someday fish in waters not yet dreamed of. Meanwhile, I have arranged a tour of some well known but interesting streams.

Part I: Basics

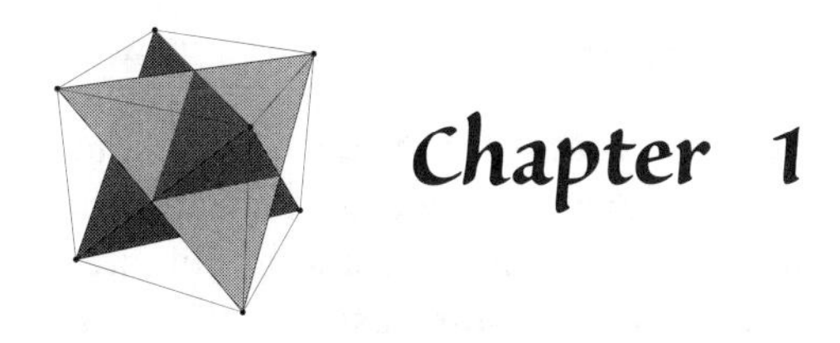

Chapter 1

Symmetry

1.1. What is Symmetry?

Imagine some symmetric objects and some non-symmetric objects. What makes a symmetric object symmetric? Are different symmetric objects symmetric in different ways?

Exercise

Exercise 1.1.1. The goal of this book is to encourage you to think things through for yourself. Take some time, and consider these questions for yourself. Start by making a list of symmetric objects: a sphere, a circle, a cube, a square, a rectangle, a rectangular box, etc. What do we mean when we say that these objects are symmetric? How is symmetry a common feature of these objects? How do the symmetries of the different objects differ?

1.2. What are Symmetries?

As an example of a symmetric object, let us take a (non-square, blank, undecorated) rectangular card. We have to choose whether to idealize the card as a two-dimensional object or whether to think of it as a thin three-dimensional object. I choose to regard it as three dimensional; that is, its motions are not restricted to a plane.

What makes the card symmetric? My answer is that there are motions of the card which leave its appearance unchanged. For example, if I left the room, you could secretly rotate the card by π about the axis through two opposite edges, as shown, and when I returned, I could not tell that you had moved the card. *A symmetry is an undetectable motion. An object is symmetric if it has symmetries.*

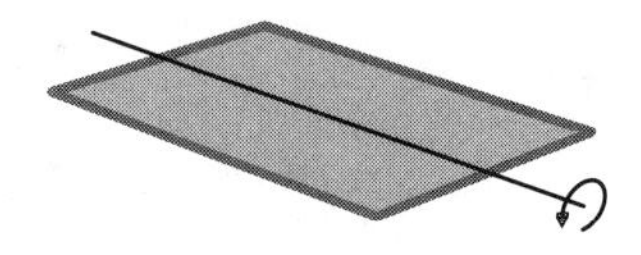

Figure 1.2.1. A Symmetry

Exercises

Exercise 1.2.1. Catalogue all the symmetries of a rectangular card. Get a card and look at it. Turn it about. Mark its parts as you need. Write out your observations and conclusions.

Exercise 1.2.2. Do the same for a square card.

Exercise 1.2.3. Do the same for a brick, i.e., a rectangular solid with three unequal edges.

1.3. Symmetries of the Rectangle and the Square

How many motions did you find for the rectangular card? Perhaps you found three motions: two rotations of π about axes through centers of opposite edges, and one rotation of π about an axis perpendicular to the faces of the card and passing through their centroids. (See Figure 1.3.1.)

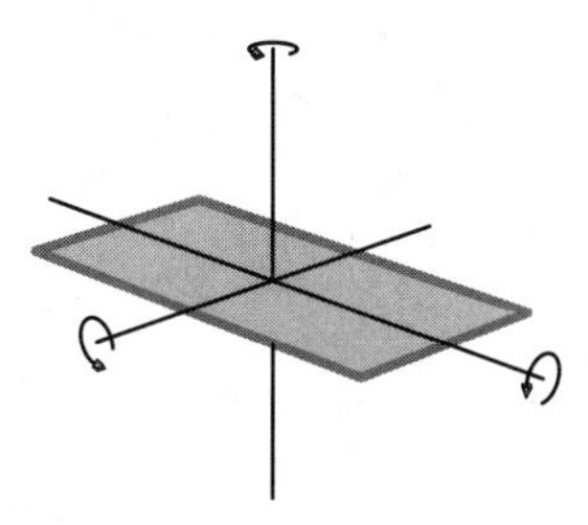

Figure 1.3.1. Symmetries of the Rectangle

It turns out to be a good idea to also include the non-motion, the rotation through 0 radians about any axis of your choice. One of the things which you could do to the card while I am out of the room is: *nothing*. When I returned I could not tell that you had done nothing rather than something; nothing is also undetectable!

If we include the non-motion, there are four different symmetries of the rectangular card.

However, another quite sensible answer is that there are infinitely many symmetries. As well as rotating by π about one of the axes, you could rotate by $-\pi$, $\pm 2\pi$, $\pm 3\pi$, $\dots$.

So which answer is right? Are there four symmetries of the rectangular card, or are there infinitely many symmetries? I will not insist that one answer is preferable, but I will assert that it is profitable to distinguish only the four symmetries. With this choice, we have to consider rotation by 2π about one of the axes the same as the non-motion, and rotation by 3π the same as rotation by π.

Another issue is whether to include reflection symmetries as well as rotation symmetries. I propose to exclude reflections for now, in order to simplify matters. I will bring reflections into the picture later in the course.

Making the same choices regarding the square card (to include the non-motion, but to distinguish only finitely many symmetries), you find that there are eight symmetries of the square: the non-motion, two rotations by

π about axes through centers of opposite edges, two rotations by π about axes through opposite corners, and rotation by $\pi/2$, π, or $3\pi/2$ about the axis perpendicular to the faces and passing through their centroids. (See Figure 1.3.2.)

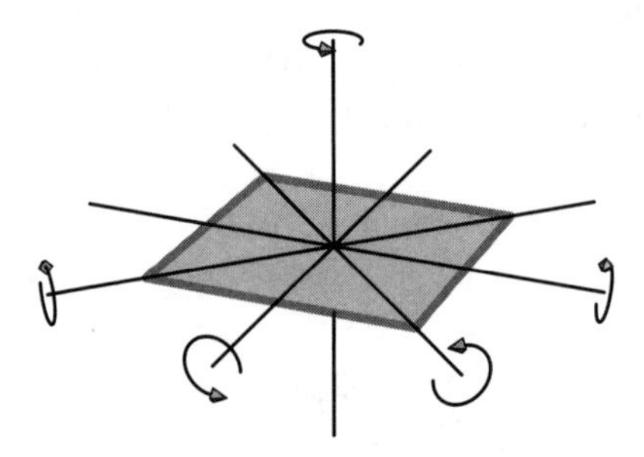

Figure 1.3.2. Symmetries of the Square

Here is an essential observation: if I leave the room and you perform two undetectable motions one after the other, I will not be able to detect the result. The result of two symmetries one after the other is also a symmetry.

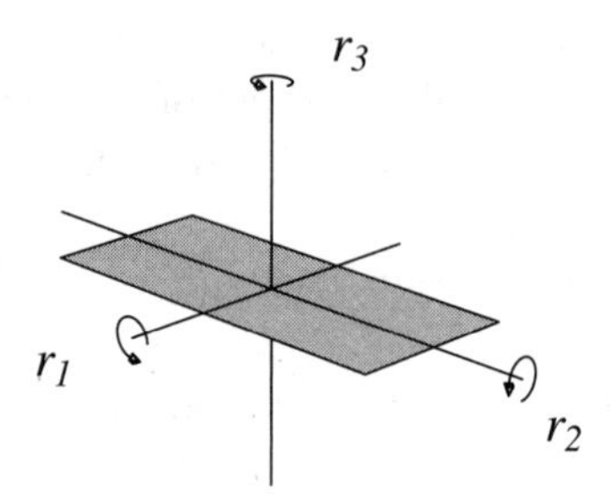

Figure 1.3.3. Labeling Symmetries of the Rectangle

Let's label the three non-trivial rotations of the rectangular card by r_1, r_2, and r_3, as shown in Figure 1.3.3, and let's call the non-motion e. If you perform first r_1, and then r_2, the result must be one of r_1, r_2, r_3, or e. Which is it? I claim that it is r_3. Likewise, if you perform first r_2 and then r_3, the result is r_1. Take your rectangular card in your hands and verify these assertions.

So we have a sort of multiplication by composition of symmetries: the product xy of symmetries x and y is the symmetry "first do y and then do x." (The order is a matter of convention; the other convention is also possible.)

Your next investigation is to work out all the products of all the symmetries of the rectangular card and of the square card. A good way to record

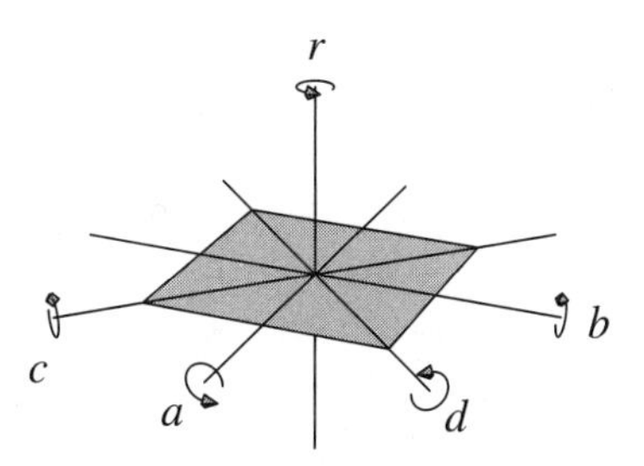

Figure 1.3.4. Labeling Symmetries of the Square

the results of your investigation is in a multiplication table: label rows and columns of a square table by the various symmetries; for the rectangle you will have four rows and columns, for the square eight rows and columns. In the cell of the table in row x and column y record the product xy. For example in the cell in row r_1 and column r_2 in the table for the rectangle, you will record the symmetry r_3.

You will have to choose some labeling for the eight symmetries of the square card in order to begin to work out the multiplication table. It will be helpful for comparing our results if we agree upon a labeling beforehand. Call the rotation by $\pi/2$ around the axis through the centroid of the faces r. The other rotations around this same axis are r^2 and r^3; we don't need other names for them. Call the non-motion e. Call the rotations by π about axes through centers of opposite edges a and b, and the rotations by π about axes through opposite vertices c and d. Also, to make comparing our results easier, let's agree to list the symmetries in the order $e, r, r^2, r^3, a, b, c, d$ in our tables. See Figure 1.3.4.

Before going on with your reading, stop here and work out the multiplication tables for the symmetries of the rectangular and square cards. For learning mathematics, it is essential to work things out for yourself.

1.4. Multiplication Tables

We label the four symmetries of the rectangular card as before; r_1 and r_2 are rotations by π about axes through center of opposite edges, and r_3 is the rotation by π through the axis through the centroid of the faces.

The multiplication table for the symmetries of the rectangle is:

	e	r_1	r_2	r_3
e	e	r_1	r_2	r_3
r_1	r_1	e	r_3	r_2
r_2	r_2	r_3	e	r_1
r_3	r_3	r_2	r_1	e

It is quite easy to give a rule for computing all of the products: the square of any element is the non-motion e. The product of any two elements other than e is the third such element.

Note that order doesn't matter. The product of two elements in either order is the same.

Here is the multiplication table for the symmetries of the square card:

	e	r	r^2	r^3	a	b	c	d
e	e	r	r^2	r^3	a	b	c	d
r	r	r^2	r^3	e	d	c	a	b
r^2	r^2	r^3	e	r	b	a	d	c
r^3	r^3	e	r	r^2	c	d	b	a
a	a	c	b	d	e	r^2	r	r^3
b	b	d	a	c	r^2	e	r^3	r
c	c	b	d	a	r^3	r	e	r^2
d	d	a	c	b	r	r^3	r^2	e

This table has the following properties, which I have emphasized by choosing the order in which to write the symmetries: The product of two powers of r (i.e., of two rotations around the axis through the centroid of the faces) is again a power of r. The square of any of the elements $\{a, b, c, d\}$ is the non-motion e. The product of any two of $\{a, b, c, d\}$ is a power of r, while the product of a power of r and one of $\{a, b, c, d\}$ (in either order) is again one of $\{a, b, c, d\}$.

Actually this last property is obvious, without doing any close computation of the products, if one thinks as follows: The symmetries $\{a, b, c, d\}$

exchange the two faces of the square card, while the powers of r do not. So, for example, the product of two symmetries which exchange the faces leaves the upper face above and the lower face below, so it has to be a power of r.

Another property is that order matters. For example, $ra = d$, whereas $ar = c$.

In case your results don't agree with mine, you may have had some bookkeeping difficulties. In order to work out the products, one has to devise some sort of bookkeeping device to keep track of the results. One possibility is to "break the symmetry" by labeling parts of the square. At the risk of over doing it somewhat, I'm going to number both the *locations* of the four vertices of the square and the corners of the square card. The numbers on the corners will travel with the symmetries of the card; those on the locations stay put. See Figure 1.4.1.

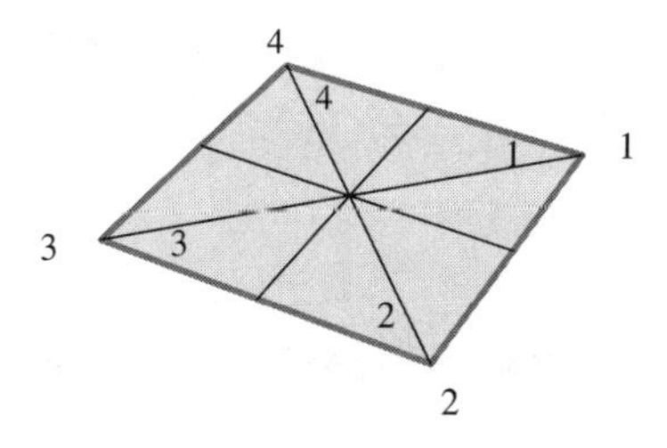

Figure 1.4.1. Breaking of Symmetry

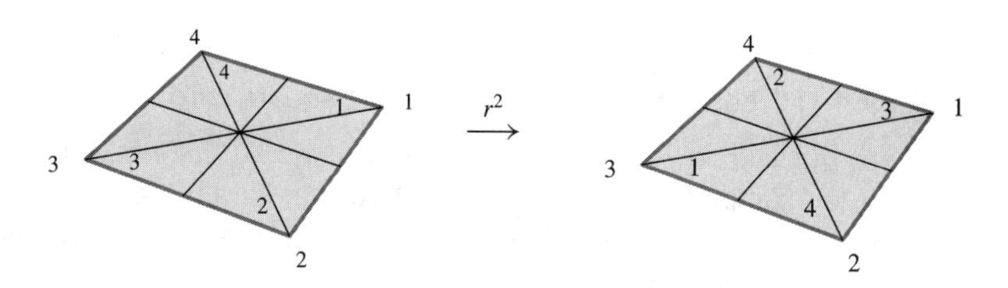

Figure 1.4.2. The Symmetry r^2

Then the various symmetries can be distinguished by the location of the numbered corners after performing the symmetry, as in Figure 1.4.2

You can make a list of where each of the eight symmetries send the numbered vertices, and then you can compute products by diagrams as in Figure 1.4.3. Comparing Figures 1.4.3 and 1.4.2, you see that $cd = r^2$.

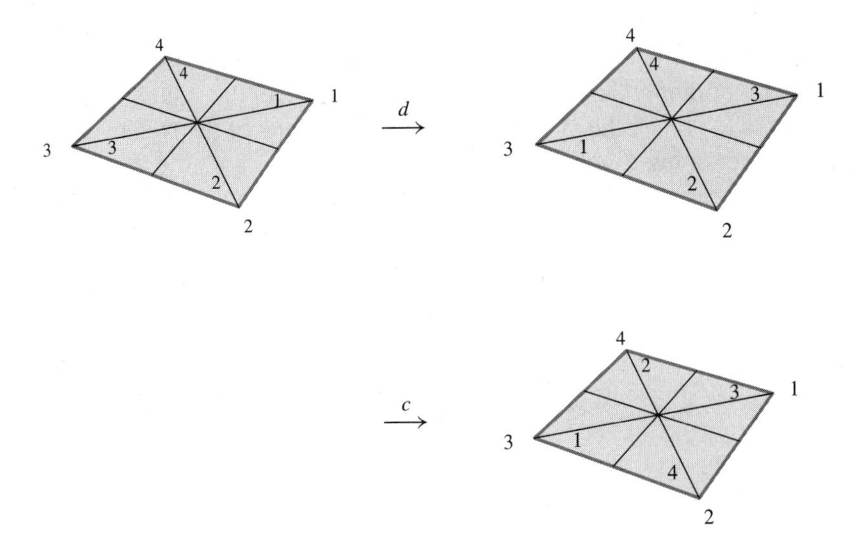

Figure 1.4.3. Computation of a Product

Exercises

Exercise 1.4.1. Consider the symmetries of the square card.

(a) Any positive power of r must be one of $\{e, r, r^2, r^3\}$. First work out some examples, say through r^{10}. Show that for any natural number k, $r^k = r^m$, where m is the non-negative remainder after division of k by 4.

(b) Observe that r^3 is the same symmetry as the rotation by $\pi/2$ about the axis through the centroid of the faces of the square, *in the clockwise sense*, looking from the top of the square; that is, r^3 is the opposite motion to r. Also $r^3 r = r r^3 = e$. So it makes sense to regard r^3 as an inverse to r and to write $r^{-1} = r^3$. Define $r^{-k} = (r^{-1})^k$ for any positive integer k. Show that $r^{-k} = r^{3k} = r^m$, where m is the unique element of $\{0, 1, 2, 3\}$ such that $m + k$ is divisible by 4.

Exercise 1.4.2. Here is another way to list the symmetries of the square card which makes it easy to compute the products of symmetries quickly.

(a) Verify that the four symmetries a, b, c, and d which exchange the top and bottom faces of the card are a, ra, r^2a, and r^3a, in some order. Which is which? Thus a complete list of the symmetries is

$$\{e, r, r^2, r^3, a, ra, r^2a, r^3a\}.$$

(b) Verify that $ar = r^{-1}a = r^3a$.

(c) Conclude that $ar^k = r^{-k}a$ for all integers k.

(d) Show that these relations suffice to compute any product.

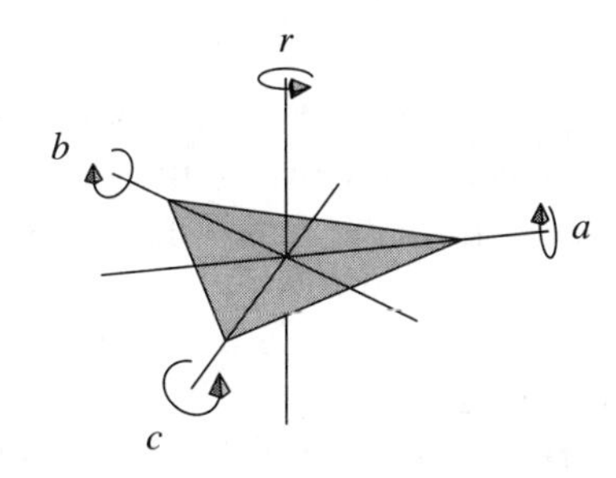

Figure 1.4.4. Symmetries of Equilateral Triangle

Exercise 1.4.3. List the symmetries of an equilateral triangular plate (there are six) and work out the multiplication table for the symmetries. (See Figure 1.4.4.)

1.5. Matrices

There is another way in which we can do the bookkeeping involved in working out the multiplication tables for symmetries of the square or the rectangle. The important idea is that, for each of the symmetries, there is a matrix A such that the linear transformation $\mathbf{v} \mapsto A\mathbf{v}$ maps the figure to itself and realizes the symmetry.

Let's arrange that the figure (square or rectangle) lies in the x-y plane with sides parallel to the coordinate axes and centroid at the origin of coordinates. Then certain axes of symmetry will coincide with the coordinate axes.

For example, we can orient the rectangle in the plane so that the axis of rotation for r_1 coincides with the x-axis, the axis of rotation for r_2 coincides with the y-axis, and the axis of rotation for r_3 coincides with the z-axis. Then the rotation r_1 leaves the x-coordinate of a point in space unchanged and changes the sign of the y and z coordinates. Similarly one can trace through what the rotations r_2 and r_3 do in terms of coordinates. The result is that the matrices

$$R_1 = \begin{bmatrix} 1 & 0 & 0 \\ 0 & -1 & 0 \\ 0 & 0 & -1 \end{bmatrix} \quad R_2 = \begin{bmatrix} -1 & 0 & 0 \\ 0 & 1 & 0 \\ 0 & 0 & -1 \end{bmatrix} \quad \text{and} \quad R_3 = \begin{bmatrix} -1 & 0 & 0 \\ 0 & -1 & 0 \\ 0 & 0 & 1 \end{bmatrix}$$

implement the three rotations r_1, r_2, and r_3. Of course the identity matrix

$$E = \begin{bmatrix} 1 & 0 & 0 \\ 0 & 1 & 0 \\ 0 & 0 & 1 \end{bmatrix}$$

implements the non-motion. Now you can check that the square of any of the R_i's is E and the product of any two of the R_i's is the third. Thus the matrices R_1, R_2, R_3, and E have the same multiplication table (using matrix multiplication) as do the symmetries r_1, r_2, r_3, and e of the rectangle. So we could do our bookkeeping by matrix multiplication, although here there isn't much bookkeeping to do.

Let us similarly work out the matrices for the symmetries of the square: Choose the orientation of the square in space so that the axes of symmetry for the rotations a, b, and r coincide with the x-, y-, and z- axes, respectively.

Then the symmetries a and b are implemented by the matrices

$$A = \begin{bmatrix} 1 & 0 & 0 \\ 0 & -1 & 0 \\ 0 & 0 & -1 \end{bmatrix} \quad B = \begin{bmatrix} -1 & 0 & 0 \\ 0 & 1 & 0 \\ 0 & 0 & -1 \end{bmatrix}.$$

The rotation r is implemented by the matrix

$$R = \begin{bmatrix} 0 & -1 & 0 \\ 1 & 0 & 0 \\ 0 & 0 & 1 \end{bmatrix},$$

and powers of r by powers of this matrix

$$R^2 = \begin{bmatrix} -1 & 0 & 0 \\ 0 & -1 & 0 \\ 0 & 0 & 1 \end{bmatrix} \quad \text{and} \quad R^3 = \begin{bmatrix} 0 & 1 & 0 \\ -1 & 0 & 0 \\ 0 & 0 & 1 \end{bmatrix}.$$

The symmetries c and d are implemented by matrices

$$C = \begin{bmatrix} 0 & -1 & 0 \\ -1 & 0 & 0 \\ 0 & 0 & -1 \end{bmatrix} \quad \text{and} \quad D = \begin{bmatrix} 0 & 1 & 0 \\ 1 & 0 & 0 \\ 0 & 0 & -1 \end{bmatrix}.$$

Here is an important idea: if two linear transformations T_1 and T_2 of $\mathbb{R}^3$, restricted to a geometric figure, give symmetries τ_1 and τ_2 of the figure, then the composed linear transformation $T_1 T_2$ gives the composed symmetry $\tau_1 \tau_2$.

But the composition of linear transformations can be computed by the matrix product of the corresponding matrices. Therefore the set of matrices $\{E, R, R^2, R^3, A, B, C, D\}$ *must* have the same multiplication table (under matrix multiplication) as does the corresponding set of symmetries $\{e, r, r^2, r^3, a, b, c, d\}$.

So we could have worked out the multiplication table for the symmetries of the square by computing products of the corresponding matrices. The computation, which previously required a certain amount of concentration, now becomes more or less automatic when we do it by matrix multiplication.

Exercises

Exercise 1.5.1. Work out the products of the matrices E, R, R^2,R^3, A, B, C, D, and verify that these products reproduce the multiplication table for the symmetries of the square.

Exercise 1.5.2. Find matrices implementing the six symmetries of the equilateral triangle. (Compare Exercise 1.4.3.) In order to standardize our notation and our co-ordinates, let's agree to put the vertices of the triangle at $(1, 0, 0)$, $(-1/2, \sqrt{3}/2, 0)$, and $(-1/2, -\sqrt{3}/2, 0)$. You may have to review some linear algebra in order to compute the matrices of the symmetries. Verify that the products of the matrices reproduce the multiplication table for the symmetries of the equilateral triangle.

1.6. Groups: Definition and Examples

Composition of symmetries of a geometric figure (such as a rectangular or square plate) evidently satisfies the following properties:

1. The product of three symmetries is independent of how the three are associated: the product of two symmetries followed by a third gives the same result as the first symmetry followed by the product of the second and third. This is the associative law for multiplication.
2. The non-motion e composed with any other symmetry (in either order) is the second symmetry.
3. For each symmetry there is an inverse, such that the composition of the symmetry with its inverse (in either order) is the non-motion e. (The inverse is just the reversed motion; the inverse of a rotation about a certain axis is the rotation about the same axis by the same angle but in the opposite sense.)

These are the axioms for a mathematical structure known as a group.

Definition 1.6.1. A group is a set G with a product, denoted here simply by juxtaposition, satisfying the following properties:

1. The product is associative: for all $a, b, c \in G$, one has $(ab)c = a(bc)$.
2. There is an identity element $e \in G$ with the property that for all $a \in G$, $ea = ae = a$.
3. For each element $a \in G$ there is an element $a^{-1} \in G$ satisfying $aa^{-1} = a^{-1}a = e$.

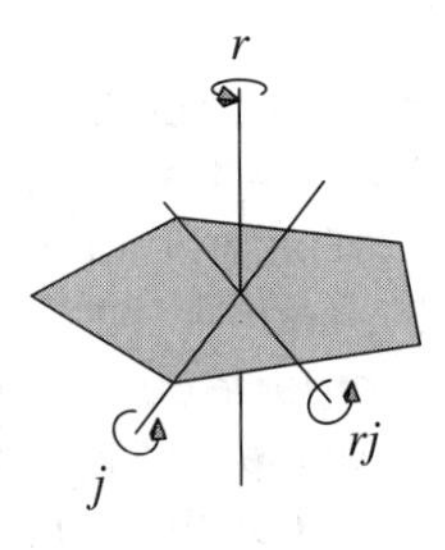

Figure 1.6.1. Pentagon

I would like to comment on the role of axioms in mathematics. You might have heard that mathematics is a deductive science in which one starts with axioms and proceeds to deduce consequences. But axioms are seldom the starting point in the exploration of a mathematical subject; that is,

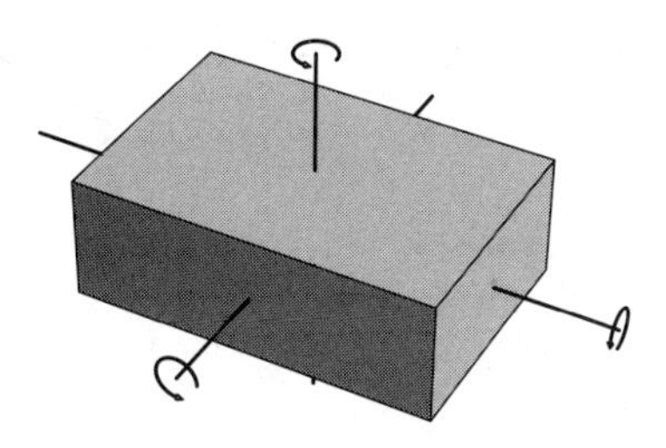

Figure 1.6.2. Brick

a subject is not created by setting up axioms for it and then working out the consequences. Rather, axioms are the distillation of experience. We have a concept of a group because groups appear of their own accord everywhere in mathematics.

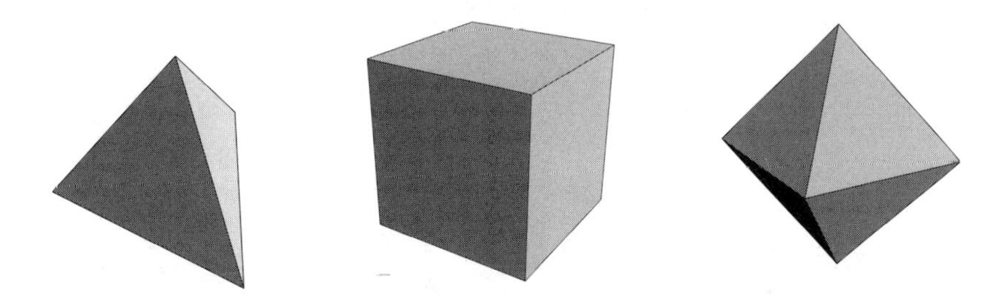

Figure 1.6.3. Regular Polyhedra

Example 1.6.2. Our primary motivating examples for the concept of a group are the groups of symmetries of geometric figures. So far, we have analyzed two of the smallest and simplest of such symmetry groups: the groups of the rectangle and the square. Eventually we will work out the symmetry groups of regular polygons (Figure 1.6.1), of rectangular solids (Figure 1.6.2), of the five regular polyhedra (Figures 1.6.3 and 1.6.4), and of "crystals," infinitely repeating geometric arrays (Figure 1.6.5). We will include reflections as well as rotations in our study of geometric symmetry (Figure 1.6.6).

Example 1.6.3. Here is an example of a group which does not seem to have to do with the symmetries of a geometric object. Consider a deck of playing cards. The cards are distinct, but I put them face down so I can't tell one from another. Let us take a very small deck, say of five cards, in order

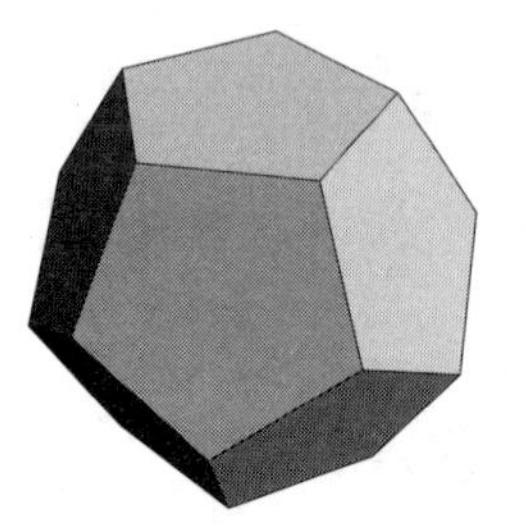
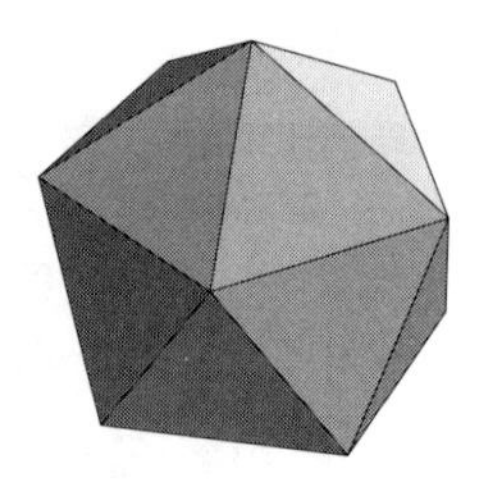

Figure 1.6.4. More Regular Polyhedra

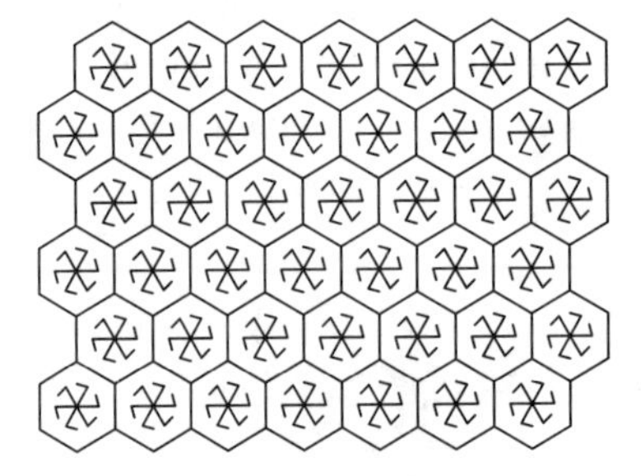

Figure 1.6.5. Crystal

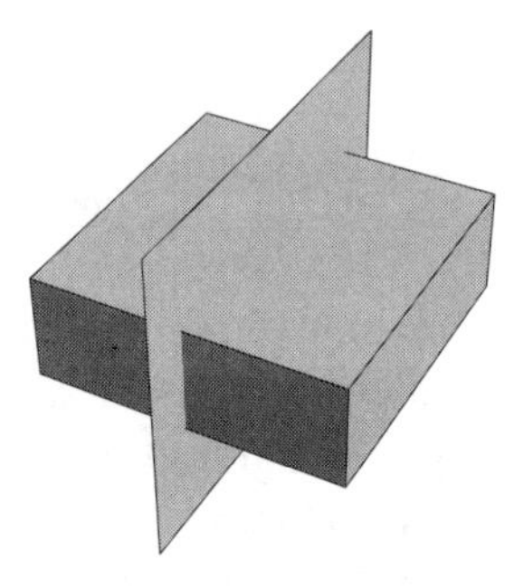

Figure 1.6.6. Reflection

to clarify our discussion. A *permutation* of the deck is rearrangement according to position; for example a permutation might put the first (top) card in the fourth position, the second card in the fifth position, the third card in the second position, the fourth card in the third position and the fifth (bottom) card in the first position. We represent this permutation by a diagram (Figure 1.6.7).

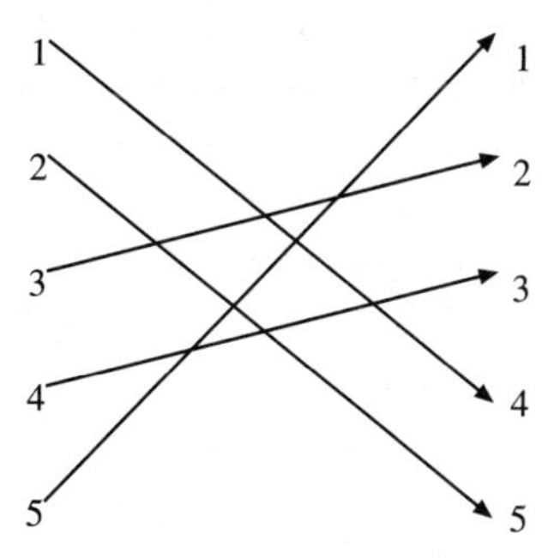

Figure 1.6.7. A Permutation

It is possible to compose two permutations as shown in Figure 1.6.8, and composition of permutations is associative.

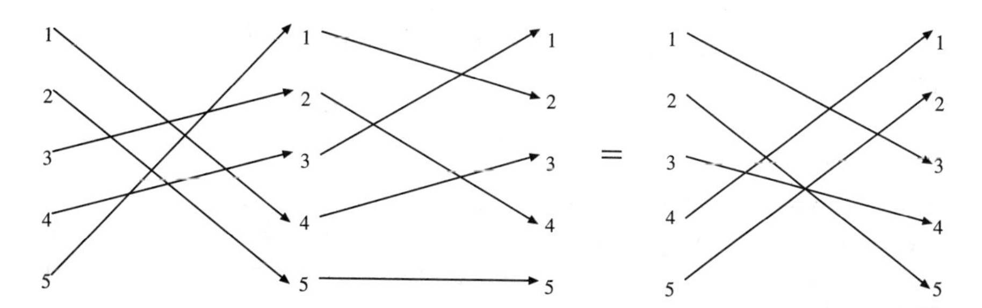

Figure 1.6.8. Composition of Permutations

There is a non-permutation e which simply leaves all the cards in place. The composition of the non-permutation with any other permutation, in either order, is the second permutation.

Finally, for any permutation there is an inverse permutation which undoes the first permutation; for example, the inverse of the permutation in Figure 1.6.7 is the permutation which puts the first card in the fifth position, the second card in the third position, the third card in the fourth position, the fourth card in the first position, and the fifth card in the second position. The diagram is obtained by reversing the arrows in Figure 1.6.7.

Composition of a permutation with its inverse leaves the cards in their original position.

Thus the set of permutations of a deck of cards forms a group under composition of permutations. We will begin to look more closely at the permutation group in Section 2.2.

Example 1.6.4. In the preface, I commented about the "concrete" and the "abstract" in mathematics. Let me ask you to think about the issue of concreteness and abstraction in the light of the following example.

A deck of cards is a very familiar and concrete (if peculiar) object. Permuting a deck so that participants in a game cannot predict the order of the cards in the deck is a familiar and concrete physical process.

Here is a particular question about this concrete process. Take the standard deck of playing cards and "cut" it perfectly; put the first 26 cards in one half-deck on the left and the second 26 cards in a second half-deck on the right. Then shuffle the two half-decks back together by alternating first one card from the left half-deck, then one card from the right, then one from the left, and so forth. If one repeats this perfect shuffle some relatively small number of times, the deck will return to its original arrangement. How many times?

Do you think this is a concrete question or an abstract question? Do you think the answer is likely to be concrete or abstract?

Although a standard deck of cards is not a very large collection of objects, the number of permutations of 52 cards is unbelievably large, namely $(52)(51)(50)\cdots(3)(2)(1) =$

$$80658175170943878571660636856403766975289505440883277824000000000000.$$

We can't possibly write a multiplication table for such a large group. Does the mere size of this group make it an abstract object, or is it no more or less abstract than the group of permutations of 5 objects, which has 120 elements?

Example 1.6.5. Look again at the diagrams in Example 1.6.3. The diagrams display *bijective maps* from the set $\{1, 2, 3, 4, 5\}$ to itself.

Recall that a function (or map) $f : X \to Y$ is *one to one* (or *injective*) if $f(x_1) \neq f(x_2)$ whenever $x_1 \neq x_2$ are distinct elements of X. A map $f : X \to Y$ is said to be *onto* (or *surjective*) if the range of f is all of Y; that is, for each $y \in Y$, there is an $x \in X$ such that $f(x) = y$. A map $f : X \to Y$ is called *invertible* (or *bijective*) if it is both injective and surjective. For each invertible map $f : X \to Y$, there is a map $f^{-1} : Y \to X$ called the inverse of f and satisfying $f \circ f^{-1} = \mathrm{id}_Y$ and $f^{-1} \circ f = \mathrm{id}_X$. (Here id_X denotes the identity map $\mathrm{id}_X : x \mapsto x$ on X and similarly id_Y denotes the identity map on Y.) For $y \in Y$, $f^{-1}(y)$ is the unique element of X such that $f(x) = y$.

Maps from a set X to itself can be composed, and composition of maps is associative. If f and g are invertible maps on X, then the composition $f \circ g$ is also invertible (with inverse $g^{-1} \circ f^{-1}$). So composition of maps defines an associative product on the set $\mathrm{Sym}(X)$ of invertible maps on X. The identity map is the identity element for this product, and the inverse of a map is the inverse for this product. Thus $\mathrm{Sym}(X)$ with the composition product is an example of a group.

This is really the same example as Example 1.6.3. Permutations of a (finite) set X are the same as bijective maps of X.

In mathematics, we often deal with sets with additional structure, for example, the set $\mathbb{R}^3$ with its structure as a vector space, or the set $\mathbb{R}^3$ with its distance function, which measures the distance between pairs of points. Whenever one has a set with an additional structure, it is germain to consider bijective maps of the set which preserve the structure, and the family of such maps *always* forms a group. Let us look at particular examples:

Example 1.6.6. Consider the set $\mathbb{R}^3$, or more generally $\mathbb{R}^n$ with its structure as a vector space. The vector space structure involves two operations: addition of vectors and multiplication of vectors by real numbers. These operations satisfy various familiar properties which we need not list here. A map or transformation $T : \mathbb{R}^n \to \mathbb{R}^n$ preserves the vector space structure if $T(\boldsymbol{a}+\boldsymbol{b}) = T(\boldsymbol{a}) + T(\boldsymbol{b})$ and $T(s\boldsymbol{a}) = sT(\boldsymbol{a})$ for all vectors $\boldsymbol{a}, \boldsymbol{b} \in \mathbb{R}^n$ and all $s \in \mathbb{R}$. That is, the maps which preserve the vector space structure are precisely the linear maps, which are the subject of linear algebra.

In linear algebra, you learn that the composition $S \circ T$ of invertible linear maps is also invertible and linear, with inverse $T^{-1} \circ S^{-1}$. Composition is always associative, so defines an associative product on the set of invertible linear maps on $\mathbb{R}^n$. The identity transformation $E(\boldsymbol{x}) = \boldsymbol{x}$ is invertible and linear, and is the identity for this product. The inverse of an invertible linear map is invertible and linear, and is the inverse for this product. Thus the set $\mathrm{GL}(\mathbb{R}^n)$ of invertible linear maps forms a group with composition as the product. (The notation "GL" stands for "general linear group".)

Example 1.6.7. This is basically the same as the previous example. You learn in linear algebra that each linear transformation of $\mathbb{R}^n$ is given by multiplication by a matrix, and that invertible transformations are given by invertible matrices. Products of matrices correspond to composition of linear transformations.

The product AB of invertible n-by-n matrices A and B is invertible with inverse $B^{-1}A^{-1}$. The product of matrices is associative, so matrix multiplication defines an associative product on the set of invertible n-by-n matrices. The identity matrix E, with 1's on the diagonal and 0's off the diagonal, is the identity element for matrix multiplication, and the inverse of a matrix is the inverse for matrix multiplication. Thus the set $\mathrm{GL}(n, \mathbb{R})$ of invertible n-by-n matrices with real entries forms a group under matrix multiplication.

Example 1.6.8. Now consider $\mathbb{R}^3$ with its distance structure. An invertible transformation of $\mathbb{R}^3$ preserves the distance structure if $d(T(\boldsymbol{a}), T(\boldsymbol{b})) = d(\boldsymbol{a}, \boldsymbol{b})$ for all elements $\boldsymbol{a}, \boldsymbol{b} \in \mathbb{R}^n$. Such maps are called *isometries.*

We will eventually learn that the group of all isometries is very closely related to the group of linear isometries, and we will obtain a very complete picture of the group of isometries.

Example 1.6.9. There are many more examples of this general type which may be more or less familiar to you, depending on which mathematical subjects you have already studied. You are not likely to have studied all of these subjects yet, so just skim over this example quickly.

In topology, the structure preserving maps of a space are the bijections which are continuous and which have continuous inverse; these are known as *homeomorphisms.* The set of homeomorphisms forms a group under composition of maps.

In multivariable calculus or differential geometry, structure preserving maps are bijective maps which are differentiable; these are known as *diffeomorphisms.* The set of diffeomorphisms of a set form a group under composition of maps.

In complex analysis, the structure preserving maps are bijective maps which are complex differentiable or *holomorphic.* The set of bijective holomorphic maps of an open set in the complex plane forms a group under composition of maps.

All the examples which I have mentioned thus far are *active*; the groups consist of *motions* or *maps* or *transformations.* But there are some other examples with which you are familiar and which do not appear naturally as groups of transformations.

Example 1.6.10. One family of examples consists of familiar *number systems*: The integers $\mathbb{Z}$, the rational numbers $\mathbb{Q}$, the real numbers $\mathbb{R}$, and the complex numbers $\mathbb{C}$ are all groups with addition as the group operation, 0 as the identity element and with the *additive* inverse serving as the inverse of an element. (The natural numbers $\mathbb{N}$ are a non-example; they do not form a group under addition, since inverses are missing.)

Example 1.6.11. In the *fields* $\mathbb{R}$, $\mathbb{Q}$, or $\mathbb{C}$, every non-zero element has a multiplicative inverse. The set of non-zero elements in these number systems are groups, with multiplication as the group operation. The element 1 is the identity element, and the inverse of an element is the *multiplicative* inverse. The notation F^* is often used for the multiplicative group of non-zero elements ($F = \mathbb{R}$, $\mathbb{Q}$, or $\mathbb{C}$).

Example 1.6.12. $\mathbb{R}^n$ and, more generally, any *vector space* is a group with vector addition as the group operation, the zero vector as the identity element, and the additive inverse as the inverse. In fact, the first few axioms for a vector space say precisely that a vector space with the operation of vector addition is a group in which moreover $\boldsymbol{a} + \boldsymbol{b} = \boldsymbol{b} + \boldsymbol{a}$ for all $\boldsymbol{a}, \boldsymbol{b}$. Vector spaces are groups whose role is not to act but rather to be acted upon; consequently the real subject of linear algebra is not vector spaces but rather the linear operators which act upon them.

Exercises

Exercise 1.6.1. Convince yourself, by means of diagrams perhaps, that the composition of permutations is associative.

Exercise 1.6.2. Draw diagrams for the non-permutation e and for the inverse of the permutation in Figure 1.6.7. Confirm by diagrams that the composition of a permutation π with e is π and that the composition of a permutation with its inverse is e.

Exercise 1.6.3.

(a) Show that the composition of isometries is an isometry.
(b) Observe that composition defines an associative product on the set of isometries.
(c) The identity transformation is an isometry.
(d) Show that the inverse of an isometry is also an isometry.
(e) Conclude that the set of isometries of $\mathbb{R}^3$ forms a group under composition of maps.

Exercise 1.6.4. Show that the set of invertible maps of $\mathbb{R}^3$ which are both isometric and linear, i.e., which preserve both the distance structure and the vector space structure, form a group under composition of maps.

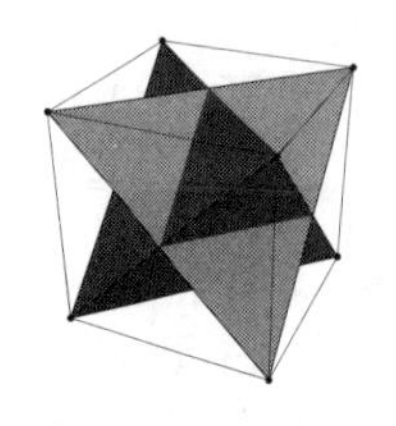

Chapter 2

A First Look at Groups

2.1. First Results

In the previous chapter, we saw many examples of groups and finally arrived at a definition, or collection of axioms, for groups. In this section we will try our hand at obtaining some first theorems about groups. For many students, this will be the first experience with constructing proofs concerning an algebraic object described by axioms. I would like to urge both students and instructors to take time with this material and not to go on before mastering it.

Our first results concern the uniqueness of the identity element in a group, and the uniqueness of the inverse of an element in a group.

Proposition 2.1.1. *Let G be a group with identity element e.*

(a) *Suppose e' and g are elements of G and $e'g = g$. Then $e' = e$. Similarly, if $e'' \in G$ and $ge'' = g$, then $e'' = e$.*

(b) *Suppose h' and g are elements of G such that $h'g = e$. Then $h' = g^{-1}$. Similarly, if $h'' \in G$ and $gh'' = e$, then $h'' = g^{-1}$.*

(c) $(g^{-1})^{-1} = g$.

Note that part (a) says that if a group element acts like the identity when multiplied by *one* element on *one* side, then it *is* the identity. Part (b) says, similarly, that if an element acts like the inverse of g on one side, then it is the inverse of g.

Proof. The idea for part (a) is to write e' as $e'e$ and to write e as gg^{-1} and then to use associativity: Since $e'g = g$, it follows that $e' = e'e = e'(gg^{-1}) = (e'g)g^{-1} = gg^{-1} = e$. The proof for e'' is similar.

To prove (b), assume that $h'g = e$. Then $h' = h'e = h'(gg^{-1}) = (h'g)g^{-1} = eg^{-1} = g^{-1}$. The proof for h'' is similar.

Since $gg^{-1} = e$, it follows from (b) that $g = (g^{-1})^{-1}$. □

Proposition 2.1.2. *Let G be a group and let $a, b \in G$. Then $(ab)^{-1} = b^{-1}a^{-1}$*

Proof. It suffices to show that $ab(b^{-1}a^{-1}) = e$. But by associativity, $ab(b^{-1}a^{-1}) = a(bb^{-1})a^{-1} = aea^{-1} = aa^{-1} = e$. □

Perhaps you have already noticed in the group multiplication tables that you have computed that each row and column contains each group element exactly once. A smart conjecture is that this is always true.

Proposition 2.1.3. *Let G be a group and let a and b be elements of G.*

(a) *The equation $ax = b$ has a unique solution in G, and likewise the equation $xa = b$ has a unique solution in G.*

(b) *The map $L_a : G \longrightarrow G$ defined by $L_a(x) = ax$ is a bijection. Similarly, the map $R_a : G \longrightarrow G$ defined by $R_a(x) = xa$ is a bijection.*

Proof. You are asked to write out a proof of this proposition in Exercise. 2.1.1. You should try the exercise before going on. Here are some hints:

In part (a), you have to show that there is a solution to the equation, and that there is only one solution. There may be several *psychological* stages to solving such a problem. Analogy is often helpful: How would you solve the equation if it involved real numbers instead of group elements? What does division mean? How would you solve the problem if it involved matrices instead of group elements? Once you have a candidate for the solution to the equation (which must be expressed in terms of a and b), show that your candidate is in fact a solution, using the axioms for a group. In order to show that your candidate is the only possible solution, assume x is a solution to the equation and show x must be equal to the group element which you have already picked out as the solution. Again, your steps must be justified by use of the group axioms.

In part (b), you have to show that a certain map is *bijective*. It would be a good idea to review the common terminology concerning functions in Appendix B; constant use is made of this terminology in this text and in all of mathematics. To show a map is bijective, you have to check that it is both *surjective* and *injective*. But surjectivity involves the existence of solutions to the equation in part (a), and injectivity involves the uniqueness of solutions!

For example, to establish injectivity of the map L_a, you have to show: if x_1 and x_2 are elements of G and $L_a(x_1) = L_a(x_2)$, then $x_1 = x_2$. So you assume that $x_1, x_2 \in G$ and $L_a(x_1) = L_a(x_2)$; that is $ax_1 = ax_2$. Now you have to show that $x_1 = x_2$. If you want to use the result of part (a), write

$b = ax_1 = ax_2$; thus, x_1 and x_2 are both solutions to the equation $ax = b$. What do you conclude? If you want to do this part independently of part (a), notice that you have to "cancel" the a on the left; what does canceling mean, and how can you justify it using the group axioms? □

Corollary 2.1.4. *If G is a finite group, each row and each column of the multiplication table of G contains each element of G exactly once.*

Proof. You are asked to prove this in Exercise 2.1.2, using the previous proposition. □

Example 2.1.5. The conclusions of Proposition 2.1.3 would be false if we were not working in a group. For example, let G be the set of non-zero integers. The equation $2x = 3$ has no solution in G; one cannot in general divide in the integers. This shows that the set of non-zero integers, with multiplication as the operation, is not a group.

With this much work done, we can figure out what are all possible groups with 1, 2, 3 or 4 elements.

Definition 2.1.6. The *order* of a group is its size or cardinality. We will denote the order of a group G by $\#G$.

First let's produce some groups of order 1, 2, 3, and 4.

Example 2.1.7. *A group of order 1:* Take the set consisting of the number 1, with the operation multiplication of numbers. This is a group.

Example 2.1.8. *A group of order 2:* Take the set consisting of the numbers 1 and -1, again with the operation multiplication of numbers. This is a group with 1 serving as the group identity. Each element is its own inverse.

Example 2.1.9. *A group of order 3:* This example uses complex numbers; a brief account of complex numbers can be found in Appendix D.

Take the set of complex numbers $C_3 = \{e^{2\pi i/3}, e^{-2\pi i/3}, 1\}$. For any real numbers s and t, one has $e^{is}e^{it} = e^{i(s+t)}$; in particular $e^{2\pi i/3}e^{-2\pi i/3} = e^0 = 1$. It follows from this that product of any two elements of C is contained in C and also that the inverse of any element of C is contained in C. Multiplication of complex numbers is associative, so in particular multiplication of elements of C is associative. This establishes that C is a group.

Write $\xi = e^{2\pi i/3}$. Then $\xi^2 = e^{-2\pi i/3}$ and $\xi^3 = 1$. So the group consists of the three distinct powers of ξ.

Example 2.1.10. *A group of order 4:* Take the set of complex numbers $C_4 = \{i, -1, -i, 1\}$ consisting of the four distinct powers of $i = e^{2\pi i/4}$. Verify that this is a group under complex multiplication, with 1 the identity element. (See Exercise 2.1.3.)

Example 2.1.11. *Another group of order 4:* The set of rotational symmetries of the rectangular card is a group of order 4.

We need a notion of two groups being essentially the same. Let me first give you an example of this idea. Inside the symmetry group of the square card, consider the set $\mathcal{R} = \{e, r, r^2, r^3\}$. Verify that this subset of the symmetry group is a group under composition of symmetries. (Exercise 2.1.4.)

The group $\mathcal{R}$ is essentially the same as the group $H = \{i, -1, -i, 1\}$ of fourth roots of 1 in $\mathbb{C}$. Define a bijection between these two groups by

$$\begin{aligned} e &\leftrightarrow 1 \\ r &\leftrightarrow i \\ r^2 &\leftrightarrow -1 \\ r^3 &\leftrightarrow -i. \end{aligned}$$

Under this bijection, the multiplication tables of the two groups match up: If one applies the bijection to each entry in the multiplication table of $\mathcal{R}$, one obtains the multiplication table for H. Verify this statement; see exercise 2.1.5. So although the groups seem to come from different contexts, they really are essentially the same and differ only in the names given to the elements.

Definition 2.1.12. We say that two groups G and H are *isomorphic* if there is a bijection $\varphi : G \longrightarrow H$ such that for all $g_1, g_2 \in G$ $\varphi(g_1 g_2) = \varphi(g_1)\varphi(g_2)$. The map φ is called an *isomorphism.*

You are asked to show in Exercise 2.1.6 that the two groups of order four given in examples above are *non-isomorphic.*

The definition says that under the bijection, the multiplication tables of the two groups match up, so the two groups differ only by a renaming of elements. Since the multiplication tables match up, one could also expect that the identity elements and inverses of elements match up, and in fact this is so:

Proposition 2.1.13. *If $\varphi : G \to H$ is an isomorphism, then $\varphi(e_G) = e_H$, and for each $g \in G$, $\varphi(g^{-1}) = \varphi(g)^{-1}$.*

Proof. For any $h \in H$, there is a $g \in G$ such that $\varphi(g) = h$. Then $\varphi(e_G)h = \varphi(e_G)\varphi(g) = \varphi(e_G g) = \varphi(g) = h$. Hence by the uniqueness of the identity in H, $\varphi(e_G) = e_H$. Likewise $\varphi(g^{-1})\varphi(g) = \varphi(gg^{-1}) = \varphi(e_G) = e_H$. This shows that $\varphi(g^{-1}) = \varphi(g)^{-1}$. □

Proposition 2.1.14.

(a) *Up to isomorphism, there is exactly one group of order 1.*
(b) *Up to isomorphism, there is exactly one group of order 2.*
(c) *Up to isomorphism, there is exactly one group of order 3.*
(d) *Up to isomorphism, there are exactly two groups of order 4.*
(e) *Up to isomorphism, there is exactly one group of order 5.*

The statement (c) means, for example, that there is a group of order 3, and any other group of the same order is isomorphic to that one. Statement (d) means that there are two distinct (non-isomorphic) groups of order 4, and any group of order 4 must be isomorphic to one of them.

Proof. The reader is guided through the proof of these statements in the exercises. The idea is to try to write down the group multiplication table, observing the constraint that each group element must appear exactly once in each row and column. □

Exercises

Exercise 2.1.1. Prove Proposition 2.1.3

Exercise 2.1.2. Show that each row and each column of the multiplication table of a finite group contains each group element exactly once. Use Proposition 2.1.3.

Exercise 2.1.3. Show that $C_4 = \{i, -1, -i, 1\}$ is a group under complex multiplication, with 1 the identity element.

Exercise 2.1.4. Show that set of symmetries $\mathcal{R} = \{e, r, r^2, r^3\}$ of the square card is a group under composition of symmetries.

Exercise 2.1.5. . Consider the group $C_4 = \{i, -1, -i, 1\}$ of fourth roots of unity in the complex numbers and the group $\mathcal{R} = \{e, r, r^2, r^3\}$ contained in the group of rotations of the square card. Show that the bijection

$$\begin{aligned} e &\leftrightarrow 1 \\ r &\leftrightarrow i \\ r^2 &\leftrightarrow -1 \\ r^3 &\leftrightarrow -i. \end{aligned}$$

produces a matching of the multiplication tables of the two groups. That is if one applies the bijection to each entry of the multiplication table of H, one produces the multiplication table of $\mathcal{R}$. Thus, the two groups are isomorphic.

Exercise 2.1.6. Show that the groups C_4 and the group of rotational symmetries of the rectangle are not isomorphic, although each group has four elements. *Hint:* In one of the groups, but not the other, each element has square equal to the identity. Show that if two groups G and H are isomorphic, and G has the property that each element has square equal to the identity, then H also has this property.

Exercise 2.1.7. Suppose that $\varphi : G \to H$ is an isomorphism of groups. Show that for all $g \in G$ and $n \in \mathbb{N}$, $\varphi(g^n) = (\varphi(g))^n$. Show that if $g^n = e$ then also $(\varphi(g))^n = e$.

The following several exercises investigate groups with a small number of elements by means of their multiplication tables. The requirements $ea = a$ and $ae = a$ for all a determine one row and one column of the multiplication table. The other constraint on the multiplication table which we know is that each row and each column must contain every group element exactly once. When the size of the group is small, these constraints suffice to determine the possible tables.

Exercise 2.1.8. Show that there is up to isomorphism only one group of order 2. *Hint:* Call the elements $\{e, a\}$. Show that there is only one possible multiplication table. Since the row and the column labeled by e are known, there is only one entry of the table which is not known. But that entry is determined by the requirement that each row and column contain each group element.

Exercise 2.1.9. Show that there is up to isomorphism only one group of order 3. *Hint:* Call the elements $\{e, a, b\}$. Show that there is only one possible multiplication table. Since the row and column labeled by e are known, there are four table entries left to determine. Show that there is only one way to fill in these entries which is consistent with the requirement that each row and column contain each group element exactly once.

Exercise 2.1.10. Show that a group with 4 elements must have a non-identity element whose square is the identity. That is some non-identity element must be its own inverse. *Hint:* You must start by assuming that you have a group G with 4 elements. *You cannot assume anything else about G except that it is a group and that it has four elements.* You must show that one of the three non-identity elements has square equal to e. Call the elements of the group $\{e, a, b, c\}$. There are two possibilities: each non-identity element has square equal to e, in which case there is nothing more to show, or some element does not have square equal to e. Suppose that $a^2 \neq e$. Thus, $a \neq a^{-1}$. Without loss of generality, write $b = a^{-1}$. Then also $a = b^{-1}$. Then what is the inverse of c?

Exercise 2.1.11. Show that there are only two distinct groups with four elements, as follows. Call the elements of the group e, a, b, c. Let a denote a non-identity element whose square is the identity. The row and column labeled by e are known. Show that the row labeled by a is determined by the requirement that each group element must appear exactly once in each row and column; similarly, the column labeled by a is determined. There are now four table entries left to determine. Show that there are exactly two possible ways to complete the multiplication table which are consistent with the constraints on multiplication tables. Show that these two ways of completing the table yield the multiplication tables of the two groups with four elements which we have already encountered.

Exercise 2.1.12. Generalizing Exercise 2.1.10, show that any group with an even number of elements must have a non-identity element whose square is the identity, that is, a non-identity element which is its own inverse. *Hint:* The non-identity elements which are not their own inverse match up in pairs $\{a, b\}$, where $a^{-1} = b$ and $b^{-1} = a$. Observe that some element cannot be part of such a pair, so it must be its own inverse.

Exercise 2.1.13. It's a fact that there is exactly one group of order 5 up to isomorphism. Find one group of order 5. *Hint:* Roots of unity in the complex numbers.

It is possible to show the uniqueness of the group of order 5 with the techniques which have at hand. Try it. *Hint:* Suppose G is a group of order 5. First show it is not possible for a non-identity element $a \in G$ to satisfy $a^2 = e$, because there is no way to complete the multiplication table which respects the constraints on group multiplication tables.

Next, show there is no non-identity element a such that e, a, a^2 are distinct elements but $a^3 = e$. Finally show that there is no non-identity element a such that e, a, a^2, a^3 are distinct elements but $a^4 = e$. Consequently for any non-identity element a, the elements e, a, a^2, a^3, a^4 are distinct, and necessarily $a^5 = e$.

Later it will be possible to obtain the uniqueness of the groups of order 2, 3, and 5 as an immediate corollary of a general result.

Definition 2.1.15. A group G is called *abelian* (or *commutative*) if for all elements $a, b \in G$, the products in the two orders are equal: $ab = ba$.

Exercise 2.1.14. The following conditions are equivalent for a group G:

(a) G is abelian.
(b) For all $a, b \in G$, $(ab)^{-1} = a^{-1}b^{-1}$.
(c) For all $a, b \in G$, $aba^{-1}b^{-1} = e$.
(d) For all $a, b \in G$, $(ab)^2 = a^2b^2$.
(e) For all $a, b \in G$ and natural numbers n, $(ab)^n = a^nb^n$. (Use induction).

Exercise 2.1.15. Show that all groups of order no more than 5 are abelian. Show that the group of symmetries of the triangular plate is non-abelian. (See Exercise 1.4.3.)

2.2. Permutation Groups

I put three identical objects in front of you on the table:

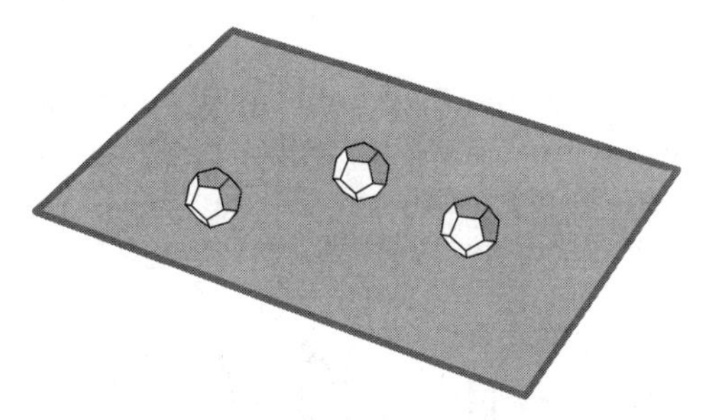

This configuration has symmetry, regardless of the nature of the objects or their relative position, just because the objects are identical. If you glance away, I could switch the objects around, and when you look back you could not tell whether I had moved them. This is a sort of discovery: *symmetry is not intrinsically a geometric concept.*

What are all the symmetries of the configuration of three objects? Any two objects can be switched while the third is left in place; there are three such symmetries. One object can be put in the place of a second, the second in the place of the third, and the third in the place of the first; There are two possibilities for such a rearrangement (corresponding to the two ways to traverse the vertices of a triangle). And there is the non-rearrangement, in which all the objects are left in place. So there are six symmetries in all.

The symmetries of a configuration of identical objects are called *permutations*.

What is the multiplication table for the group of six permutations of three objects? Before we can work this out, we have to devise some sort of bookkeeping system. Let's number not the objects but the three positions which they occupy. Then we can describe each symmetry by recording for each i, $1 \leq i \leq 3$, the final position of the object which starts in position i. For example the permutation which switches the objects in positions 1 and 3 and leaves the object in position 2 in place will be described by

$$\begin{pmatrix} 1 & 2 & 3 \\ 3 & 2 & 1 \end{pmatrix}.$$

The permutation which moves the object in position 1 to position 2, that in position 2 to position 3 and that in position 3 to position 1 is denoted by

$$\begin{pmatrix} 1 & 2 & 3 \\ 2 & 3 & 1 \end{pmatrix}.$$

With this notation, the six permutations of three objects are

$$\begin{pmatrix} 1 & 2 & 3 \\ 1 & 2 & 3 \end{pmatrix} \qquad \begin{pmatrix} 1 & 2 & 3 \\ 2 & 3 & 1 \end{pmatrix} \qquad \begin{pmatrix} 1 & 2 & 3 \\ 3 & 1 & 2 \end{pmatrix}$$
$$\begin{pmatrix} 1 & 2 & 3 \\ 2 & 1 & 3 \end{pmatrix} \qquad \begin{pmatrix} 1 & 2 & 3 \\ 1 & 3 & 2 \end{pmatrix} \qquad \begin{pmatrix} 1 & 2 & 3 \\ 3 & 2 & 1 \end{pmatrix}.$$

The product of permutations is computed by following each object as it is moved by the two permutations. If the first permutation moves an object from position i to position j and the second moves an object from position j to position k, then the composition moves an object from i to k. For example:

$$\begin{pmatrix} 1 & 2 & 3 \\ 2 & 3 & 1 \end{pmatrix}\begin{pmatrix} 1 & 2 & 3 \\ 1 & 3 & 2 \end{pmatrix} = \begin{pmatrix} 1 & 2 & 3 \\ 2 & 1 & 3 \end{pmatrix}$$

Recall our convention that the element on the right in the product is the first permutation and that on the left is the second.

Our notation for permutations suggests a slightly different point of view: a permutation of three objects is just a bijective function on the set $\{1, 2, 3\}$; the permutation moves an object from position i to position j if the function maps i to j. It should be clear, upon reflection, that the composition of permutations is the same as the composition of bijective maps.

In general the permutation group of a set of n identical objects is the same as the group of all bijective maps on the set $\{1, 2, \dots, n\}$. We generally write S_n for this group rather than $\mathrm{Sym}(\{1, 2, \dots, n\})$. It is not difficult to see that the size of this group is $n! = n(n-1)\cdots(2)(1)$. For there are n possible images for 1; for each of these possibilities, there are $n-1$ possible images for 2, and so forth. For n of moderate size, these groups are quite enormous.

There is an alternative notation for permutations which is convenient for many purposes. I explain it by example. Consider the permutation

$$\pi = \begin{pmatrix} 1 & 2 & 3 & 4 & 5 & 6 & 7 \\ 4 & 3 & 1 & 2 & 6 & 5 & 7 \end{pmatrix}$$

in S_7. This permutation takes 1 to 4, 4 to 2, 2 to 3, and 3 back to 1; it takes 5 to 6 and 6 back to 5; and it fixes (doesn't move) 7. Correspondingly we write $\pi = (1423)(56)$.

A permutation such as (1423) which permutes several numbers cyclically (1 to 4, 4 to 2, 2 to 3, and 3 to 1) and leaves all other numbers fixed is called a *cycle*. Two cycles are called *disjoint* if each leaves fixed the numbers moved by the other. The expression $\pi = (1423)(56)$ for π as a product of disjoint cycles is called *cycle notation*.

Here is (the outline of) an algorithm for writing a permutation $\pi \in S_n$ in cycle notation. Let a_1 be the first number $(1 \le a_1 \le n)$ which is not fixed by π. Write

$$\begin{aligned} a_2 &= \pi(a_1) \\ a_3 &= \pi(a_2) = \pi(\pi(a_1)) \\ a_4 &= \pi(a_3) = \pi(\pi(\pi(a_1))), \end{aligned}$$

and so forth. The numbers

$$a_1, a_2, \dots$$

cannot be all distinct since each is in $\{1, 2, \dots, n\}$. It follows that there is a number k such that $a_1, a_2, \dots, a_k$ are all distinct, and $\pi(a_k) = a_1$. (Exercise 2.2.8.) The permutation π permutes the numbers $\{a_1, a_2, \dots, a_k\}$ among themselves, and the remaining numbers

$$\{1, 2, \dots, n\} \setminus \{a_1, a_2, \dots, a_k\}$$

among themselves, and the restriction of π to $\{a_1, a_2, \dots, a_k\}$ is the cycle $(a_1, a_2, \dots, a_k)$. (Exercise 2.2.7.) If π fixes all numbers in $\{1, 2, \dots, n\} \setminus \{a_1, a_2, \dots, a_k\}$, then

$$\pi = (a_1, a_2, \dots, a_k).$$

Otherwise, consider the first number $b_1 \notin \{a_1, a_2, \dots, a_k\}$ which is not fixed by π. Write

$$\begin{aligned} b_2 &= \pi(b_1) \\ b_3 &= \pi(b_2) = \pi(\pi(b_1)) \\ b_4 &= \pi(b_3) = \pi(\pi(\pi(b_1))), \end{aligned}$$

and so forth; as before, there is an integer l such that $b_1, \dots, b_l$ are all distinct and $\pi(b_l) = b_1$. Now π permutes the numbers

$$\{a_1, a_2, \dots, a_k\} \cup \{b_1, \dots, b_l\}$$

among themselves, and the remaining numbers

$$\{1, 2, \dots, n\} \setminus (\{a_1, a_2, \dots, a_k\} \cup \{b_1, \dots, b_l\})$$

among themselves; furthermore the restriction of π to

$$\{a_1, a_2, \dots, a_k\} \cup \{b_1, \dots, b_l\}$$

is the product of disjoint cycles

$$(a_1, a_2, \dots, a_k)(b_1, \dots, b_l).$$

Continue in this way until π has been written as a product of disjoint cycles.

Let me show you how to express the idea of the algorithm a little more formally and also more concisely, using mathematical induction. In the explanation above, the phrase "continue in this way" is a signal that to formalize the argument it is necessary to use induction.

Because disjoint cycles π_1 and π_2 commute ($\pi_1\pi_2 = \pi_2\pi_1$, exercise), uniqueness in the following statement means uniqueness up to order; the factors are unique, and the order in which the factors are written is irrelevant. Also note that $(a_1, a_2, \dots, a_k)$ is the same cyclic permutation as $(a_2, \dots, a_k, a_1)$, and there is no preferred first entry in the cycle notation. Finally, in order not to have to make an exception for the identity element e, we regard e as the product of the empty collection of cycles.

Theorem 2.2.1. *Every permutation can be written uniquely as a product of disjoint cycles.*

Proof. We prove by induction on the cardinality of a finite set X that every permutation in $\mathrm{Sym}(X)$ can be written uniquely as a product of disjoint cycles. If $\#X = 1$, there is nothing to do, since the only permutation of X is the identity e. Suppose, therefore, that the result holds for all finite sets of cardinality less than $\#X$. Let π be a non-identity permutation of X. Choose $x_0 \in X$ such that $\pi(x_0) \neq x_0$. Denote $x_1 = \pi(x_0)$, $x_2 = \pi(x_1)$, and so forth. Since $\#X$ is finite, there is a number k such that $x_0, x_1, \dots, x_k$ are all distinct and $\pi(x_k) = x_0$. See Exercise 2.2.8. The sets $X_1 = \{x_0, x_1, \dots, x_k\}$ and $X_2 = X \setminus X_1$ are each invariant under π; that is, $\pi(X_i) = X_i$ for $i = 1, 2$, and therefore π is the product of $\pi_1 = \pi_{|X_1}$ and $\pi_2 = \pi_{|X_2}$. See Exercise 2.2.7. But π_1 is the cycle $(x_0, x_1, \dots, x_k)$, and by the induction hypothesis π_2 is a product of disjoint cycles. Hence π is also a product of disjoint cycles.

The uniqueness statement follows from a variation on the same argument: The cycle containing x_0 is uniquely determined by the list $x_0, x_1, \dots$. Any expression of π as a product of disjoint cycles must contain this cycle. The product of the remaining cycles in the expression yields π_2; but by the induction hypothesis, the decomposition of π_2 as a product of disjoint cycles is unique. Hence, the cycle decomposition of π is unique. □

Exercises

Exercise 2.2.1. Work out the full multiplication table for the group of permutations of three objects.

Exercise 2.2.2. Compare the multiplication table of S_3 with that for the group of symmetries of an equilateral triangular card. (See Figure 1.4.4 and compare Exercise 1.4.3.) Make the following identifications:

$$\begin{pmatrix} 1 & 2 & 3 \\ 2 & 3 & 1 \end{pmatrix} \longleftrightarrow r$$

$$\begin{pmatrix} 1 & 2 & 3 \\ 3 & 1 & 2 \end{pmatrix} \longleftrightarrow r^{-1}$$

$$\begin{pmatrix} 1 & 2 & 3 \\ 1 & 3 & 2 \end{pmatrix} \longleftrightarrow a$$

$$\begin{pmatrix} 1 & 2 & 3 \\ 3 & 2 & 1 \end{pmatrix} \longleftrightarrow b$$

$$\begin{pmatrix} 1 & 2 & 3 \\ 2 & 1 & 3 \end{pmatrix} \longleftrightarrow c$$

Show that the multiplication tables are then identical.

Exercise 2.2.3. Work out the decomposition in disjoint cycles for:

(a) $\begin{pmatrix} 1 & 2 & 3 & 4 & 5 & 6 & 7 \\ 2 & 5 & 6 & 3 & 7 & 4 & 1 \end{pmatrix}$

(b) $(12)(1234)$

(c) $(12)(234)$

(d) $(12)(23)(34)$
(e) $(13)(1234)(13)$
(f) $(12)(13)(14)$

Exercise 2.2.4.

(a) Explain how to compute the inverse of a permutation which is given in two-line notation. Compute the inverse of

$$\begin{pmatrix} 1 & 2 & 3 & 4 & 5 & 6 & 7 \\ 2 & 5 & 6 & 3 & 7 & 4 & 1 \end{pmatrix}.$$

(b) Explain how to compute the inverse of a permutation which is given as a product of cycles (disjoint or not). One trick of problem solving is to simplify the problem by considering special cases. First you should consider the case of a single cycle, and it will probably be helpful to begin with a *short* cycle. A 2-cycle is its own inverse, so the first interesting case is that of a 3-cycle. Once you have figure out the inverse for a 3-cycle and a 4-cycle, you will probably be able to guess the general pattern. Now you can begin work on a product of several cycles.

Exercise 2.2.5. On the basis of your computations in Problem 2.2.3, make some conjectures about patterns for certain products of two-cycles, and for certain products of two-cycles and other cycles.

Exercise 2.2.6. Show that any k-cycle $(a_1, \dots, a_k)$ can be written as a product of $(k-1)$ 2-cycles. Conclude that any permutation can be written as a product of some number of 2-cycles. *Hint:* For the first part, look at your computations in problem 2.2.3 to discover the right pattern. Then do a proper proof by induction.

The following two exercises supply important details for the proof of the existence and uniqueness of the disjoint cycle decomposition for a permutation of a finite set:

Exercise 2.2.7. Suppose X is the union of disjoint sets X_1 and X_2, $X = X_1 \cup X_2$ and $X_1 \cap X_2 = \emptyset$. Suppose X_1 and X_2 are invariant for a permutation $\pi \in \text{Sym}(X)$. Write π_i for the permutation $\pi_{|X_i} \in \text{Sym}(X_i)$ for $i = 1, 2$, and (noticing the abuse of notation) also write π_i for the permutation of X which is π_i on X_i and the identity on $X \setminus X_i$. Show that $\pi = \pi_1\pi_2 = \pi_2\pi_1$.

Exercise 2.2.8.

(a) Let π be a non-identity permutation in $\text{Sym}(X)$, where X is a finite set. Let x_0 be some element of X which is not fixed by π. Denote $x_1 = \pi(x_0)$, $x_2 = \pi(x_1)$, and so forth. Show that there is a number k such that $x_0, x_1, \dots, x_k$ are all distinct and $\pi(x_k) = x_0$. *Hint:* Let

k be the least integer such that $\pi(x_k) = x_{k+1} \in \{x_0, x_1, \dots, x_k\}$. Show that $\pi(x_k) = x_0$. To do this, show that the assumption $\pi(x_k) = x_l$ for some l, $1 \le l \le k$ leads to a contradiction.

(b) Show that $X_1 = \{x_0, x_1, \dots, x_k\}$ and $X_2 = X \setminus X_1$ are both invariant under π.

Exercise 2.2.9. Abelian groups are something of a rarity among symmetry groups. Show that S_n is non-abelian for all $n \ge 3$. *Hint:* Find a pair of 2-cycles which do not commute.

2.3. Divisibility in the Integers

This section is about the integers, the group which is probably most familiar to you. The set of integers $\mathbb{Z}$ is a group under addition, with 0 serving as the identity element and $-a$ the inverse of any element a. Of course, the integers also have a second operation, multiplication, but as you learned in elementary school, multiplication in the integers can be interpreted in terms of repeated addition: For $a \in \mathbb{Z}$ and $n \in \mathbb{Z}$, $n > 0$, $na = a + \cdots + a$ (n times), and $(-n)a = n(-a)$. Finally, $0a = 0$. Thus both operations in $\mathbb{Z}$ can be interpreted in terms of the additive group structure, and we needn't make any great fuss about the second operation for now.

The main theme of this section is the elementary theory of *divisibility* in the integers. The fundamental fact is the following familiar result (division with remainder).

Proposition 2.3.1. *For natural numbers a and d, there exist integers q and r such $a = qd + r$ and $0 \le r < d$.*

Proof. The set $S = \{k \in \mathbb{N} : kd > a\}$ is non-empty, since $d\mathbb{N}$ is infinite, while $\{1, 2, \dots, a\}$ is finite. By the well ordering principle, S has a least element $q + 1$. See Appendix C. Then $q \ge 0$ and $qd \le a$, since $q \notin S$. Set $r = a - qd \ge 0$. Since $(q + 1)d > a$, it follows that $d > a - qd = r$. □

For integers a and d, with $d \ne 0$, we say that *d divides a* if $a/d \in \mathbb{Z}$.

Proposition 2.3.2. *For integers n and m, let*

$$I(m, n) = \{am + bn : a, b \in \mathbb{Z}\}.$$

(a) *For $x, y \in I(m, n)$, $x + y \in I(m, n)$ and $-x \in I(m, n)$*

(b) *For all $x \in \mathbb{Z}$, $xI(m, n) \subseteq I(m, n)$.*

(c) *If $\beta \in \mathbb{N}$ divides m and n, then β divides all elements of $I(m, n)$.*

Proof. Exercise 2.3.2. □

Definition 2.3.3. A natural number α is a *greatest common divisor* of non-zero integers m and n if

1. α divides m and n and
2. whenever $\beta \in \mathbb{N}$ divides m and n, then β also divides α.

Notice that if a greatest common divisor exists, it is unique. We denote the greatest common divisor by g.c.d.(m, n).

Next, we show that the greatest common divisor of two natural numbers m and n is an element of $I(m, n)$, and can be found by repeated use of division with remainder (Proposition 2.3.1).

Suppose without loss of generality that $m \geq n$. Define sequences $n = n_0 > n_1 > n_2 \cdots \geq 0$ and $q_1, q_2, \ldots$ by induction: Define n_1 and q_1 by:

$$m = q_1 n + n_1 \quad \text{and} \quad 0 \leq n_1 < n_0$$

If $n_1, \ldots, n_{k-1}$ and $q_1, \ldots, q_{k-1}$ have been defined and $n_{k-1} > 0$, then define n_k and q_k by

$$n_{k-2} = q_k n_{k-1} + n_k \quad \text{and} \quad 0 \leq n_k < n_{k-1}.$$

This process must stop after no more than n steps with some remainder $n_{r+1} = 0$. Then we have the following system of relations:

$$\begin{aligned}
m &= q_1 n + n_1 \\
n &= q_2 n_1 + n_2 \\
&\ldots \\
n_{k-2} &= q_k n_{k-1} + n_k \\
&\ldots \\
n_{r-1} &= q_{r+1} n_r.
\end{aligned}$$

Proposition 2.3.4. *The natural number n_r is the greatest common divisor of m and n.*

Proof. Exercise 2.3.4. □

Definition 2.3.5. Non-zero integers m and n are *relatively prime* if g.c.d.$(m, n) = 1$.

Proposition 2.3.6. *Two integers m and n are relatively prime if, and only if,* $1 \in I(m, n)$.

Proof. Exercise 2.3.8. □

Proposition 2.3.7. *Let p be a prime number, and a and b non-zero integers. If p divides ab, then p divides a or p divides b.*

Proof. If p does not divide a, then a and p are relatively prime, so $1 = \alpha a + \beta p$ for some integers α and β. Multiplying by b gives $b = \alpha ab + \beta pb$, which shows that b is divisible by p. □

Corollary 2.3.8. *Suppose that a prime number p divides a product $a_1 a_2 \dots a_r$ of non-zero integers. Then p divides one of the factors.*

Proof. Exercise 2.3.9. □

Theorem 2.3.9. *The prime factorization of a natural number is unique.*

Proof. We have to show that for all natural numbers n, if n has factorizations:

$$n = q_1 q_2 \dots q_r,$$
$$n = p_1 p_2 \dots p_s,$$

where the q_i's and p_j's are prime and $q_1 \le q_2 \le \dots \le q_r$ and $p_1 \le p_2 \le \dots \le p_s$, then $r = s$ and $q_i = p_i$ for all i. We do this by induction on n. First check the case $n = 1$; 1 cannot be written as the product of any non-empty collection of prime numbers. So consider a natural number $n \ge 2$ and assume inductively that the assertion of unique factorization holds for all natural numbers less than n. Consider two factorizations of n as above, and assume without loss of generality that $q_1 \le p_1$. Since q_1 divides $n = p_1 p_2 \dots p_s$, it follows from Exercise 2.3.8 that q_1 divides, and hence is equal to one of the p_i. Since also $q_1 \le p_1 \le p_k$ for all k, it follows that $p_1 = q_1$. Now dividing both sides by q_1, we get:

$$n/q_1 = q_2 \dots q_r,$$
$$n/q_1 = p_2 \dots p_s,$$

(Note that n/q_1 could be 1 and one or both of $r-1$ and $s-1$ could be 0.) Since $n/q_1 < n_1$, it follows from the induction hypothesis that $r = s$ and $q_i = p_i$ for all $i \geq 2$. □

Exercises

Exercise 2.3.1. Show that for integers a and d, with $d \neq 0$, there exist integers q and r such $a = qd + r$ and $|r| < |d|$. Show that it is even possible to find q and r such $a = qd + r$ and $0 \leq r < |d|$.

Exercise 2.3.2. Prove Proposition 2.3.2.

Exercise 2.3.3. Try the algorithm for computing the greatest common divisor of integers m and n on an example, say $m = 60$ and $n = 8$.

Exercise 2.3.4.

(a) Verify that in the algorithm preceding Proposition 2.3.4, all the numbers n_k are elements of $I(m, n)$. In particular $n_r \in I(m, n)$.

(b) Verify that all the n_k are multiples of n_r, and in particular m and n are multiples of n_r.

(c) Conclude from parts (a) and (b) that n_r is the greatest common divisor of m and n.

Exercise 2.3.5. Show that for non-zero integers m and n, $\text{g.c.d.}(m, n) = \text{g.c.d.}(|m|, |n|)$.

Exercise 2.3.6. Show that for non-zero integers m and n, $\text{g.c.d.}(m, n)$ is the largest natural number dividing m and n.

Exercise 2.3.7. Show that for non-zero integers m and n, $\text{g.c.d.}(m, n)$ is the smallest element of $I(m, n) \cap \mathbb{N}$. (The existence of a smallest element of $I(m, n) \cap \mathbb{N}$ follows from the well ordering principle for the natural numbers, see Appendix C.)

Exercise 2.3.8. Show that two integers m and n are relatively prime if, and only if, $1 \in I(m, n)$.

Exercise 2.3.9. Show that if a prime number p divides a product $a_1 a_2 \dots a_r$ of non-zero integers, then p divides one of the factors.

Exercise 2.3.10.

(a) Write a program in your favorite programming language to compute the greatest common divisor of two non-zero integers, using the approach of repeated division with remainders.

(b) Another method of finding the greatest common divisor would be to compute the prime factorizations of the the two integers and then to take the largest collection of prime factors common to the two factorizations. How do the two methods compare in computational efficiency?

Exercise 2.3.11. Explore the idea of the greatest common divisor of *several* integers, $n_1, n_2, \dots, n_k$.

(a) Make a reasonable definition of $\text{g.c.d}(n_1, n_2, \dots, n_k)$.

(b) Let $I = I(n_1, n_2, \dots, n_k) =$

$$\{m_1 n_1 + m_2 n_2 + \dots m_k n_k : m_1, \dots, m_k \in \mathbb{Z}\}.$$

Show that if $x, y \in I$, then $x + y \in I$ and $-x \in I$. Show that if $x \in \mathbb{Z}$ and $a \in I$, then $xa \in I$.

(c) Show that $\text{g.c.d}(n_1, n_2, \dots, n_k)$ is the smallest element of $I \cap \mathbb{N}$.

(d) Develop an algorithm to compute $\text{g.c.d}(n_1, n_2, \dots, n_k)$.

(e) Develop a computer program to compute the greatest common divisor of any finite collection of non-zero integers.

2.4. Subgroups and Cyclic Groups

Definition 2.4.1. A non-empty subset H of a group G is called a *subgroup* if H is itself a group *with the group operation inherited from G.*

For H to be a subgroup of G, it is necessary that

1. For all elements h_1 and h_2 of H, the product $h_1 h_2$ is also an element of H.
2. For all $h \in H$, the inverse h^{-1} is an element of H.

It is not necessary to check more than this. Associativity of the product is inherited from G, so it need not be checked. Also, if conditions (1) and (2) are satisfied, then e is automatically in H; indeed, H is non-empty, so contains some element h; according to (2), $h^{-1} \in H$ as well, and then according to (1), $e = hh^{-1} \in H$.

These observations are a great labor saving device. Very often when one needs to check that some set H with a binary operation is a group, H is already contained in some known group, so one needs only check points (1) and (2).

One says that a subset H of a group G is *closed under multiplication* if condition (1) is satisfied. One says that H is *closed under inverses* if condition (2) is satisfied.

Example 2.4.2. An n-by-n matrix A is said to be *orthogonal* if $A^t A = E$. Show that the set $\mathrm{O}(n, \mathbb{R})$ of n-by-n real valued orthogonal matrices is a group.

Proof. If $A \in \mathrm{O}(n, \mathbb{R})$, then A has a left inverse A^t, so A is invertible with inverse A^t. Thus $\mathrm{O}(n, \mathbb{R}) \subseteq \mathrm{GL}(n, \mathbb{R})$. Therefore it suffices to check that the product of orthogonal matrices is orthogonal and that the inverse of an orthogonal matrix is orthogonal. But if A and B are orthogonal, then $(AB)^t = B^t A^t = B^{-1} A^{-1} = (AB)^{-1}$; hence AB is orthogonal. If $A \in \mathrm{O}(n, \mathbb{R})$, then $(A^{-1})^t = (A^t)^t = A = (A^{-1})^{-1}$, so $A^{-1} \in \mathrm{O}(n, \mathbb{R})$. □

Here are some additional examples of subgroups:

Example 2.4.3. In any group G, G itself and $\{e\}$ are subgroups.

Example 2.4.4. The set of all complex numbers of modulus (absolute value) equal to 1 is a subgroup of the group of all non-zero complex numbers, with multiplication as the group operation. See Appendix D.

Proof. For any non-zero complex numbers a and b, $|ab| = |a||b|$, and $|a^{-1}| = |a|^{-1}$. It follows that the set of complex number of modulus 1 is closed under multiplication and under inverses. □

Example 2.4.5. In the group of symmetries of the square, the subset $\{e, r, r^2, r^3\}$ is a subgroup. Also the subset $\{e, r^2, a, b\}$ is a subgroup; the latter subgroup is isomorphic to the symmetry group of the rectangle, since each non-identity element has square equal to the identity, and the product of any two non-identity elements is the third.

Example 2.4.6. In the permutation group S_4, the set of permutations π satisfying $\pi(4) = 4$ is a subgroup. This subgroup, since it permutes the numbers $\{1, 2, 3\}$, and leaves 4 fixed, is isomorphic to S_3.

Example 2.4.7. In S_4, there are eight 3-cycles. There are three elements which are products of disjoint 2-cycles, namely $(12)(34)$, $(13)(24)$, and $(14)(23)$. These eleven elements, together with the identity, form a subgroup of S_4.

Proof. At the moment we have no theory to explain this fact, so we have to verify by computation that the set is closed under multiplication. The amount of computation required can be reduced substantially by observing some patterns in products of cycles, as in Exercise 2.4.1.

The set is clearly closed under inverses.

Eventually we will have a theory which will make this result transparent. □

I now discuss a certain type of subgroup which appears in all groups. Take any group G and any element $a \in G$. Consider all powers of a: Define $a^0 = e$, $a^1 = a$, and for $k > 1$, define a^k to be the product of k factors of a. (A little more properly, a^k is defined inductively by declaring $a^k = aa^{k-1}$.) For $k > 1$ define $a^{-k} = (a^{-1})^k$.

We now have a^k defined for all integers k, and it is a fact that $a^k a^l = a^{k+l}$ for all integers k and l. It's not hard to convince yourself that this must be true by looking at a few examples for k and l of the same sign and of the opposite sign; a proper proof has to be done byinduction. Likewise, one can show that for all integers k, $(a^k)^{-1} = a^{-k}$. It is also useful to observe, and to prove by induction, that $a^{kl} = (a^k)^l$ for all integers k and l. You are asked in Exercise 2.4.5 to verify these facts by inductive proofs.

Definition 2.4.8. Let a be an element of a group G. The set of powers of a $\langle a \rangle = \{a^k : k \in \mathbb{Z}\}$ is a subgroup of G, called the *cyclic subgroup generated by a*. If there is an element $a \in G$ such that $\langle a \rangle = G$, one says that G is a *cyclic group*. We say that a is a *generator* of the cyclic group.

Example 2.4.9. Take $G = \mathbb{Z}$, with addition as the group operation, and take any element $d \in \mathbb{Z}$. Because the group operation is addition, the set of powers of d with respect to this operation is the set of multiples of d in the ordinary sense. For example, the third power of d is $d + d + d = 3d$. Thus, we obtain Example 2.4.2.

Example 2.4.10. As a second example, the set of all powers of r in the symmetries of the square is $\{e, r, r^2, r^3\}$.

There are two possibilities for $\langle a \rangle$, as we are reminded by the two examples. One possibility is that all the powers a^k are distinct, in which case, of course, the subgroup $\langle a \rangle$ is infinite; if this is so, we say that a has infinite order.

The other possibility is that two powers of a coincide. Suppose $k < l$ and $a^k = a^l$. Then $e = (a^k)^{-1}a^l = a^{l-k}$, so some positive power of a is the identity. Let n be the least positive integer such that $a^n = e$. Then $e, a, a^2, \dots, a^{n-1}$ are all distinct (Exercise 2.4.6.) and $a^n = e$. Now any integer k (positive or negative) can be written as $k = mn + r$, where the remainder r satisfies $0 \le r \le n-1$. Hence $a^k = a^{mn+r} = a^{mn}a^r = e^m a^r = ea^r = a^r$. Thus $\langle a \rangle = \{e, a, a^2, \ldots, a^{n-1}\}$.

Definition 2.4.11. The *order* of the cyclic subgroup generated by a is called *the order of a*. If the order of a is finite, then it is the least positive integer n such that $a^n = e$. We denote the order of a by $o(a)$.

Proposition 2.4.12. *Let G be a cyclic group of finite order n. For all $b \in G$, the order of b divides n.*

Proof. Let $b = a^\ell$. Then $b^n = (a^\ell)^n = (a^n)^\ell = e^\ell = e$. If k denotes the order of b, then $n = qk + r$, where $0 \le r < k$. But then $e = b^n = b^{qk}b^r = eb^r = b^r$. Since $r < k$, this is only possible if $r = 0$, by definition of the order of an element. Thus the order k of b divides n. □

As a model for a cyclic subgroup of order $n > 1$, we can take the set of nth roots of 1 in the complex numbers. The nth roots of 1 are the powers of $\xi = e^{2\pi i/n}$, which has order n. Refer to Appendix D for a discussion of the complex numbers.

Proposition 2.4.13. *Let a be an element of a group G.*
(a) *If a has infinite, order then $\langle a \rangle$ is isomorphic to $\mathbb{Z}$.*
(b) *If a has finite order n, then $\langle a \rangle$ is isomorphic to the group C_n of nth roots of 1.*

Proof. For part (a), define a map $\varphi : \mathbb{Z} \longrightarrow \langle a \rangle$ by $\varphi(k) = a^k$. This map is surjective by definition of $\langle a \rangle$, and it is injective because all powers of a are distinct. Furthermore $\varphi(k + l) = a^{k+l} = a^k a^l$. So φ is an isomorphism between $\mathbb{Z}$ and $\langle a \rangle$.

For part (b), since C_n has n elements $1, \xi, \xi^2, \ldots, \xi^{n-1}$ and $\langle a \rangle$ has n elements $1, a, a^2, \ldots, a^{n-1}$, we can define a bijection $\varphi : C_n \longrightarrow \langle a \rangle$ by $\varphi(\xi^k) = a^k$ for $0 \le k \le n-1$. The multiplication in C_n is given by $\xi^k \xi^l = \xi^r$, where r is the remainder after division of $k + l$ by n, and the multiplication in $\langle a \rangle$ is given by the analogous rule. So φ is an isomorphism. □

Proposition 2.4.14.
(a) *Any subgroup of $\mathbb{Z}$ is cyclic and isomorphic to $\mathbb{Z}$.*
(b) *Let $G = \langle a \rangle$ be a finite cyclic group. Any subgroup of G is also cyclic.*

Proof. Exercise 2.4.8. □

Proposition 2.4.15. *Let G be a group and let $H_1, H_2, \dots, H_n$ be subgroups of G. Then $H_1 \cap H_2 \cap \cdots \cap H_n$ is a subgroup of G. More generally, if $\{H_\alpha\}$ is any collection of subgroups, then $\bigcap_\alpha H_\alpha$ is a subgroup.*

Proof. Exercise 2.4.9. □

We have seen that any cyclic group of finite order n is isomorphic to the group C_n of roots of n-roots of unity in $\mathbb{C}$. Thus there is, up to isomorphism, exactly one cyclic group of order n. We are now going to construct a new model for the cyclic group of order n. For the remainder of this discussion, n is a fixed positive integer. For each integer k, let $[k] = \{k + jn : j \in \mathbb{Z}\}$; that is $[k]$ is the subset of $\mathbb{Z}$ consisting of all integers differing from k by some multiple of n.

Lemma 2.4.16. *For integers r and k the following are equivalent:*

(a) $r \in [k]$
(b) $k \in [r]$
(c) $[r] = [k]$
(d) $(k - r)$ *is divisible by* n

Proof. If $r \in [k]$ then there is a $j \in \mathbb{Z}$ such that $r = k + jn$. Hence $k = r - nj \in [r]$. Thus (a) implies (b), and similarly (b) implies (a). Furthermore $r \in [k]$ if, and only if, $r - k = jn$ for some $j \in \mathbb{Z}$, so (a) and (d) are equivalent. Finally, if $r = k + jn$, then for all integers ℓ, one has $r + \ell n = k + (j + \ell)n \in [k]$, so $[r] \subseteq [k]$; thus (a) implies that $[r] \subseteq [k]$ and similarly (b) implies $[k] \subseteq [r]$. Since (a) and (b) are equivalent, they each imply that $[r] = [k]$. On the other hand, if $[r] = [k]$, then $r \in [r] = [k]$, so (c) implies (a). □

How many different subsets $[k]$ are there? According to Proposition 2.3.1, for each $k \in \mathbb{Z}$, there is a unique $r \in \{0, 1, \dots, n-1\}$ such that $k - r$ is divisible by n. This means that $[0], [1], \dots, [n-1]$ are distinct and that for any $k \in \mathbb{Z}$, $[k]$ is equal to one of $[0], [1], \dots, [n-1]$. It also means that $\mathbb{Z}$ is the disjoint union of the sets $[0], [1], \dots, [n-1]$.

Let us write $a \equiv b \mod n$ if $a - b$ is divisible by n; one reads this as "a is congruent to b modulo n." Note that $a \equiv b \mod n$ if, and only if, $[a] = [b]$, by the previous lemma. The relation $a \equiv b \mod n$ has the following properties:

Lemma 2.4.17.

(a) *For all* $a \in \mathbb{Z}$, $a \equiv a \mod n$.

(b) *For all* $a, b \in \mathbb{Z}$, $a \equiv b \mod n$ *if, and only if,* $b \equiv a \mod n$.

(c) *For all* $a, b, c \in \mathbb{Z}$, *if* $a \equiv b \mod n$ *and* $b \equiv c \mod n$, *then* $a \equiv c \mod n$.

Proof. For (a), $a - a = 0$ is divisible by n. For (b), $a - b$ is divisible by n if, and only if, $b - a$ is divisible by n. Finally, if $a - b$ and $b - c$ are both divisible by n, then also $a - c = (a - b) + (b - c)$ is divisible by n. □

We let $\mathbb{Z}_n$ be the set whose elements are $[0], [1], \ldots, [n-1]$, i.e. all the distinct subsets of $\mathbb{Z}$ of the form $[k]$. We define an operation $+$ on $\mathbb{Z}_n$ as follows. For any subsets A and B of $\mathbb{Z}$ we define $A + B$ to be $\{a + b : a \in A \text{ and } b \in B\}$. Now since the elements of $\mathbb{Z}_n$ are subsets of $\mathbb{Z}$ we can define addition on $\mathbb{Z}_n$ precisely in the sense of addition of subsets. What remains to check is that the sum $[a] + [b]$ of two elements of $\mathbb{Z}_n$ is again an element of $\mathbb{Z}_n$. More precisely:

Lemma 2.4.18. $[a] + [b] = [a + b]$.

Proof. This is an equality of subsets of $\mathbb{Z}$. As is often the case, the most efficient way to prove equality of sets is to prove that each is a subset of the other, and this is what we do.

Take arbitrary elements $a + kn \in [a]$ and $b + \ell n \in [b]$. Then their sum $a + b + (k + \ell)n$ is in $[a + b]$. Thus $[a] + [b] \subseteq [a + b]$. Conversely, an arbitrary element $a + b + kn$ in $[a + b]$ can be written as $(a + kn) + b$, evidently an element of $[a] + [b]$. Thus $[a + b] \subseteq [a] + [b]$. □

Proposition 2.4.19. $\mathbb{Z}_n$ *is a cyclic group of order* n.

Proof. Everything follows from the previous lemma. In fact $[0]$ is an identity element because $[0] + [k] = [0 + k] = [k]$. Furthermore, $[-k] = [n - k]$ is the inverse of $[k]$ because $[-k] + [k] = [-k + k] = [0]$. Finally associativity holds because $([a] + [b]) + [c] = [a + b] + c = [a + b + c] = [a] + [b + c] = [a] + ([b] + [c])$. This proves that $\mathbb{Z}_n$ is a group. This group is cyclic with generator $[1]$ because for any $[k]$, $[1] + [1] + \cdots + [1]$ (k times) equals $[k]$. (Properly, one should show this by induction.) □

Exercises

Exercise 2.4.1.

(a) Show that $\{e, (12)(34), (13)(24), (14)(23)\}$ is a subgroup of S_4.

(b) Now examine products of two 3-cycles in S_4. Notice that the two three cycles have either all three digits in common, or they have two out of three digits in common. If they have three digits in common, they are either the same or inverses. If they have two digits in common, then they can be written as $(a_1a_2a_3)$ and $(a_1a_2a_4)$, or as $(a_1a_2a_3)$ and $(a_2a_1a_4)$. Show that in all cases the product is either the identity, another 3-cycle, or a product of two disjoint 2-cycles.

(c) Finally show that the product of a 3-cycle and an element of the form $(ab)(cd)$ is again a 3-cycle.

(d) Let H be the subset of S_4 consisting of all three-cycles, all products of disjoint two-cycles, and the identity. Show that H is a subgroup.

Exercise 2.4.2. In the group of integers $\mathbb{Z}$ (with addition), the set of multiples of any number d is a subgroup; that is, for any $d \in \mathbb{Z}$, $d\mathbb{Z} = \{nd : n \in \mathbb{Z}\}$ is a subgroup of $\mathbb{Z}$.

Exercise 2.4.3. Let H be a subgroup of $\mathbb{Z}$.

(a) Show that $H \cap \mathbb{N} \neq \emptyset$

(b) Let d be the smallest element of $H \cap \mathbb{N}$, which exists by the well ordering principle for the natural numbers, see Appendix C. Show that $d\mathbb{Z} = H$. *Hint:* First show that $d\mathbb{Z} \subseteq H$. To show the opposite containment, let $h \in H$, and consider division with remainder of h by d; that is, apply Proposition 2.3.2 to write $h = qd + r$, where $0 \leq r < d$. What can you conclude about r?

Exercise 2.4.4.

(a) Let $m, n \in \mathbb{Z}$, $m, n \neq 0$. Show that $I(m, n) = \{am + bn : a, b \in \mathbb{Z}\}$ is a subgroup of $\mathbb{Z}$, and is equal to $d\mathbb{Z}$, where $d = \text{g.c.d.}(m, n)$.

(b) Let $n_1, n_2, \dots, n_r \in \mathbb{Z}$, $n_i \neq 0$. Show that

$$I = \{m_1n_1 + \cdots + m_rn_r : m_i \in \mathbb{Z}\}$$

is a subgroup of $\mathbb{Z}$, and $I = d\mathbb{Z}$, where $d = \text{g.c.d.}(n_1, n_2, \dots, n_r)$.

Exercise 2.4.5. Prove by induction the following facts about powers of elements in a group: $a^ka^l = a^{k+l}$ and $(a^k)^{-1} = a^{-k}$, for all integers k and l.

Refer to Appendix C for a discussion of induction and multiple induction.

Exercise 2.4.6. Let a be an element of a group. Let n be the least positive integer such that $a^n = e$. Show that $e, a, a^2, \dots, a^{n-1}$ are all distinct. Conclude that the order of the subgroup generated by a is n.

Exercise 2.4.7. Consider the set of n-th roots of unity in the complex numbers, $C_n = \{e^{k2\pi i/n} : 0 \le k \le n-1\}$. Show that this set forms a group under multiplication of complex numbers. Show that the group is cyclic with generator $\zeta = e^{2\pi i/n}$.

Exercise 2.4.8. Prove Proposition 2.4.14. *Hint:* The first part is contained in Exercise 2.4.3. For the second part, use an idea similar to that of Exercise 2.4.3.

Exercise 2.4.9. Prove Proposition 2.4.15. Refer to Appendix B for a discussion of intersections of arbitrary collections of sets.

Exercise 2.4.10. List all the subgroups of S_3.

Exercise 2.4.11. List all the subgroups of the group of symmetries of the square card.

Exercise 2.4.12. Show that the order of a cycle in S_n is the length of the cycle. For example, the order of (1234) is 4. What is the order of a product of two disjoint cycles? Begin (of course!) by considering some examples. Note for instance that the product of a 2-cycle and a 3-cycle (disjoint) is 6,while the order of the product of 2 disjoint 2-cycles is 2.

Exercise 2.4.13. Show that any n-th root of 1 in $\mathbb{C}$ is a power of $e^{2\pi i/n}$.

Exercise 2.4.14.

(a) Let R_θ denote the rotation matrix

$$\begin{bmatrix} \cos\theta & -\sin\theta \\ \sin\theta & \cos\theta \end{bmatrix}.$$

Show that the set of R_θ, where θ varies through the real numbers, forms a group under matrix multiplication. In particular, $R_\theta R_\mu = R_{\theta+\mu}$, and $R_\theta^{-1} = R_{-\theta}$.

(b) Let J denote the matrix of reflection in the x-axis,

$$J = \begin{bmatrix} 1 & 0 \\ 0 & -1 \end{bmatrix}.$$

Show $JR_\theta = R_{-\theta}J$.

(c) Let J_θ be the matrix of reflection in the line containing the origin and the point $(\cos\theta, \sin\theta)$. Compute J_θ and show that

$$J_\theta = R_\theta J R_{-\theta} = R_{2\theta} J.$$

(d) Let $R = R_{\pi/2}$. Show that the eight matrices

$$\{R^k J^l : 0 \leq k \leq 3 \text{ and } 0 \leq l \leq 1\}$$

form a group, which can be identified with the group of symmetries of the square.

For any group G and any subset $S \subseteq G$ there is a smallest subgroup of G which contains S, which is called the *subgroup generated by* S. A "constructive" view of this subgroup is that it consists of all possible products $g_1 g_2 \cdots g_n$, where $g_i \in S$ or $g_i^{-1} \in S$. Another view of the subgroup generated by S, which is sometimes useful, is that it is the intersection of the family of all subgroups of G which contain S; this family is non-empty since G itself is such a subgroup. One says that *G is generated by S* if G is the only subgroup which contains S.

Exercise 2.4.15. The symmetric group S_n (for $n \geq 2$) is generated by the 2-cycles $(12), (23), \ldots, (n-1\ n)$.

Exercise 2.4.16. The symmetric group S_n (for $n \geq 2$) is generated by the 2-cycle (12) and the n-cycle $(12 \ldots n)$.

Exercise 2.4.17. The subgroup of $\mathbb{Z}$ generated by any finite set of non-zero integers $n_1, \ldots, n_k$ is $\mathbb{Z}d$, where d is the greatest common divisor of $\{n_1, \ldots, n_k\}$.

2.5. The Dihedral Groups

In this section, we will work out the symmetry groups of regular polygons and of the disk, which might be thought of as a "limit" of regular polygons as the number of sides increases. We regard these figures as thin plates, capable of rotations in three dimensions. Their symmetry groups are known collectively as the *dihedral groups*.

We have already found the symmetry group of the equilateral triangle (regular 3-gon) in Exercise 1.4.3 and of the square (regular 4-gon) in Sections 1.3 and 1.4. For now, it will be convenient to work first with the disk

$$\{\begin{bmatrix} x \\ y \\ 0 \end{bmatrix} : x^2 + y^2 \leq 1\},$$

whose symmetry group we denote by D.

Observe that the rotation r_t through any angle t around the z-axis is a symmetry of the disk. Such rotations satisfy $r_t r_s = r_{t+s}$ and in particular

$r_t r_{-t} = r_0 = e$, where e is the non-motion. It follows that $N = \{r_t : t \in \mathbb{R}\}$ is a subgroup of D.

For any line in the $x - y$ plane through the origin, the rotation by π about that line is a symmetry of the disk (which interchanges the top and bottom faces of the disk). Denote by j_t the rotation about the line ℓ_t which passes through the origin and the point

$$\begin{bmatrix} \cos(t) \\ \sin(t) \\ 0 \end{bmatrix},$$

and write $j = j_0$ for the rotation about the x axis. Each j_t generates a subgroup of D of order 2.

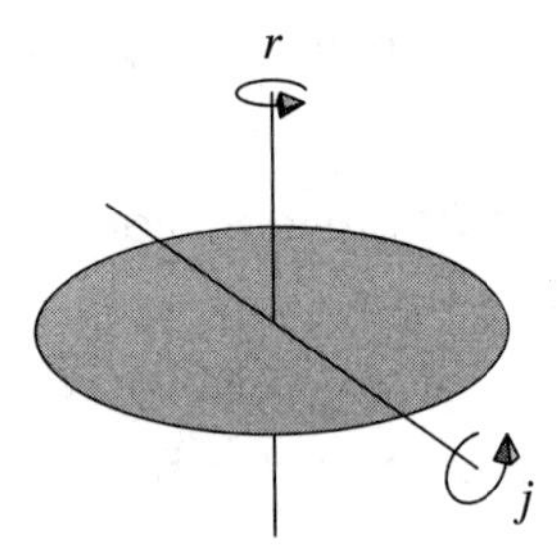

Figure 2.5.1. Symmetries of the Disk

Next, we observe that each j_t can be expressed in terms of j and the rotation r_t. To perform the rotation j_t about the line ℓ_t, one can rotate the disk until the line ℓ_t overlays the x-axis, then perform the rotation j about the x-axis, and finally rotate the disk so that ℓ_t is returned to its original position. Thus $j_t = r_t j r_{-t}$, or $j_t r_t = r_t j$. Therefore, we need only work out how to compute products involving j and the rotations r_t.

Note that j applied to a point

$$\begin{bmatrix} \rho\cos(s) \\ \rho\sin(s) \\ 0 \end{bmatrix}$$

in the disk is

$$\begin{bmatrix} \rho\cos(-s) \\ \rho\sin(-s) \\ 0 \end{bmatrix},$$

and r_t applied to

$$\begin{bmatrix} \rho\cos(s) \\ \rho\sin(s) \\ 0 \end{bmatrix}$$

is

$$\begin{bmatrix} \rho\cos(s+t) \\ \rho\sin(s+t) \\ 0 \end{bmatrix}.$$

In the exercises, you are asked to verify the following facts about the group D:

1. $jr_t = r_{-t}j$, and $j_t = r_{2t}j = jr_{-2t}$.
2. All products in D can be computed using these relations.
3. The symmetry group D of the disk consists of the rotations r_t for $t \in \mathbb{R}$ and the "flips" $j_t = r_{2t}j$. Writing $N = \{r_t : t \in \mathbb{R}\}$, we have $D = N \cup Nj$.
4. The subgroup N of D satisfies $aNa^{-1} = N$ for all $a \in D$.

Next, we turn to the symmetries of the regular polygons. Consider a regular n-gon with vertices at

$$\begin{bmatrix} \cos(k\pi/n) \\ \sin(k\pi/n) \\ 0 \end{bmatrix}$$

for $k = 0, 1, \ldots, n-1$. Denote the symmetry group of the n-gon by D_n.

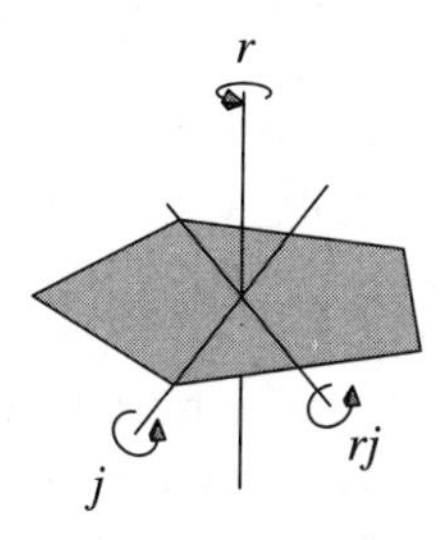

Figure 2.5.2. Symmetries of the Pentagon

In the exercises, you are asked to verify the following facts about the symmetries of the n-gon:

1. The rotation $r = r_{2\pi/n}$ through an angle of $2\pi/n$ about the z-axis generates a cyclic subgroup of D_n of order n.
2. The "flips" $j_{k\pi/n} = r_{k2\pi/n}\, j = r^k\, j$, for $k \in \mathbb{Z}$, are symmetries of the n-gon.

3. The distinct flip symmetries of the n-gon are $r^k\, j$ for $k = 0, 1, \ldots, n-1$.
4. If n is odd, then the axis of each of the "flips" passes through a vertex of the n-gon and the midpoint of the opposite edge. See Figure 2.5.2 for the case $n = 5$.
5. If n is even and k is even, then $j_{k\pi/n} = r^k\, j$ is a rotation about an axis passing through a pair of opposite vertices of the n-gon.
6. If n is even and k is odd, then $j_{k\pi/n} = r^k\, j$ is a rotation about an axis passing through the midpoints of a pair of opposite edges of the n-gon. See Figure 2.5.3 for the case $n = 6$.

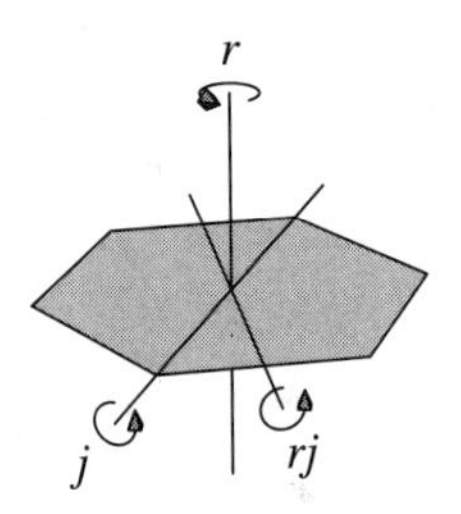

Figure 2.5.3. Symmetries of the Hexagon

The symmetry group D_n consists of the $2n$ symmetries r^k and $r^k\, j$, for $0 \le k \le n-1$. It follows from our computations for the symmetries of the disk that $jr = r^{-1} j$, so $jr^k = r^{-k} j$ for all k. This relation allows the computation of all products in D_n.

The group D_n can appear as the symmetry group of a geometric figure, or of a physical object, in a slightly different form. Think, for example, of a five-petalled flower, or a star-fish, which look quite different from the top and from the bottom. Or think of a pentagonal card with its two faces of different colors. The rotation j, which exchanges top and bottom, is not a symmetry of such an object. However, the *reflection* in the plane perpendicular to the faces of the pentagon and passing through one edge and the center of the opposite side is a symmetry of the figure. One can show that the group generated by the rotations and such reflections is isomorphic to D_n. See Exercise 2.5.6.

The plane as Figure 2.5.4 has D_9 symmetry. The figure was generated by several million iterations of a discrete dynamical system exhibiting "chaotic" behavior; the figure is shaded according to the probability of the moving "particle" entering a region of the diagram – the darker regions are visited more frequently. A beautiful book by M. Field and M. Golubitsky, *Symmetry in Chaos*, Oxford University Press, 1992, discusses symmetric

Figure 2.5.4. Object with D_9 Symmetry

Figure 2.5.5. Object with Z_5 Symmetry

figures arising from chaotic dynamical systems, and displays many beautiful figures produced by such systems. Figures 2.5.4 and 2.5.5 were produced using algorithms from the book of Field and Golubitsky.

Is it possible for a geometric figure to have Z_n symmetry but not D_n symmetry? Certainly. Figure 2.5.5 has Z_5 symmetry, but not D_5 symmetry.

Exercises

Exercise 2.5.1. Show that the elements j and r_t of the symmetry group D of the disk satisfy the relations $jr_t = r_{-t}j$, and $j_t = r_{2t}j = jr_{-2t}$.

Exercise 2.5.2. The symmetry group D of the disk consists of the rotations r_t for $t \in \mathbb{R}$ and the "flips" $j_t = r_{2t}j$.

(a) Writing $N = \{r_t : t \in \mathbb{R}\}$, we have $D = N \cup Nj$.
(b) All products in D can be computed using the relation $jr_t = r_{-t}j$.
(c) The subgroup N of D satisfies $aNa^{-1} = N$ for all $a \in D$.

Exercise 2.5.3. The symmetries of the disk are implemented by linear transformations of $\mathbb{R}^3$. Write the matrices of the symmetries r_t and j with respect to the standard basis of $\mathbb{R}^3$. Denote these matrices by R_t and J respectively. Confirm the relation $JR_t = R_{-t}J$.

Exercise 2.5.4. Consider the group D_n of symmetries of the n-gon.

(a) The rotation $r = r_{2\pi/n}$ through an angle of $2\pi/n$ about the z-axis generates a cyclic subgroup of D_n of order n.

(b) The "flips" $j_{k\pi/n} = r_{k2\pi/n}\, j = r^k\, j$, for $k \in \mathbb{Z}$, are symmetries of the n-gon.

(c) The distinct flip symmetries of the n-gon are $r^k\, j$ for $k = 0, 1, \ldots, n-1$.

Exercise 2.5.5.

(a) If n is odd, then the axis of each of the "flips" passes through a vertex of the n-gon and the midpoint of the opposite edge. See Figure 2.5.2 for the case $n = 5$.

(b) If n is even and k is even, then $j_{k\pi/n} = r^k\, j$ is a rotation about an axis passing through a pair of opposite vertices of the n-gon.

(c) If n is even and k is odd, then $j_{k\pi/n} = r^k\, j$ is a rotation about an axis passing through the midpoints of a pair of opposite edges of the n-gon. See Figure 2.5.3 for the case $n = 6$.

Exercise 2.5.6. Consider a card in the shape of an n-gon, whose two faces are distinguishable. Show that the symmetry group of this figure (including reflections) is isomorphic to D_n.

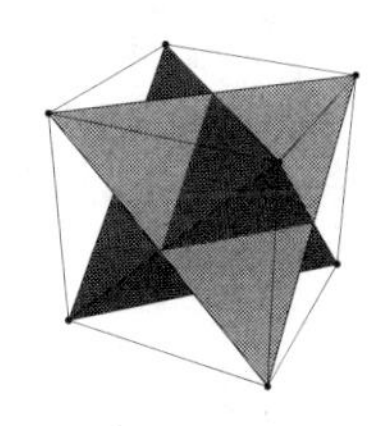

Chapter 3

Basic Theory of Groups

3.1. Homomorphisms and Isomorphisms

We have already introduced the concept of an *isomorphism* between two groups: An isomorphism $\varphi : G \longrightarrow H$ is a bijection which preserves group multiplication, i.e., $\varphi(g_1 g_2) = \varphi(g_1)\varphi(g_1)$ for all $g_1, g_2 \in G$.

For example, the set of eight 3-by-3 matrices:

$$\{E, R, R^2, R^3, A, RA, R^2A, R^3A\},$$

where E is the 3-by-3 identity matrix, and

$$A = \begin{bmatrix} 1 & 0 & 0 \\ 0 & -1 & 0 \\ 0 & 0 & -1 \end{bmatrix} \quad R = \begin{bmatrix} 0 & -1 & 0 \\ 1 & 0 & 0 \\ 0 & 0 & 1 \end{bmatrix},$$

given in Section 1.5, is a subgroup of $\mathrm{GL}(3, \mathbb{R})$. The map $\varphi : r^k a^l \mapsto R^k A^l$ $(0 \leq k \leq 3, 0 \leq l \leq 1)$ is an isomorphism from the group of symmetries of the square to this group of matrices.

Similarly, the set of eight 2-by-2 matrices

$$\{E, R, R^2, R^3, J, RJ, R^2J, R^3J\},$$

where now E is the 2-by-2 identity matrix and

$$J = \begin{bmatrix} 1 & 0 \\ 0 & -1 \end{bmatrix} \quad R = \begin{bmatrix} 0 & -1 \\ 1 & 0 \end{bmatrix}$$

is a subgroup of $\mathrm{GL}(2, \mathbb{R})$, and the map $\psi : r^k a^l \mapsto R^k J^l$ $(0 \leq k \leq 3, 0 \leq l \leq 1)$ is an isomorphism from the group of symmetries of the square to this group of matrices.

There is a more general concept which proves to be very useful:

Definition 3.1.1. A map between groups $\varphi : G \longrightarrow H$ is called a *homomorphism* if it preserves group multiplication, $\varphi(g_1 g_2) = \varphi(g_1)\varphi(g_1)$ for all $g_1, g_2 \in G$.

There is no requirement here that φ be either injective or surjective.

Example 3.1.2. We consider some homomorphisms of the symmetry group of the square into permutation groups. Place the square card in the x-y-plane so that the axes of symmetry for the rotations a, b, and r coincide with the x-, y-, and z- axes, respectively. Each symmetry of the card induces a bijective map of the space $S = \{(x, y, 0) : |x| \le 1, |y| \le 1\}$ occupied by the card. For example, the symmetry a induces the map

$$\begin{bmatrix} x \\ y \\ 0 \end{bmatrix} \mapsto \begin{bmatrix} x \\ -y \\ 0 \end{bmatrix}.$$

The map associated to each symmetry sends the set V of four vertices of S onto itself. So for each symmetry σ of the square, we get an element $\pi(\sigma)$ of $\mathrm{Sym}(V)$. Composition of symmetries corresponds to composition of maps of S and of V, so the assignment $\sigma \mapsto \pi(\sigma)$ is a homomorphism from the symmetry group of the square to $\mathrm{Sym}(V)$. This homomorphism is injective, since a symmetry of the square is entirely determined by what it does to the vertices, but it cannot be surjective, since the square has only eight symmetries while $\#\mathrm{Sym}(V) = 24$.

Example 3.1.3. To make these observations more concrete and computationally useful, we number the vertices of S. It should be emphasized that we are not numbering the corners of the card, which move along with the card, but rather *the locations* of these corners, which stay put. This is a point about which one can easily be confused.

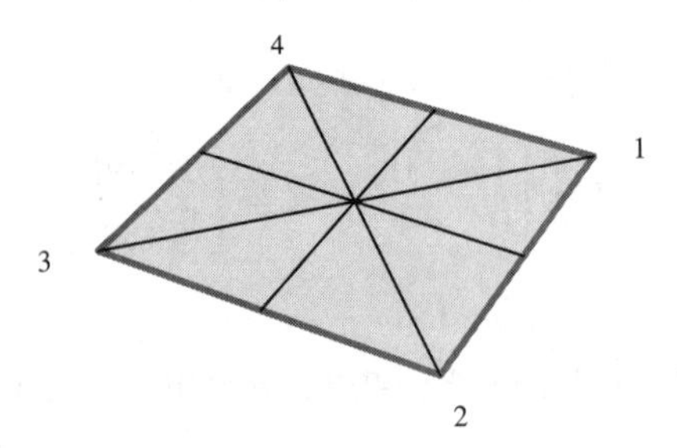

Figure 3.1.1. Labeling the Vertices of the Square

Numbering the vertices gives us a homomorphism φ from the group of symmetries of the square into S_4. Observe, for example, that $\varphi(r) = (1432)$, $\varphi(a) = (14)(23)$, and $\varphi(c) = (24)$. Now you can compute that:

$$\varphi(a)\varphi(r) = (14)(23)(1432) = (24) = \varphi(c) = \varphi(ar).$$

You are asked in Exercise 3.1.1 to complete the tabulation of the map φ from the symmetry group of the square into S_4 and to verify the homomorphism property.

Note that all of this is a formalization of the computation by pictures which was done in Section 1.4.

Example 3.1.4. There are other sets of geometric objects associated with the square which are permuted by symmetries of the square: the set of edges, the set of diagonals, the set of pairs of opposite edges. Let's consider the diagonals. Numbering the diagonals gives a homomorphism ψ from the group of symmetries of the square into S_2.

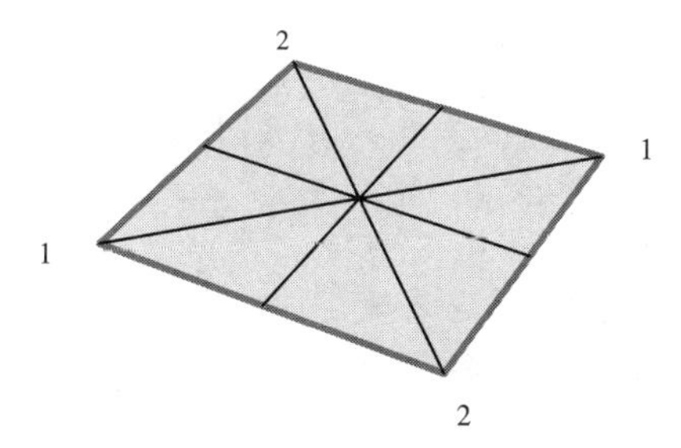

Figure 3.1.2. Labeling the Diagonals of the Square

You can compute, for example, that $\psi(r) = \psi(a) = (12)$, while $\psi(c) = e$. You are asked in Exercise 3.1.2 to complete the tabulation of the map ψ and to verify its homomorphism property.

Example 3.1.5. It is well known that the determinant of an invertible matrix is non-zero, and that the determinant satisfies the identity $\det(AB) = \det(A)\det(b)$. Therefore $\det : \mathrm{GL}(n, \mathbb{R}) \to \mathbb{R}^*$ is a homomorphism from the group of invertible matrices (see Example 1.6.7) to the group of non-zero real numbers under multiplication (see Example 1.6.11).

Example 3.1.6. Recall that a linear transformation $T : \mathbb{R}^n \to \mathbb{R}^n$ has the property that $T(\boldsymbol{a}+\boldsymbol{b}) = T(\boldsymbol{a}) + T(\boldsymbol{b})$. Thus T is a group homomorphism from the additive group $\mathbb{R}^n$ to itself. (Compare Example 1.6.12.) More concretely, for any n-by-n matrix M, one has $M(\boldsymbol{a}+\boldsymbol{b}) = M\boldsymbol{a} + M\boldsymbol{b}$. Thus multiplication by M is a group homomorphism from the additive group $\mathbb{R}^n$ to itself.

Example 3.1.7. Let G be any group and a and element of G. The the map from $\mathbb{Z}$ to G given by $k \mapsto a^k$ is a group homomorphism. This is equivalent to the statement that $a^{k+\ell} = a^k a^\ell$ for all integers k and ℓ. The image of this homomorphism is the cyclic subgroup of G generated by g.

Example 3.1.8. There is a homomorphism from $\mathbb{Z}$ to $\mathbb{Z}_n$ defined by $k \mapsto [k]$. This is precisely the content of Lemma 2.4.18. This example is a special case of the previous example, with $G = \mathbb{Z}_n$ and the chosen element $a = [1] \in G$. The map is given by $k \mapsto k[1] = [k]$.

Example 3.1.9. Let G be an *abelian* group and n a fixed integer. Then the map from G to G given by $g \mapsto g^n$ is a group homomorphism. This is equivalent to the statement that $(ab)^n = a^n b^n$ when a, b are elements in an abelian group.

Let us now turn from the examples to some general observations. Our first observation is that the composition of homomorphisms is a homomorphism.

Proposition 3.1.10. *Let $\varphi : G \longrightarrow H$ and $\psi : H \longrightarrow K$ be homomorphisms of groups. Then the composition $\psi \circ \varphi : G \longrightarrow K$ is also a homomorphism.*

Proof. Exercise 3.1.3. □

Next we check that homomorphisms preserve the group identity and inverses.

Proposition 3.1.11. *Let $\varphi : G \longrightarrow H$ be a homomorphism of groups.*

(a) $\varphi(e_G) = e_H$.

(b) *For each $g \in G$, $\varphi(g^{-1}) = (\varphi(g))^{-1}$*

Proof. For any $g \in G$,

$$\varphi(e_G)\varphi(g) = \varphi(e_G g) = \varphi(g).$$

It follows from Proposition 2.1.1 (a) that $\varphi(e_G) = e_H$. Similarly, for any $g \in G$,

$$\varphi(g^{-1})\varphi(g) = \varphi(g^{-1}g) = \varphi(e_G) = e_H,$$

so Proposition 2.1.1 (b) implies that $\varphi(g^{-1}) = (\varphi(g))^{-1}$. □

The next observation is that homomorphisms preserve subgroups.

Proposition 3.1.12. *Let $\varphi : G \longrightarrow H$ be a homomorphism of groups.*

(a) *For each subgroup $A \subseteq G$, $\varphi(A)$ is a subgroup of G.*

(b) *For each subgroup $B \subseteq H$,*

$$\varphi^{-1}(B) = \{g \in G : \varphi(g) \in B\}$$

is a subgroup of G.

Proof. We have to show that $\varphi(A)$ is closed under multiplication and inverses. Let h_1 and h_2 be elements of $\varphi(A)$. There exist elements $a_1, a_2 \in A$ such that $h_i = \varphi(a_i)$ for $i = 1, 2$. Then

$$h_1 h_2 = \varphi(a_1)\varphi(a_2) = \varphi(a_1 a_2) \in \varphi(A),$$

since $a_1 a_2 \in A$. Likewise, for $h \in \varphi(A)$, there is an $a \in A$ such that $\varphi(a) = h$. Then, using Proposition 3.1.11 (b) and closure of A under inverses, we compute

$$h^{-1} = (\varphi(a))^{-1} = \varphi(a^{-1}) \in \varphi(A).$$

The proof of part (b) is left as an exercise. □

You might think at first that it is not very worthwhile to look at non-injective homomorphisms $\varphi : G \to H$ since such a homomorphism loses information about G. But in fact, such a homomorphism also reveals certain information about the structure of G which otherwise might be missed. For example, consider the homomorphism ψ from the symmetry group G of the square into the symmetric group S_2 induced by the action of G on the two diagonals of the square, as discussed in Example 3.1.4. Let N denote the set of symmetries σ of the square such that $\psi(\sigma) = e$. You can compute that $N = \{e, c, d, r^2\}$. (Do it now!) From the general theory which I am about to one sees that N is a special sort of subgroup G, called a normal subgroup. Understanding such subgroups helps to understand the structure of G. For now, verify for yourself that N is, in fact, a subgroup.

Definition 3.1.13. A subgroup N of a group G is said to be *normal* if for all $g \in G$, $gNg^{-1} = N$. Here gNg^{-1} means $\{gng^{-1} : n \in N\}$.

Definition 3.1.14. Let $\varphi : G \longrightarrow H$ be a homomorphism of groups. The *kernel* of the homomorphism φ, denoted $\ker(\varphi)$ is $\varphi^{-1}(e_H) = \{g \in G : \varphi(g) = e_H\}$.

According to Proposition 3.1.12 (b), $\ker(\varphi)$ is a subgroup of G (since $\{e_H\}$ is a subgroup of H). We now observe that it is a normal subgroup.

Proposition 3.1.15. *Let $\varphi : G \longrightarrow H$ be a homomorphism of groups. Then* $\ker(\varphi)$ *is a normal subgroup of G.*

Proof. It suffices to show that $g\ker(\varphi)g^{-1} = \ker(\varphi)$ for all $g \in G$. If $x \in \ker(\varphi)$, then $\varphi(gxg^{-1}) = \varphi(g)\varphi(x)(\varphi(g))^{-1} = \varphi(g)e(\varphi(g))^{-1} = e$. Thus $gxg^{-1} \in \ker(\varphi)$. We have now shown that for all $g \in G$, $g\ker(\varphi)g^{-1} \subseteq \ker(\varphi)$. We still need to show the opposite containment. But if we replace g by g^{-1}, we obtain that for all $g \in G$, $g^{-1}\ker(\varphi)g \subseteq \ker(\varphi)$; this is equivalent to $\ker(\varphi) \subseteq g\ker(\varphi)g^{-1}$. Since we have both $g\ker(\varphi)g^{-1} \subseteq \ker(\varphi)$ and $\ker(\varphi) \subseteq g\ker(\varphi)g^{-1}$, we have equality of the two sets. □

Example 3.1.16. The kernel of the determinant $\det : \mathrm{GL}(n, \mathbb{R}) \to \mathbb{R}^*$ is the subgroup of matrices with determinant equal to 1. This subgroup is called the *special linear group* and denoted $\mathrm{SL}(n, \mathbb{R})$.

Example 3.1.17. Let G be any group and $a \in G$. If the order of a is n, then the kernel of the homomorphism $k \mapsto a^k$ from $\mathbb{Z}$ to G is the set of all multiples of n, $\{kn : k \in \mathbb{Z}\}$. If a is of infinite order, then the kernel of the homomorphism is $\{0\}$.

Example 3.1.18. In particular the kernel of the homomorphism from $\mathbb{Z}$ to $\mathbb{Z}_n$ defined by $k \mapsto [k]$ is $[0] = \{kn : k \in \mathbb{Z}\}$.

Example 3.1.19. If G is an abelian group and n is a fixed integer, then the kernel of the homomorphism $g \mapsto g^n$ from G to G is the set of elements whose order divides n.

Additional examples of homomorphisms are explored in the exercises. In particular, it is shown in the exercises that there is a homomorphisms $\epsilon : S_n \to \{\pm 1\}$ with the property that $\epsilon(\tau) = -1$ for any 2-cycle τ. This is an example of a homomorphism which is very far from being injective and which picks out an essential structural feature of the symmetric group.

Definition 3.1.20. The homomorphism ϵ is called the *sign* (or *parity*) homomorphism. A permutation π is said to be *even* if $\epsilon(\pi) = 1$, that is, if π is in the kernel of the sign homomorphism. Otherwise π is said to be *odd*. The subgroup of even permutations (that is, the kernel of ϵ) is generally denoted A_n. This subgroup is also referred to as the *alternating group*.

The following statement about even and odd permutations is implicit in the exercises:

Proposition 3.1.21. *A permutation π is even if, and only if, π can be written as a product of an even number of 2-cycles.*

Even and odd permutations have the following property: The product of two even permutations is even; the product of an even and an odd permutation is odd, and the product of two odd permutations is even. Hence

Corollary 3.1.22. *The set of odd permutations in S_n is $(12)A_n$, where A_n denotes the subgroup of even permutations.*

Proof. $(12)A_n$ is contained in the set of odd permutations. But if σ is any odd permutation, then $(12)\sigma$ is even, so $\sigma = (12)((12)\sigma) \in (12)A_n$. □

Corollary 3.1.23. *A k-cycle is even if k is odd and odd if k is even.*

Proof. According to Exercise 2.2.6, a k cycle can be written as a product of $(k-1)$ 2-cycles. □

Exercises

Exercise 3.1.1. Let φ be the map from symmetries of the square into S_4 induced by the numbering of the vertices of the square in Figure 3.1.1. Complete the tabulation of $\varphi(\sigma)$ for σ in the symmetry group of the square, and verify the homomorphism property of φ by computation.

Exercise 3.1.2. Let ψ be the map from the symmetry group of the square into S_2 induced by the labeling of the diagonals of the square as in Figure 3.1.2. Complete the tabulation of ψ and verify the homomorphism property by computation. Identify the kernel of ψ.

Exercise 3.1.3. Prove Proposition 3.1.10.

Exercise 3.1.4. Prove part (b) of Proposition 3.1.12.

Exercise 3.1.5. For any subgroup A of a group G, and $g \in G$, show that gAg^{-1} is a subgroup of G.

Exercise 3.1.6. Show that every subgroup of an abelian group is normal.

Exercise 3.1.7. Let $\varphi : G \longrightarrow H$ be a homomorphism of groups with kernel N. For $a, x \in G$, show that

$$\varphi(a) = \varphi(x) \Leftrightarrow a^{-1}x \in N \Leftrightarrow aN = xN.$$

Exercise 3.1.8. Let $\varphi : G \longrightarrow H$ be a homomorphism of G *onto* H. If A is a normal subgroup of G, show that $\varphi(A)$ is a normal subgroup of H.

The following exercises examine an important homomorphism from S_n to $C_2 = \{1, -1\}$ (for any n).

Exercise 3.1.9. Let $\{x_1, x_2, \dots, x_n\}$ be variables. For any polynomial p in n variables and for $\sigma \in S_n$, define

$$\sigma(p)(x_1, \dots, x_n) = p(x_{\sigma(1)}, \dots, x_{\sigma(n)}).$$

Check that $\sigma(\tau(p)) = (\sigma\tau)(p)$ for all σ and $\tau \in S_n$

Exercise 3.1.10. Now fix $n \in \mathbb{N}$, and define

$$\Delta = \prod_{1 \le i < j \le n} (x_i - x_j).$$

For any $\sigma \in S_n$, check that $\sigma(\Delta) = \pm\Delta$. Show that the map $\epsilon : \sigma \mapsto \sigma(\Delta)/\Delta$ is a homomorphism from S_n to $\{1, -1\}$.

Exercise 3.1.11. Show that for any 2-cycle (a, b), $\epsilon((a, b)) = -1$; hence if a permutation π is a product of k 2-cycles , then $\epsilon(\pi) = (-1)^k$. Now any permutation can be written as a product of 2-cycles. (Exercise 2.2.6). If a permutation π can be written as a product of k 2-cycles and also as a product of l 2-cycles, then $\epsilon(\pi) = (-1)^k = (-1)^l$, so the parity of k and of l is the same. The parity is even if, and only if, $\epsilon(\pi) = 1$.

Exercise 3.1.12. For each permutation $\pi \in S_n$, define an n-by-n matrix $T(\pi)$ as follows. Let $\hat{\boldsymbol{e}}_1, \hat{\boldsymbol{e}}_2, \dots, \hat{\boldsymbol{e}}_n$ be the standard basis of $\mathbb{R}^n$; $\hat{\boldsymbol{e}}_k$ has a 1 in the k-th coordinate and zeros elsewhere. Define

$$T(\pi) = [\hat{\boldsymbol{e}}_{\pi(1)}, \hat{\boldsymbol{e}}_{\pi(2)}, \dots, \hat{\boldsymbol{e}}_{\pi(n)}];$$

that is, the k-th column of $T(\pi)$ is the basis vector $\hat{\boldsymbol{e}}_{\pi(k)}$. Show that the map $\pi \mapsto T(\pi)$ is a homomorphism from S_n into $\mathrm{GL}(n, \mathbb{R})$. What is the range of T?

Exercise 3.1.13.

(a) Show that for any k cycle $(a_1, a_2, \dots, a_k) \in S_n$, and for any permutation $\pi \in S_n$, one has

$$\pi(a_1, a_2, \dots, a_k)\pi^{-1} = (\pi(a_1), \pi(a_2), \dots, \pi(a_k)).$$

Hint: As always, first look at some examples for small n and k. Both sides are permutations, i.e. maps defined on $\{1, 2, \dots, n\}$. Show that they are the same maps by showing that they do the same thing.

(b) Show that for any two k cycles, $(a_1, a_2, \dots, a_k)$ and $(b_1, b_2, \dots, b_k)$ in S_n there is a permutation $\pi \in S_n$ such that:

$$\pi(a_1, a_2, \dots, a_k)\pi^{-1} = (b_1, b_2, \dots, b_k).$$

(c) Suppose that α and β are elements of S_n and that $\beta = g\alpha g^{-1}$ for some $g \in S_n$. Show that when α and β are written as a product of disjoint cycles, they both have exactly the same number of cycles of each length. (For example if $\alpha \in S_{10}$ is a product of two 3-cycles, one 2-cycle, and four 1-cycles, then so is β.) One says that α and β *have the same cycle structure.*

(d) Conversely, suppose α and β are elements of S_n and they have the same cycle structure. Show that there is an element $g \in S_n$ such that $\beta = g\alpha g^{-1}$.

Exercise 3.1.14. Show that ϵ is the unique homomorphism from S_n onto $\{1, -1\}$. *Hint:* Let $\varphi : S_n \longrightarrow \{\pm 1\}$ be a homomorphism. If $\varphi((12)) = -1$, show, using the results of Exercise 3.1.13, that $\varphi = \epsilon$. If $\varphi((12)) = +1$, show that φ is the trivial homomorphism, $\varphi(\pi) = 1$ for all π.

Exercise 3.1.15. For $m < n$, one can consider S_m as a subgroup of S_n. Namely, S_m is the subgroup of S_n which leaves fixed the numbers from $m + 1$ to n. The parity of an element of S_m can be computed in two ways: as an element of S_m and as an element of S_n. Show that two answers always agree.

The following concept is used in the next exercises:

Definition 3.1.24. An *automorphism* of a group G is an isomorphism from G onto G.

Exercise 3.1.16. Fix an element g in a group G. Show that the map $c_g : G \to G$ defined by $c_g(a) = gag^{-1}$ is an automorphism of G.

Exercise 3.1.17. Show that conjugate elements of S_n have the same parity. More generally, if $\phi : S_n \to S_n$ is an automorphism, then ϕ preserves parity.

Exercise 3.1.18.

(a) Show that the set of matrices with positive determinant is a normal subgroup of $\mathrm{GL}(n, \mathbb{R})$.

(b) Show that $\epsilon = \det \circ T$, where T is the homomorphism defined in Exercise 3.1.12 and ϵ is the sign homomorphism. *Hint:* Determine the range of $\det \circ T$ and use the uniqueness of the sign homomorphism from Exercise 3.1.14.

Exercise 3.1.19.

(a) For $A \in \mathrm{GL}(n, \mathbb{R})$ and $\boldsymbol{b} \in \mathbb{R}^n$, define the transformation $T_{A,\boldsymbol{b}} : \mathbb{R}^n \to \mathbb{R}^n$ by $T_{A,\boldsymbol{b}}(\boldsymbol{x}) = A\boldsymbol{x} + \boldsymbol{b}$. Show that the set of all such transformations forms a group G.

(b) Consider the set of matrices

$$\begin{bmatrix} A & b \\ 0 & 1 \end{bmatrix},$$

where $A \in \mathrm{GL}(n, \mathbb{R})$ and $\boldsymbol{b} \in \mathbb{R}^n$, and where the 0 denotes a 1-by-n row of zeros. Show that this is a subgroup of $\mathrm{GL}(n+1, \mathbb{R})$, and that it is isomorphic to the group described in part (a).

(c) Show that the map $T_{A,\boldsymbol{b}} \mapsto A$ is a homomorphism from G to $\mathrm{GL}(n, \mathbb{R})$, and that the kernel K of this homomorphism is isomorphic to $\mathbb{R}^n$, considered as an abelian group under vector addition.

3.2. Cosets and Lagrange's Theorem

Consider the subgroup $H = \{e, (1\ 2)\} \subseteq S_3$. For each of the six elements $\pi \in S_3$ you can compute the set $\pi H = \{\pi\sigma : \sigma \in H\}$. For example, $(23)H = \{(23), (132)\}$. Do the computation now, and check that you get the following results:

$$\begin{aligned} eH = (1\ 2)H &= H \\ (2\ 3)H = (1\ 2\ 3)H &= \{(2\ 3), (1\ 3\ 2)\} \\ (1\ 3)H = (1\ 2\ 3)H &= \{(1\ 3), (1\ 2\ 3)\} \end{aligned}$$

As π varies through S_3, only three different sets πH are obtained, each occurring twice.

Definition 3.2.1. Let H be subgroup of a group G. A subset of the form gH, where $g \in G$ is called a *left coset of H in G.*

Example 3.2.2. S_3 may be identified with a subgroup of S_4 consisting of permutations which leave 4 fixed and permute $\{1, 2, 3\}$. For each of the 24 elements $\pi \in S_4$ you can compute the set πS_3.

This computation requires a little labor. If you want, you can get a computer to do some of the repetitive work; for example, programs for computations in the symmetric group are distributed with the symbolic mathematics program *Mathematica*.

With the notation $H = \{\sigma \in S_4 : \sigma(4) = 4\}$, the results are:

$$
\begin{aligned}
eH &= (1\,2)H = (1\,3)H = (2\,3)H = (1\,2\,3)H = (1\,3\,2)H = H \\
(4\,3)H &= (4\,3\,2)H = (2\,1)(4\,3)H \\
&= (2\,4\,3\,1)H = (4\,3\,2\,1)H = (4\,3\,1)H \\
&= \{(4\,3),\ (4\,3\,2),\ (2\,1)(4\,3),\ (2\,4\,3\,1),\ (4\,3\,2\,1),\ (4\,3\,1)\} \\
(4\,2)H &= (3\,4\,2)H = (4\,2\,1)H \\
&= (4\,2\,3\,1)H = (3\,4\,2\,1)H = (3\,1)(4\,2)H \\
&= \{(4\,2),\ (3\,4\,2),\ (4\,2\,1),\ (4\,2\,3\,1),\ (3\,4\,2\,1),\ (3\,1)(4\,2)\}
\end{aligned}
$$

$$
\begin{aligned}
(4\,1)H &= (4\,1)\,(3\,2)H = (2\,4\,1)H \\
&= (2\,3\,4\,1)H = (3\,2\,4\,1)H = (3\,4\,1)H \\
&= \{(4\,1),\ (4\,1)\,(3\,2),\ (2\,4\,1),\ (2\,3\,4\,1),\ (3\,2\,4\,1),\ (3\,4\,1)\}
\end{aligned}
$$

The regularity of the data above for left cosets of subgroups of symmetric groups is so striking that it can't be missed. Make some conjectures about cosets of a subgroup H in a group G before you read further.

Proposition 3.2.3. *Let H be a subgroup of a group G, and let a and b be elements of G. The following conditions are equivalent:*

(a) $a \in bH$.
(b) $b \in aH$.
(c) $aH = bH$.
(d) $b^{-1}a \in H$.
(e) $a^{-1}b \in H$.

Proof. If condition (a) is satisfied, then there is an element $h \in H$ such that $a = bh$; but then $b = ah^{-1} \in aH$. Thus, (a) implies (b), and similarly (b) implies (a). Now suppose that (a) holds and choose $h \in H$ such that $a = bh$. Then for all $h_1 \in H$, $ah_1 = bhh_1 \in bH$; thus $aH \subseteq bH$. Similarly (b) implies that $bH \subseteq aH$. Since (a) is equivalent to (b), each implies (c). Because $a \in aH$ and $b \in bH$, (c) implies (a) and (b). Finally, (d) and (e) are equivalent by taking inverses, and $a = bh \in bH \Leftrightarrow b^{-1}a = h \in H$, so (a) and (d) are equivalent. □

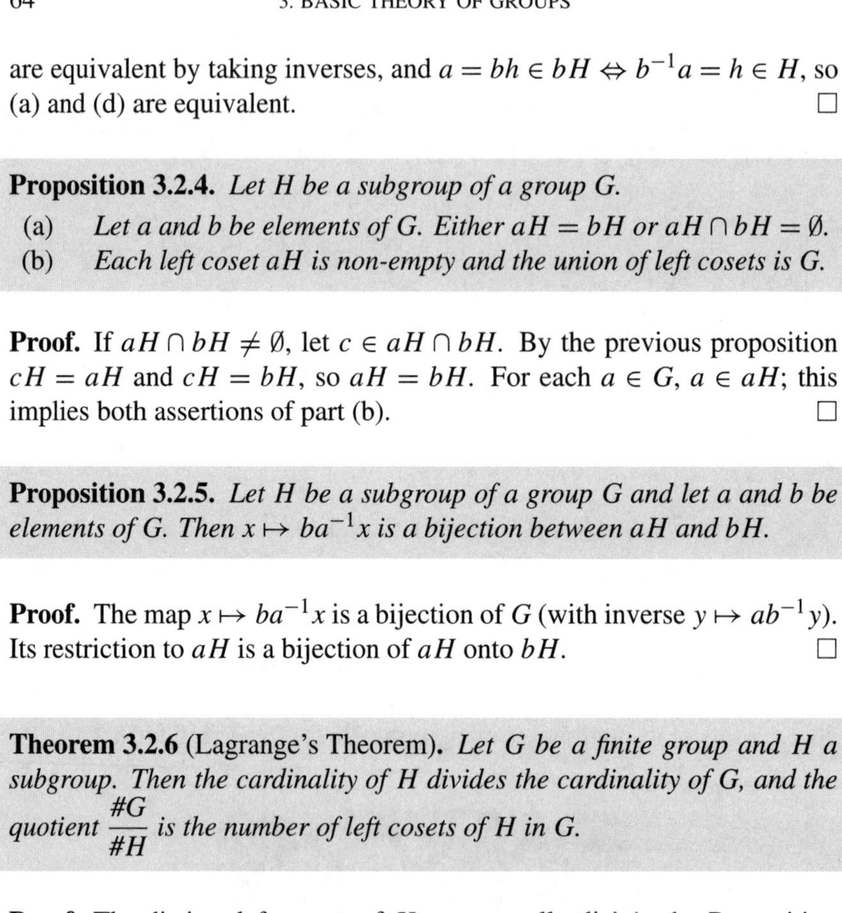

Proposition 3.2.4. *Let H be a subgroup of a group G.*

(a) *Let a and b be elements of G. Either $aH = bH$ or $aH \cap bH = \emptyset$.*

(b) *Each left coset aH is non-empty and the union of left cosets is G.*

Proof. If $aH \cap bH \neq \emptyset$, let $c \in aH \cap bH$. By the previous proposition $cH = aH$ and $cH = bH$, so $aH = bH$. For each $a \in G$, $a \in aH$; this implies both assertions of part (b). □

Proposition 3.2.5. *Let H be a subgroup of a group G and let a and b be elements of G. Then $x \mapsto ba^{-1}x$ is a bijection between aH and bH.*

Proof. The map $x \mapsto ba^{-1}x$ is a bijection of G (with inverse $y \mapsto ab^{-1}y$). Its restriction to aH is a bijection of aH onto bH. □

Theorem 3.2.6 (Lagrange's Theorem). *Let G be a finite group and H a subgroup. Then the cardinality of H divides the cardinality of G, and the quotient $\dfrac{\#G}{\#H}$ is the number of left cosets of H in G.*

Proof. The distinct left cosets of H are mutually disjoint by Proposition 3.2.4 and each has the same size (namely $\#H = \#eH$) by Proposition 3.2.5. Since the union of the left cosets is G, the cardinality of G is the cardinality of H times the number of distinct left cosets of H. □

Definition 3.2.7. For a subgroup H of a group G, the *index* of H in G is the number of left cosets of H in G. The index is denoted $[G : H]$.

Index also makes sense for infinite groups. For example, take the larger group to be $\mathbb{Z}$ and the subgroup to be $n\mathbb{Z}$. Then $[\mathbb{Z} : n\mathbb{Z}] = n$, because there are n cosets of $n\mathbb{Z}$ in $\mathbb{Z}$. Every subgroup of $\mathbb{Z}$ has the form $n\mathbb{Z}$ for some n, so every subgroup has finite index. But it is also possible for a subgroup of an infinite group to have infinite index. For example, the cosets of $\mathbb{Z}$ in $\mathbb{R}$ are in bijective correspondence with the elements of the half-open interval $[0, 1)$; there are uncountably many cosets!

Corollary 3.2.8. *Let p be a prime number and suppose G is a group of order p. Then*

(a) *G has no subgroups other than G and $\{e\}$.*
(b) *G is cyclic, and in fact, for any non-identity element $a \in G$, $G = \langle a \rangle$.*
(c) *Every homomorphism from G into another group is either trivial (i.e., every element of G is sent to the identity) or injective.*

Proof. The first assertion follows immediately from Lagrange's theorem, since the size of a subgroup can only be p or 1. If $a \neq e$, then the subgroup $\langle a \rangle$ is not $\{e\}$, so must be G. The last assertion also follows from the first, since the kernel of a homomorphism is a subgroup. □

Any two groups of prime order p are isomorphic, since each is cyclic. This generalizes (substantially) the results which we obtained before on the uniqueness of the groups of orders 2, 3, and 5.

Corollary 3.2.9. *Let G be any finite group, and let $a \in G$. Then the order $o(a)$ divides the order of G.*

Proof. The order of a is the cardinality of the subgroup $\langle a \rangle$. □

The index of subgroups satisfies a multiplicativity property:

Proposition 3.2.10. *Suppose $K \subseteq H \subseteq G$ are subgroups. Then*

$$[G : K] = [G : H][H : K].$$

Proof. If the groups are finite, then by Lagrange's Theorem,

$$[G : K] = \frac{\#G}{\#K} = \frac{\#G}{\#H}\frac{\#H}{\#K} = [G : H][H : K].$$

If the groups are infinite, one has to use another approach, which is discussed in the exercises. □

Definition 3.2.11. For any group G, the *center* $Z(G)$ of G is the set of elements which commute with all elements of G,

$$Z(G) = \{a \in G : ag = ga \text{ for all } g \in G\}.$$

You are asked in the exercises to show that the center of a group is a normal subgroup, and to compute the center of several particular groups.

Exercises

Exercise 3.2.1. Check that the left cosets of the subgroup

$$K = \{e, (123), (132)\}$$

in S_3 are

$$eK = (123)K = (132)K = K$$
$$(12)K = (13)K = (23)K = \{(12), (13), (23)\}.$$

and that each occurs three times in the list $(gK)_{g \in S_3}$. Note that K is the subgroup of even permutations and the other coset of K is the set of odd permutations.

Exercise 3.2.2. Suppose $K \subseteq H \subseteq G$ are subgroups. Suppose $h_1K, \dots, h_RK$ is a list of the distinct cosets of K in H, and $g_1H, \dots, g_SH$ is a list of the distinct cosets of H in G. Show that $\{g_sh_rH : 1 \le s \le S, 1 \le r \le R\}$ is the set of distinct cosets of H in G. *Hint:* There are two things to show. First, you have to show that if $(r, s) \ne (r', s')$, then $g_sh_rK \ne g_{s'}h_{r'}K$. Second, you have to show that if $g \in G$, then for some (r, s), $gK = g_sh_rK$.

Exercise 3.2.3. Try to extend the idea of the previous exercise to the case where at least one of the pairs $K \subseteq H$ and $H \subseteq G$ has infinite index.

Exercise 3.2.4. Consider the group S_3.

(a) Find all the left cosets and all the right cosets of the subgroup $H = \{e, (12)\}$ of S_3, and observe that not every left coset is also a right coset.

(b) Find all the left cosets and all the right cosets of the subgroup $K = \{e, (123), (132)\}$ of S_3, and observe that every left coset is also a right coset.

Exercise 3.2.5. For a subgroup N of a group G, prove that the following are equivalent:

(a) N is normal.

(b) Each left coset of N is also a right coset. That is, for each $a \in G$, there is a $b \in G$ such that $aN = Nb$.

(c) For each $a \in G$, $aN = Na$.

Exercise 3.2.6. Suppose N is a subgroup of a group G and $[G : N] = 2$. Show that N is normal using the criterion of the previous exercise.

Exercise 3.2.7. Show that if G is a finite group and N is a subgroup of index 2, then for elements a and b of G, the product ab is an element of N if, and only if, either both of a and b are in N or neither of a and b is in N.

Exercise 3.2.8. For two subgroups H and K of a group G and an element $a \in G$, the double coset HaK is the set of all products hak where $h \in H$ and $k \in K$. Show that two double cosets HaK and HbK are either equal or disjoint.

Exercise 3.2.9. Consider the additive group $\mathbb{R}$ and its subgroup $\mathbb{Z}$. Describe a coset $t + \mathbb{Z}$ geometrically. Show that the set of all cosets of $\mathbb{Z}$ in $\mathbb{R}$ is $\{t + \mathbb{Z} : 0 \le t < 1\}$. What are the analogous results for $\mathbb{Z}^2 \subseteq \mathbb{R}^2$?

Exercise 3.2.10. Consider the additive group $\mathbb{Z}$ and its subgroup $n\mathbb{Z}$ consisting of all integers divisible by n. Show that the distinct cosets of $n\mathbb{Z}$ in $\mathbb{Z}$ are $\{n\mathbb{Z}, 1 + n\mathbb{Z}, 2 + n\mathbb{Z}, \dots, n - 1 + n\mathbb{Z}\}$.

Exercise 3.2.11.

(a) Show that the center of a group G is a normal subgroup of G.
(b) What is the center of the group S_3?

Exercise 3.2.12.

(a) What is the center of the group D_4 of symmetries of the square?
(b) What is the center of the dihedral group D_n?

Exercise 3.2.13. Find the center of the group $\mathrm{GL}(2, \mathbb{R})$ of 2-by-2 invertible real matrices. Do the same for $\mathrm{GL}(3, \mathbb{R})$. *Hint:* It is a good idea to go outside of the realm of invertible matrices in considering this problem. This is because it is useful to exploit linearity, but the group of invertible matrices is not a linear space. So first explore the condition for a matrix A to commute with *all* matrices B, not just invertible matrices. Note that the condition $AB = BA$ is linear in B, so a matrix A commutes with all matrices if, and only if, it commutes with each member of a linear spanning set of matrices. So now consider the so-called matrix units E_{ij}, which have a 1 in the i, j position and zeros elsewhere. The set of matrix units is a basis of the linear space of matrices. Find the condition for a matrix to commute with all of the E_{ij}'s. It remains to show that if a matrix commutes with all *invertible* matrices, then it also commutes with all E_{ij}'s. (The results of this problem hold just as well with the real numbers $\mathbb{R}$ replaced by the complex numbers $\mathbb{C}$ or the rational numbers $\mathbb{Q}$.)

Exercise 3.2.14. Show that the symmetric group S_n has a unique subgroup of index 2, namely the subgroup A_n of even permutations. *Hint:* Such subgroup N is normal. Hence if it contains one element of a certain cycle structure, then it contains all elements of that cycle structure, according to Exercise 3.3.10 and Exercise 3.1.13. Can N contain any 2-cycles? Show that N contains a product of k 2-cycles if, and only if, k is even, by using Exercise 3.2.7. Conclude $N = A_n$ by Proposition 3.1.21.

Exercise 3.2.15. Let G be an abelian group. For any integer $n > 0$ show that the map $\varphi : a \mapsto a^n$ is a homomorphism from G into G. Characterize the kernel of φ. Show that if n is relatively prime to the order of G, then φ is an isomorphism; hence for each element $g \in G$ there is a unique $a \in G$ such that $g = a^n$.

3.3. Equivalence Relations and Set Partitions

Consider a group G and a subgroup H. Associated to this data, one has the family of left cosets of H in G. The content of Proposition 3.2.4 is that distinct left cosets are disjoint, and the union of all left cosets is G. This is an example of a set partition:

Definition 3.3.1. A *partition* of a set X (into non-empty subsets) is a collection of mutually disjoint non-empty subsets whose union is X.

Mutually disjoint means that the intersection of any two of the sets is empty.

Associated to the data of a group G and a subgroup H, one can also define a binary relation on G by $a \sim b \pmod{H}$ or $a \sim_H b$ if a and b are in the same left coset of H, that is, if $b^{-1}a \in H$. This relation rather evidently has the following properties:

1. $a \sim_H a$.
2. $a \sim_H b \Leftrightarrow b \sim_H a$.
3. If $a \sim_H b$ and $b \sim_H c$, then also $a \sim_H c$.

This is an example of an *equivalence relation.*

Definition 3.3.2. An equivalence relation $\sim$ on a set X is a binary relation with the properties:

1. *Reflexivity:* For each $x \in X$, $x \sim x$.
2. *Symmetry:* For $x, y \in X$, $x \sim y \Leftrightarrow y \sim x$.
3. *Transitivity:* For $x, y, z \in X$, if $x \sim y$ and $y \sim z$, then $x \sim z$.

Notice that the equivalence relation $a \sim_H b$ determines the partition of G into left cosets of H, and vice versa. This is a general feature:

Definition 3.3.3. If $\sim$ is an equivalence relation on X, then for each $x \in X$, the *equivalence class* of x is the set

$$[x] = \{y \in X : x \sim y\}.$$

Note that $x \in [x]$ because of reflexivity.

Proposition 3.3.4. *Let $\sim$ be an equivalence relation on X. For $x, y \in X$, the following conditions are equivalent:*

(a) $x \in [y]$.
(b) $y \in [x]$.
(c) $[x] = [y]$.

Proof. Exercise. □

Corollary 3.3.5. *Let $\sim$ be an equivalence relation on X. Then the collection of equivalence classes is a partition of X into non-empty subsets.*

Proof. If $z \in [x] \cap [y]$, then $[x] = [z] = [y]$, by the proposition. Thus, different equivalence classes are disjoint. The classes are non-empty with union X since $x \in [x]$. □

Conversely, given a partition of X into non-empty sets, one can define an equivalence relation on X by $x \sim y$ if x and y are in the same subset of the partition. It is straightforward to check that this is an equivalence relation.

You are asked to show in an exercise that if one starts with a partition and forms the associated equivalence relation, and then takes the family of equivalence classes for this equivalence relation, one ends up with exactly the partition one started with. Likewise, if one starts with an equivalence relation, forms the partition into equivalence classes, and then builds the equivalence relation related to this partition, then one just gets back the equivalence relation one started with.

Thus, equivalence relations and partitions are just two sides of one phenomenon. This is a phenomenon with a third side as well: Consider again the example determined by a subgroup H of a group G.

Definition 3.3.6. The set of left cosets of H in G is usually denoted G/H. The surjective map $\pi : G \longrightarrow G/H$ defined by $\pi(a) = aH$ is called *the canonical projection* or *quotient map* of G onto G/H.

Note that the *fibers* of the canonical projection π, namely the sets $\pi^{-1}(aH)$, where $aH \in G/H$, are exactly the left cosets of H.

Now consider any surjective map f from a set X onto another set Y; then one can define a relation on X by $x_1 \sim x_2$ if $f(x_1) = f(x_2)$. It is left as an exercise to show that this is an equivalence relation. Moreover, the collection of fibers of the map f, that is, the collection of $f^{-1}(y)$, for $y \in Y$ is a partition of X, and in fact, is the partition determined by the equivalence relation.

Conversely, if $\sim$ is an equivalence relation on X, define $X/\sim$ to be the set of equivalence classes and define a surjection f of X onto $X/\sim$ by $f(x) = [x]$. You are asked to show in Exercise 3.3.4 that if one starts with an equivalence relation $\sim$ on X, builds the corresponding surjective map $f : X \longrightarrow X/\sim$, and then the equivalence relation associated with this map, one gets back the original equivalence relation.

The converse assertion is slightly more subtle. Suppose one starts with a surjection $f : X \longrightarrow Y$, and forms the corresponding equivalence relation $\sim$ and then the surjection $\pi : X \longrightarrow X/\sim$. Now one cannot expect that this is the map f which one started with. However, it is equivalent to the original f in the following sense: There is a bijection $\tilde{f} : X/\sim \longrightarrow Y$ such that $\tilde{f} \circ \pi = f$.

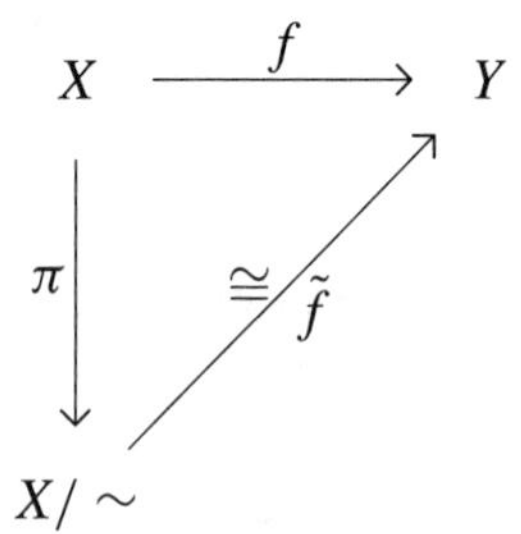

There is only one possible way to define $\tilde{f}$; one has to define $\tilde{f}([x]) = f(x)$. It is then straightforward to check that $\tilde{f}$ has the desired property, $\tilde{f} \circ \pi = f$

I summarize this discussion by a proposition.

Proposition 3.3.7. *Let X be any set. There is a one-to-one correspondence between:*

(a) *Equivalence relations on X,*
(b) *Partitions of X by non-empty subsets, and*
(c) *Surjections of X onto another set Y (up to equivalence).*

Here is an equivalence relation which is useful for studying the structure of groups:

Definition 3.3.8. Elements a and b of a group G are said to be *conjugate* if there is a $g \in G$ such that $b = gag^{-1}$.

You are asked to show in the exercises that conjugacy is an equivalence relation, and to find all the conjugacy equivalence classes in several groups of small order.

Definition 3.3.9. The equivalence classes for conjugacy are called *conjugacy classes.*

Note that the center of a group is related to the notion of conjugacy in the following way: The center of consists of all elements whose conjugacy class is a singleton. That is, $g \in Z(G) \Leftrightarrow$ the conjugacy class of g is $\{g\}$.

Exercises

Exercise 3.3.1. Prove 3.3.4.

Exercise 3.3.2. Show that if one starts with a partition and forms the associated equivalence relation, and then takes the family of equivalence classes for this equivalence relation, one ends up with exactly the partition one started with.

Likewise, if one starts with an equivalence relation, forms the partition into equivalence classes, and then builds the equivalence relation related to this partition, then one just gets back the equivalence relation one started with.

Exercise 3.3.3. Consider any surjective map f from a set X onto another set Y; then one can define an relation on X by $x_1 \sim x_2$ if $f(x_1) = f(x_2)$. Check that this is an equivalence relation. Show that the associated partition of X is the partition into "fibers" $f^{-1}(y)$ for $y \in Y$.

Exercise 3.3.4. If one starts with an equivalence relation $\sim$ on X, builds the corresponding surjective map $f : X \longrightarrow X/\sim$, and then the equivalence relation associated with this map, one gets back the original equivalence relation.

Exercise 3.3.5. Let $f : X \longrightarrow Y$ be a surjective map and define the corresponding equivalence relation on X by $x \sim y$ if, and only if, $f(x) = f(y)$. Show that there is a bijection $\tilde{f} : X/\sim \to Y$ with the property $\tilde{f} \circ \pi = f$.

The next several exercises concern conjugacy classes in a group.

Exercise 3.3.6. Show that conjugacy of group elements is an equivalence relation.

Exercise 3.3.7. What are the conjugacy classes in S_3?

Exercise 3.3.8. What are the conjugacy classes in the symmetry group of the square D_4?

Exercise 3.3.9. What are the conjugacy classes in the dihedral group D_5?

Exercise 3.3.10. Show that a subgroup is normal if, and only if, it is a union of conjugacy classes.

3.4. Homomorphism Theorems

Consider the permutation group S_n with its normal subgroup of even permutations. For the moment write $\mathcal{E}$ for the subgroup of even permutations and $\mathcal{O}$ for the coset $\mathcal{O} = (12)\mathcal{E} = \mathcal{E}(12)$ consisting of odd permutations. The subgroup $\mathcal{E}$ is the kernel of the sign homomorphism $\epsilon : S_n \longrightarrow \{1, -1\}$.

Since the product of two permutations is even if, and only if, both are even or both are odd, one has the following multiplication table for the two cosets of $\mathcal{E}$:

	$\mathcal{E}$	$\mathcal{O}$
$\mathcal{E}$	$\mathcal{E}$	$\mathcal{O}$
$\mathcal{O}$	$\mathcal{O}$	$\mathcal{E}$

The products are taken in the following sense: The product AB of two subsets of a group is $\{ab : a \in A \text{ and } b \in B\}$.

Thus the multiplication on the cosets of $\mathcal{E}$ reproduces the multiplication on the group $\{1, -1\}$. This is a general phenomenon:

Theorem 3.4.1. *Let N be a normal subgroup of a group G. The set of cosets G/N has a unique product which makes G/N a group, and which makes the quotient map $\pi : G \longrightarrow G/N$ a group homomorphism.*

Proof. Define the product of aN and bN to be their product as subsets of G, in the sense explained above. This product is associative, because products of subsets of G is associative. Since N is normal,

$$\begin{aligned} aNbN &= \{an_1bn_2 : n_1, n_2 \in N\} \\ &= \{abn_1'n_2 : n_1', n_2 \in N\} = \{abn : n \in N\} = abN. \end{aligned}$$

It is clear that N itself serves as the identity for this multiplication and that $a^{-1}N$ is the inverse of aN. Thus G/N with this multiplication is a group. Furthermore, π is a homomorphism because

$$\pi(ab) = abN = aNbN = \pi(a)\pi(b).$$

The uniqueness of the product follows simply from the surjectivity of π: in order for π to be a homomorphism, it is necessary that $aNbN = abN$. □

Example 3.4.2 (Finite Cyclic Groups as quotients of $\mathbb{Z}$)**.** The construction of $\mathbb{Z}_n$ in Section 2.4 is an example of the quotient group construction. The (normal) subgroup in the construction is $n\mathbb{Z} = \{\ell n : \ell \in \mathbb{Z}\}$. The cosets of $n\mathbb{Z}$ in $\mathbb{Z}$ are of the form $k + n\mathbb{Z} = [k]$; the distinct cosets are $[0] = n\mathbb{Z}$, $[1] = 1 + n\mathbb{Z}, \dots, [n-1] = n - 1 + n\mathbb{Z}$. The product (sum) of two cosets is $[a] + [b] = [a + b]$. So the group we called $\mathbb{Z}_n$ is precisely $\mathbb{Z}/n\mathbb{Z}$. The quotient homomorphism $\mathbb{Z} \to \mathbb{Z}_n$ is given by $k \mapsto [k]$.

This model for the finite cyclic groups reveals an additional structure. You are asked to show in Exercise 3.4.3 that $\mathbb{Z}_n$ has a multiplicative structure which is compatible with its (additive) group structure.

Example 3.4.3. Now consider a cyclic group G of order n with generator a. There is a homomorphism $\varphi : \mathbb{Z} \to G$ of $\mathbb{Z}$ *onto* G defined by $\varphi(k) = a^k$. The kernel of this homomorphism is precisely all multiples of n, the order of a; $\ker(\varphi) = n\mathbb{Z}$. I claim that φ "induces" an isomorphism $\tilde{\varphi} : \mathbb{Z}_n \to G$, defined by $\tilde{\varphi}([k]) = a^k = \varphi(k)$. It is necessary to check that this makes sense, i.e. that $\tilde{\varphi}$ is "well defined," because we have attempted to define the value of $\tilde{\varphi}$ on a coset $[k]$ in terms of a particular representative of the coset. Would we get the same result if we took another representative, say $k + 17n$ instead of k? In fact, we would get the same answer: if $[a] = [b]$, then $a - b \in n\mathbb{Z} = \ker(\varphi)$, and therefore, $\varphi(a) - \varphi(b) = \varphi(a - b) = 0$. Thus $\varphi(a) = \varphi(b)$. This shows that the map $\tilde{\varphi}$ is well defined.

Next we have to check the homomorphism property of $\tilde{\varphi}$. This property is valid because $\tilde{\varphi}([a][b]) = \tilde{\varphi}([ab]) = \varphi(ab) = \varphi(a)\varphi(b) = \tilde{\varphi}([a])\tilde{\varphi}([b])$.

The homomorphism $\tilde{\varphi}$ has the same range as φ, so it is surjective. It is also has trivial kernel: if $\tilde{\varphi}([k]) = 0$, then $\varphi(k) = 0$, so $k \in n\mathbb{Z} = [0]$, so $[k] = [0]$. Thus $\tilde{\varphi}$ is an isomorphism.

Example 3.4.4. Take the additive abelian group $\mathbb{R}$ as G and the subgroup $\mathbb{Z}$ as N. Since $\mathbb{R}$ is abelian, all of its subgroups are normal, and, in particular, $\mathbb{Z}$ is a normal subgroup.

The cosets of $\mathbb{Z}$ in $\mathbb{R}$ were considered in Exercise 3.2.9, where you were asked to verify that the cosets are parameterized by the set of real numbers t such that $0 \leq t < 1$. In fact, two real numbers are in the same coset modulo $\mathbb{Z}$ precisely if they differ by an integer, $s \equiv t \mod \mathbb{Z} \Leftrightarrow s - t \in \mathbb{Z}$. For any real number t, let $[[t]]$ denote the greatest integer less than or equal to t. Then $t - [[t]] \in [0, 1)$ and $t \equiv (t - [[t]]) \mod \mathbb{Z}$. On the other hand, no two real numbers in $[0, 1)$ are congruent modulo $\mathbb{Z}$. Thus we have a bijection between $\mathbb{R}/\mathbb{Z}$ and $[0, 1)$ which is given by $[t] \mapsto t - [[t]]$.

We get a more instructive geometric picture of the set $\mathbb{R}/\mathbb{Z}$ of cosets of $\mathbb{R}$ modulo $\mathbb{Z}$ if we take, instead of the half open interval $[0, 1)$, the closed interval $[0, 1]$, but *identify* the endpoints 0 and 1: The picture is a circle of circumference 1. Actually we can take a circle of any convenient size, and it is more convenient to take a circle of radius 1, namely

$$\{e^{2\pi it} : t \in \mathbb{R}\} = \{e^{2\pi it} : 0 \leq t < 1\}.$$

So now we have bijections between set $\mathbb{R}/\mathbb{Z}$ of cosets of $\mathbb{R}$ modulo $\mathbb{Z}$, the set $[0, 1)$, and the unit circle $\mathbb{T}$, given by

$$[t] \mapsto t - [[t]] \mapsto e^{2\pi it} = e^{2\pi i(t-[[t]])}.$$

Let us write φ for the map $t \mapsto e^{2\pi it}$ from $\mathbb{R}$ onto the unit circle, and $\tilde{\varphi}$ for the map $[t] \mapsto \varphi(t) = e^{2\pi it}$. Our discussion shows that $\tilde{\varphi}$ is well defined. We know that the unit circle $\mathbb{T}$ is itself a group, and we recall that that the exponential map $\varphi : \mathbb{R} \to \mathbb{T}$ is a group homomorphism, namely,

$$\varphi(s + t) = e^{2\pi i(s+t)} = e^{2\pi is}e^{2\pi it} = \varphi(s)\varphi(t).$$

Furthermore, the kernel of φ is precisely $\mathbb{Z}$.

We now have a good geometric picture of the quotient group $\mathbb{R}/\mathbb{Z}$ *as a set*, but we still have to discuss the group structure of $\mathbb{R}/\mathbb{Z}$. The definition of the product (addition!) on $\mathbb{R}/\mathbb{Z}$ is $[t] + [s] = [t + s]$. But observe that

$$\tilde{\varphi}([s] + [t]) = \tilde{\varphi}([s + t]) = e^{2\pi i(s+t)} = e^{2\pi is}e^{2\pi it} = \tilde{\varphi}(s)\tilde{\varphi}(t).$$

Thus $\tilde{\varphi}$ is a group isomorphism from the quotient group $\mathbb{R}/\mathbb{Z}$ to $\mathbb{T}$.

Our work can be summarized in the following diagram, in which all of the maps are group homomorphisms, and the map π is the quotient map from $\mathbb{R}$ to $\mathbb{R}/\mathbb{Z}$.

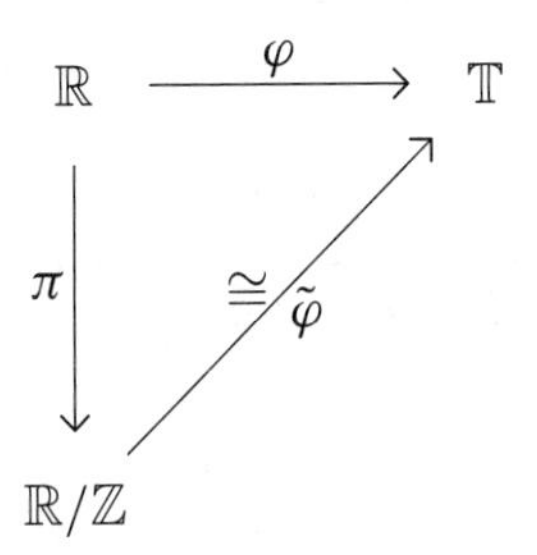

Example 3.4.5. Recall from Exercise 3.1.19 the "$Ax + b$" group or affine group $\mathrm{Aff}(n)$ consisting of transformations of $\mathbb{R}^n$ of the form

$$T_{A,\boldsymbol{b}}(\boldsymbol{x}) = A\boldsymbol{x} + \boldsymbol{b},$$

where $A \in \mathrm{GL}(n, \mathbb{R})$ and $\boldsymbol{b} \in \mathbb{R}^n$. Let N be the subset consisting of the transformations $T_{E,\boldsymbol{b}}$, where E is the identity transformation,

$$T_{E,\boldsymbol{b}}(\boldsymbol{x}) = \boldsymbol{x} + \boldsymbol{b}.$$

The composition rule in $\mathrm{Aff}(n)$ is

$$T_{A,\boldsymbol{b}} T_{A',\boldsymbol{b}'} = T_{AA',A\boldsymbol{b}'+\boldsymbol{b}}.$$

The inverse of $T_{A,\boldsymbol{b}}$ is $T_{A^{-1},-A^{-1}\boldsymbol{b}}$. N is a subgroup isomorphic to the additive group $\mathbb{R}^n$ because

$$T_{E,\boldsymbol{b}} T_{E,\boldsymbol{b}'} = T_{E,\boldsymbol{b}+\boldsymbol{b}'},$$

and N is normal. In fact,

$$T_{A,\boldsymbol{b}} T_{E,\boldsymbol{c}} T_{A,\boldsymbol{b}}^{-1} = T_{E,A\boldsymbol{c}}.$$

Let us examine the condition for two elements $T_{A,\boldsymbol{b}}$ and $T_{A',\boldsymbol{b}'}$ to be congruent modulo N. The condition is

$$T_{A,\boldsymbol{b}} T_{A',\boldsymbol{b}'}^{-1} = T_{A,\boldsymbol{b}} T_{A'^{-1},-A'^{-1}\boldsymbol{b}'} = T_{AA'^{-1},\boldsymbol{b}-AA'^{-1}\boldsymbol{b}'} \in N.$$

This is equivalent to $A = A'$. Thus the class of $T_{A,\boldsymbol{b}}$ modulo N is $[T_{A,\boldsymbol{b}}] = \{T_{A,\boldsymbol{b}'} : \boldsymbol{b}' \in \mathbb{R}^n\}$, and the cosets of N can be parameterized by $A \in \mathrm{GL}(n)$. In fact, the map $[T_{A,\boldsymbol{b}}] \mapsto A$ is a bijection between the set $\mathrm{Aff}(n)/N$ of cosets of $\mathrm{Aff}(n)$ modulo N and $\mathrm{GL}(n)$.

Let us write φ for the map $\varphi : T_{A,\boldsymbol{b}} \mapsto A$ from $\mathrm{Aff}(n)$ to $\mathrm{GL}(n)$, and $\tilde{\varphi}$ for the map $\tilde{\varphi} : [T_{A,\boldsymbol{b}}] \mapsto A$ from $\mathrm{Aff}(n)/N$ to $\mathrm{GL}(n)$. The map φ is a (surjective) homomorphism, because

$$\varphi(T_{A,\boldsymbol{b}} T_{A',\boldsymbol{b}'}) = \varphi(T_{AA',A\boldsymbol{b}'+\boldsymbol{b}}) = AA' = \varphi(T_{A,\boldsymbol{b}})\varphi(T_{A',\boldsymbol{b}'}),$$

and furthermore the kernel of φ is N.

The definition of the product in $\mathrm{Aff}(n)/N$ is

$$[T_{A,\boldsymbol{b}}][T_{A',\boldsymbol{b}'}] = [T_{A,\boldsymbol{b}} T_{A',\boldsymbol{b}'}] = [T_{AA',\boldsymbol{b}+A\boldsymbol{b}'}].$$

It follows that

$$\tilde{\varphi}([T_{A,\boldsymbol{b}}][T_{A',\boldsymbol{b}'}]) = \tilde{\varphi}([T_{AA',\boldsymbol{b}+A\boldsymbol{b}'}]) = AA' = \tilde{\varphi}([T_{A,\boldsymbol{b}}])\tilde{\varphi}([T_{A',\boldsymbol{b}'}]),$$

and therefore $\tilde{\varphi}$ is an isomorphism of groups.

We can summarize our findings in the diagram:

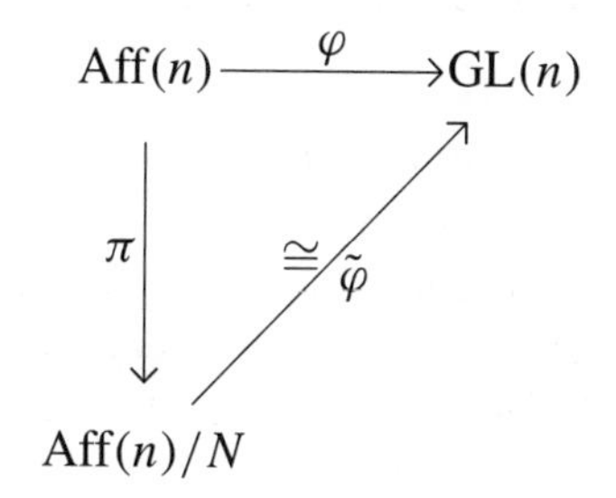

The features which we have noticed in the several examples are quite general:

Theorem 3.4.6 (Homomorphism Theorem). *Let $\varphi : G \longrightarrow \bar{G}$ be a surjective homomorphism with kernel N. Let $\pi : G \longrightarrow G/N$ be the quotient homomorphism. There is an isomorphism $\tilde{\varphi} : G/N \longrightarrow \bar{G}$ satisfying $\tilde{\varphi} \circ \pi = \varphi$. (See the following diagram.)*

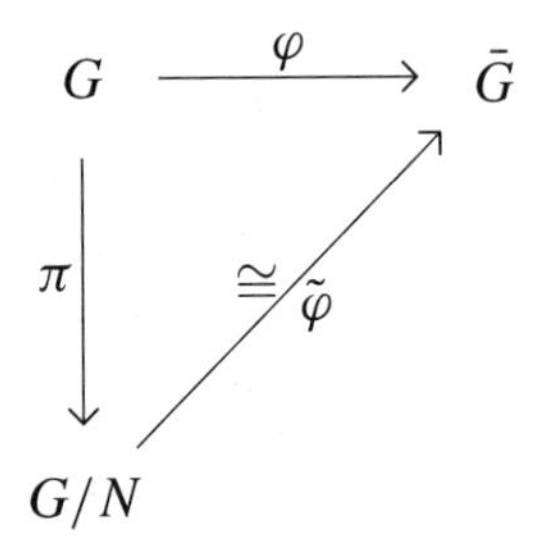

Proof. There is only one possible way to define $\tilde{\varphi}$ so that it will satisfy $\tilde{\varphi} \circ \pi = \varphi$, namely $\tilde{\varphi}(aN) = \varphi(a)$. According to Exercise 3.3.5 $\tilde{\varphi}$ is well defined and bijective. Furthermore

$$\tilde{\varphi}(aNbN) = \tilde{\varphi}(abN) = \varphi(ab) = \varphi(a)\varphi(b) = \tilde{\varphi}(aN)\tilde{\varphi}(bN).$$

□

A slightly different proof is suggested in Exercise 3.4.1.

The two theorems (Theorem 3.4.1 and Theorem 3.4.6) say that normal subgroups and (surjective) homomorphisms are two sides of one coin: Given a normal subgroup N, there is a surjective homomorphism with N as kernel, and, on the other hand, a surjective homomorphism is essentially determined by its kernel.

Theorem 3.4.6 also reveals the best way to understand a quotient group G/N. The best way is to find a natural model, namely some naturally defined group $\bar{G}$ together with a surjective homomorphism $\varphi : G \to \bar{G}$ with kernel N. Then, according to the theorem, $G/N \cong \bar{G}$. With this in mind, we take another look at the examples given above, as well as several more examples.

Example 3.4.7. Let a be an element of a group H which has order n. There is a homomorphism $\varphi : \mathbb{Z} \to H$ given by $k \mapsto a^k$. This homomorphism has range $\langle a \rangle$ and kernel $n\mathbb{Z}$. Therefore, by the Homomorphism Theorem, $\mathbb{Z}/n\mathbb{Z} \cong \langle a \rangle$. In particular, if $\zeta = e^{2\pi i/n}$, then $\varphi(k) = \zeta^k$ induces an isomorphism of $\mathbb{Z}/n\mathbb{Z}$ onto the group C_n of n-th roots of unity in $\mathbb{C}$.

Example 3.4.8. The homomorphism $\varphi : \mathbb{R} \to \mathbb{C}$ given by $\varphi(t) = e^{2\pi i t}$has range $\mathbb{T}$ and kernel $\mathbb{Z}$. Thus by the Homomorphism Theorem, $\mathbb{R}/\mathbb{Z} \cong \mathbb{T}$.

Example 3.4.9. The map $\varphi : \mathrm{Aff}(n) \to \mathrm{GL}(n)$ defined by $T_{A,\boldsymbol{b}} \mapsto A$ is a surjective homomorphism with kernel $N = \{T_{E,\boldsymbol{b}} : \boldsymbol{b} \in \mathbb{R}^n\}$. Therefore, by the Homomorphism Theorem, $\mathrm{Aff}(n)/N \cong \mathrm{GL}(n)$.

Example 3.4.10. The set $\mathrm{SL}(n, \mathbb{R})$ of matrices of determinant 1 is a normal subgroup of $\mathrm{GL}(n, \mathbb{R})$. In fact, $\mathrm{SL}(n, \mathbb{R})$ is the kernel of the homomorphism $\det : \mathrm{GL}(n, \mathbb{R}) \to \mathbb{R}^*$, and this implies that $\mathrm{SL}(n, \mathbb{R})$ is a normal subgroup. It also implies that the quotient group $\mathrm{GL}(n, \mathbb{R})/\mathrm{SL}(n, \mathbb{R})$ is naturally isomorphic to $\mathbb{R}^*$.

Example 3.4.11. Consider $G = \mathrm{GL}(n, \mathbb{R})$, the group of n-by-n invertible matrices. Set $Z = G \cap \mathbb{R}E$, the set of invertible scalar matrices. Then Z is evidently a normal subgroup of G, and is, in fact, the center of G. A coset of Z in G has the form $[A] = AZ = \{\lambda A : \lambda \in \mathbb{R}^*\}$, the set of all non-zero multiples of the invertible matrix A; two matrices A and B are equivalent modulo Z precisely if one is a scalar multiple of the other. By our general construction of quotient groups, we can form G/Z, whose elements are cosets of Z in G, with the product $[A][B] = [AB]$. G/Z is called the *projective linear group.*

The rest of this example is fairly difficult, and it might be best to skip it on the first reading. We would like to find some natural realization or model of the quotient group. Now a natural model for a group is generally as a group of transformations of something or the other, so we would have to look for some objects which are naturally transformed not by matrices but rather by matrices modulo scalar multiples.

At least two natural models are available for G/Z. One is as transformations of projective $n-1$ dimensional space $\mathbb{P}^{n-1}$, and the other is as transformations of G itself.

Projective $n-1$ dimensional space consists of n-vectors modulo scalar multiplication. More precisely, one defines an equivalence relation $\sim$ on the set $\mathbb{R}^n \setminus \{\boldsymbol{0}\}$ of non-zero vectors in $\mathbb{R}^n$ by $\boldsymbol{x} \sim \boldsymbol{y}$ if there is a non-zero scalar λ such that $\boldsymbol{x} = \lambda \boldsymbol{y}$. Then $\mathbb{P}^{n-1} = (\mathbb{R}^n \setminus \{\boldsymbol{0}\})/\sim$, the set of equivalence classes of vectors. There is another picture of $\mathbb{P}^{n-1}$ which is a little easier to visualize; every non-zero vector $\boldsymbol{x}$ is equivalent to the unit vector $\boldsymbol{x}/||\boldsymbol{x}||$, and furthermore two unit vectors $\boldsymbol{a}$ and $\boldsymbol{b}$ are equivalent if and only $\boldsymbol{a} = \pm\boldsymbol{b}$; therefore, $\mathbb{P}^{n-1}$ is also realized as $S^{n-1}/\pm$, the unit sphere in n-dimensional space, modulo equivalence of antipodal points. Write $[\boldsymbol{x}]$ for the class of a non-zero vector $\boldsymbol{x}$.

There is a homomorphism of G into $\mathrm{Sym}(\mathbb{P}^{n-1})$, the group of invertible maps from $\mathbb{P}^{n-1}$ to $\mathbb{P}^{n-1}$, defined by $\varphi(A)([x]) = [Ax]$; one has to check, as usual, that $\varphi(A)$ is a well-defined transformation of $\mathbb{P}^{n-1}$ and that φ is a homomorphism. I leave this as an exercise. What is the kernel of φ? It is precisely the invertible scalar matrices Z. Thus $\varphi(G) \cong G/Z$, by the Homomorphism Theorem, and thus G/Z has been identified as a group of transformations of projective space.

A second model for G/Z is developed in Exercise 3.4.6, as a group of transformations of G itself.

Everything in this example works in exactly the same way when $\mathbb{R}$ is replaced by $\mathbb{C}$. When moreover $n = 2$, there is a natural realization of $\mathrm{GL}(n, \mathbb{C})/Z$ as "fractional linear transformations" of $\mathbb{C}$. For this, see Exercise 3.4.5.

Proposition 3.4.12. *Let $\varphi : G \longrightarrow \bar{G}$ be a homomorphism of G onto $\bar{G}$, and let K denote the kernel of φ.*

(a) *If B is a subgroup of $\bar{G}$, then $\varphi^{-1}(B)$ is a subgroup of G containing K.*

(b) *The map $B \mapsto \varphi^{-1}(B)$ is a bijection between subgroups of $\bar{G}$ and subgroups of G containing K.*

(c) *$N \subseteq \bar{G}$ is a normal subgroup of $\bar{G}$ if, and only if, $\varphi^{-1}(N)$ is a normal subgroup of G.*

Proof. For each subgroup B of $\bar{G}$, $\varphi^{-1}(B)$ is a subgroup of G by Proposition 3.1.12, and furthermore $\varphi^{-1}(B) \supseteq \varphi^{-1}\{e\} = K$. This proves part (a).

To prove (b), I show that the map $A \mapsto \varphi(A)$ is the inverse of the map $B \mapsto \varphi^{-1}(B)$. If B is a subgroup of $\bar{G}$, then $\varphi(\varphi^{-1}(B))$ is a subgroup of $\bar{G}$, which *a priori* is contained in B. But since φ is surjective, $B = \varphi(\varphi^{-1}(B))$.

For a subgroup A of G containing $\ker(\varphi)$, $\varphi^{-1}(\varphi(A))$ is a subgroup of G which *a priori* contains A. If x is in that subgroup, then there is an $a \in A$ such that $\varphi(x) = \varphi(a)$. This is equivalent to $a^{-1}x \in \ker(\varphi) \subseteq A$ (exercise). Hence, $x \in aA = A$. This shows that $\varphi^{-1}(\varphi(A)) = A$, which completes the proof of part (b).

Suppose N is normal in $\bar{G}$. I have to show that $\varphi^{-1}(N)$ is normal in G. For all $g \in G$, $g\varphi^{-1}(N)g^{-1} \supseteq gKg^{-1} = K$, so $g\varphi^{-1}(N)g^{-1}$ is a subgroup of G containing K. Therefore, by part (b),

$$\begin{aligned} g\varphi^{-1}(N)g^{-1} &= \varphi^{-1}(\varphi(g\varphi^{-1}(N)g^{-1})) \\ &= \varphi^{-1}(\varphi(g)N\varphi(g)^{-1}) = \varphi^{-1}(N), \end{aligned}$$

since N is normal. Thus if N is normal in $\bar{G}$, then $\varphi^{-1}(N)$ is normal in G. The converse was shown in Exercise 3.1.8. $\square$

Example 3.4.13. What are all the subgroups of $\mathbb{Z}_n$? Since $\mathbb{Z}_n = \mathbb{Z}/n\mathbb{Z}$, the subgroups of $\mathbb{Z}_n$ correspond one to one with subgroups of $\mathbb{Z}$ containing the kernel of the quotient map $\varphi : \mathbb{Z} \to \mathbb{Z}/n\mathbb{Z}$, namely $n\mathbb{Z}$. But the subgroups of $\mathbb{Z}$ are cyclic and of the form $k\mathbb{Z}$ for some $k \in Z$. So when does $k\mathbb{Z}$ contain $n\mathbb{Z}$? Precisely when $n \in k\mathbb{Z}$, or when k divides n. Thus the subgroups of $\mathbb{Z}_n$ correspond one to one with *positive integer divisors of* n. The image of $k\mathbb{Z}$ in $\mathbb{Z}_n$ is cyclic with generator $[k]$ and with order n/k.

The following is a very useful generalization of the Homomorphism Theorem.

Proposition 3.4.14. *Let $\varphi : G \to \bar{G}$ be a surjective homomorphism of groups with kernel N. Let $H \subseteq N$ be a subgroup which is normal in G, and let $\pi : G \to G/H$ denote the quotient map. Then there is a surjective homomorphism $\tilde{\varphi} : G/H \to \bar{G}$ such that $\tilde{\varphi} \circ \pi = \varphi$. (See the diagram below.) The kernel of $\tilde{\varphi}$ is $N/H \subseteq G/H$.*

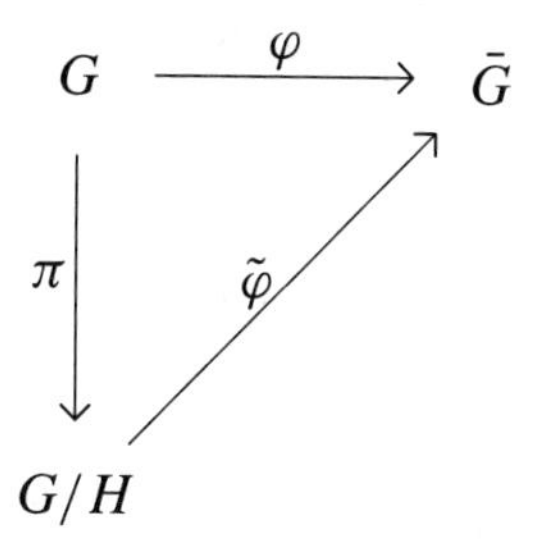

Proof. As in the proof of the Homomorphism Theorem, there is only one way to define $\tilde{\varphi}$ consistent with the requirement that $\tilde{\varphi} \circ \pi = \varphi$, namely $\varphi(aH) = \varphi(a)$. It is necessary to check that this is well defined and a homomorphism. But if $aH = bH$, then $b^{-1}a \in H \subseteq N = \ker(\varphi)$, so $\varphi(b^{-1}a) = 1$, or $\varphi(a) = \varphi(b)$. This shows that the map $\tilde{\varphi}$ is well defined. The homomorphism property follows as in the proof of the Homomorphism Theorem. □

Corollary 3.4.15.

(a) *Let $H \subseteq N \subseteq G$ be subgroups with both H and N normal in G. Then $xH \mapsto xN$ defines a homomorphism of G/H onto G/N with kernel N/H*

(b) *There is a natural isomorphism $(G/H)/(N/H) \cong G/N$*

Proof. Statement (a) is the special case of the Proposition with $\bar{G} = G/N$ and $\varphi : G \to G/N$ the quotient map. For part (b), apply the homomorphism theorem to the surjective homomorphism $G/H \to G/N$ with kernel N/H. □

Example 3.4.16. When is there a surjective homomorphism from one cyclic group $\mathbb{Z}_k$ to another cyclic group $\mathbb{Z}_\ell$?

Suppose first that $\psi : \mathbb{Z}_k \to \mathbb{Z}_\ell$ is a surjective homomorphism such that $\psi[1] = [1]$. Let φ_k and φ_ℓ be the natural quotient maps of $\mathbb{Z}$ onto $\mathbb{Z}_k$ and $\mathbb{Z}_\ell$ respectively. We have maps

$$\mathbb{Z} \xrightarrow{\varphi_k} \mathbb{Z}_k \xrightarrow{\psi} \mathbb{Z}_\ell,$$

and $\psi \circ \varphi_k$ is a surjective homomorphism of $\mathbb{Z}$ onto $\mathbb{Z}_\ell$ such that $\psi \circ \varphi_k(1) = [1]$; therefore, $\psi \circ \varphi_k = \varphi_\ell$. But then the kernel of φ_k is contained in the kernel of φ_ℓ, which is to say that every integer multiple of k is divisible by ℓ. In particular k is divisible by ℓ.

The assumption that $\psi[1] = [1]$ is not essential and can be eliminated as follows: Suppose that $\psi : \mathbb{Z}_k \to \mathbb{Z}_\ell$ is a surjective homomorphism with $\psi([1]) = [a]$. The cyclic subgroup group generated by $[a]$ is all of $\mathbb{Z}_\ell$, and in particular $[a]$ has order ℓ. Thus there is a surjective homomorphism $\mathbb{Z} \to \mathbb{Z}_\ell$ defined by $n \mapsto [na]$, with kernel $\ell\mathbb{Z}$. It follows from the Homomorphism Theorem that there is an isomorphism $\theta : \mathbb{Z}_\ell \to \mathbb{Z}_\ell$ such that $\theta([1]) = [a]$. But then $\theta^{-1} \circ \psi : \mathbb{Z}_k \to \mathbb{Z}_\ell$ is a surjective homomorphism such that $\theta^{-1} \circ \psi([1]) = \theta^{-1}([a]) = [1]$. It follows that k is divisible by ℓ.

Conversely, if k is divisible by ℓ, then $k\mathbb{Z} \subseteq \ell\mathbb{Z} \subseteq Z$. Since $\mathbb{Z}$ is abelian, all subgroups are normal, and by the Corollary, there is a surjective homomorphism $\mathbb{Z}_k \to \mathbb{Z}_\ell$ such that $[1] \mapsto [1]$.

We conclude that there is a surjective homomorphism from $\mathbb{Z}_k$ to $\mathbb{Z}_l$ if, and only if, ℓ divides k.

Proposition 3.4.17. *Let $\varphi : G \longrightarrow \bar{G}$ be a surjective homomorphism. Let $\bar{K}$ be a normal subgroup of $\bar{G}$ and let $K = \varphi^{-1}(\bar{K})$. Then*

(a) $x \mapsto \varphi(x)\bar{K}$ *is a homomorphism of G onto $\bar{G}/\bar{K}$.*

(b) $xK \mapsto \varphi(x)\bar{K}$ *is an isomorphism of G/K onto $\bar{G}/\bar{K}$.*

Proof. We know from the previous proposition that K is a normal subgroup of G containing $\ker(\varphi)$. Write ψ for the canonical projection $\psi : \bar{G} \longrightarrow \bar{G}/\bar{K}$. Then $\psi \circ \varphi : G \longrightarrow \bar{G}/\bar{K}$ is a surjective homomorphism, because it is a composition of surjective homomorphisms. The kernel of $\psi \circ \varphi$ is the set of $x \in G$ such that $\varphi(x) \in \ker(\psi) = \bar{K}$; that is $\ker(\psi \circ \varphi) = \varphi^{-1}(\bar{K}) = K$. This proves (a).

Statement (b) now follows from the Homomorphism Theorem 3.4.6. □

Proposition 3.4.18. *Let $\varphi : G \longrightarrow H$ be a surjective homomorphism with kernel N. Let A be a subgroup of G. Then $\varphi^{-1}(\varphi(A)) = AN = \{an : a \in A \text{ and } n \in N\}$, and AN is a subgroup of G containing N. Furthermore, $AN/N \cong \varphi(A) \cong A/(A \cap N)$.*

Proof. Let $x \in G$. Then

$$\begin{aligned} x \in \varphi^{-1}(\varphi(A)) &\Leftrightarrow \text{there exists } a \in A \text{ such that } \varphi(x) = \varphi(a) \\ &\Leftrightarrow \text{there exists } a \in A \text{ such that } x \in aN \\ &\Leftrightarrow x \in AN. \end{aligned}$$

Thus, $AN = \varphi^{-1}(\varphi(A))$, which, by Proposition 3.4.12, is a subgroup of G containing N. Now applying Theorem 3.4.6 to the restriction of φ to AN gives the isomorphism $AN/N \cong \varphi(AN) = \varphi(A)$. On the other hand, applying the theorem to the restriction of φ to A gives $A/(A \cap N) \cong \varphi(A)$. □

Example 3.4.19. Let G be the symmetry group of the square, which is generated by elements r and j satisfying $r^4 = e = j^2$ and $jrj = r^{-1}$. Let N be the subgroup $\{e, r^2\}$; then N is normal because $jr^2j = r^{-2} = r^2$. What is G/N? The group G/N has order 4, and is generated by two commuting elements rN and jN each of order 2. (Note that rN and jN commute because $rN = r^{-1}N$, and $jr^{-1} = rj$, so $jrN = jr^{-1}N = rjN$.) Hence, G/N is isomorphic to the group $\mathcal{V}$ of symmetries of the rectangle. Let $A = \{e, j\}$. Then AN is a 4-element subgroup of G (also isomorphic to $\mathcal{V}$) and $AN/N = \{N, jN\} \cong \mathbb{Z}_2$. On the other hand, $A \cap N = \{e\}$, so $A/(A \cap N) \cong A \cong \mathbb{Z}_2$.

Example 3.4.20. Let $G = \mathrm{GL}(n, \mathbb{C})$, the group of n-by-n invertible complex matrices. Let Z the the subgroup of invertible scalar matrices. G/Z is the complex projective linear group. Let $A = \mathrm{SL}(n, \mathbb{C})$. Then $AZ = G$ because for any invertible matrix X, one has $X = \det(X)X'$, where $X' = \det(X)^{-1}X \in A$. On the other hand, $A \cap Z$ is the group of invertible scalar matrices with determinant 1; such a matrix must have the form ζE where ζ is an n-th root of unity in $\mathbb{C}$. We have $G/Z = AZ/Z = A/(A \cap Z) = A/\{\zeta E : \zeta \text{ is a root of unity}\}$.

The same holds with $\mathbb{C}$ replaced by $\mathbb{R}$, and here the result is more striking, because $\mathbb{R}$ contains few roots of unity. If n is odd, the only n-th root of unity in $\mathbb{R}$ is 1, so we see that the projective linear group is isomorphic to $\mathrm{SL}(n, \mathbb{R})$. On the other hand, if n is even, then -1 is also an n-th root of unity and the projective linear group is isomorphic to $\mathrm{SL}(n, \mathbb{R})/\{\pm 1E\}$.

Exercises

Exercise 3.4.1. Adopt the notation of Theorem 3.4.6. Recall from Exercise 3.1.7 that for $a, x \in G$,

$$\varphi(a) = \varphi(x) \Leftrightarrow a^{-1}x \in N \Leftrightarrow aN = xN$$

Show that it follows from this that $\tilde{\varphi}$ defined by $\tilde{\varphi}(aN) = \varphi(a)$ is both well defined and injective. Observe that the surjectivity of $\tilde{\varphi}$ follows from the surjectivity of φ.

Exercise 3.4.2. Consider the affine group $\mathrm{Aff}(n)$ consisting of transformations of $\mathbb{R}^n$ of the form $T_{A,\boldsymbol{b}}(\boldsymbol{x}) = A\boldsymbol{x} + \boldsymbol{b}$ ($A \in \mathrm{GL}(n, \mathbb{R})$ and $\boldsymbol{b} \in \mathbb{R}^n$).

(a) Show that the inverse of $T_{A,\boldsymbol{b}}$ is $T_{A^{-1},-A^{-1}\boldsymbol{b}}$.

(b) Show that $T_{A,\boldsymbol{b}}T_{E,\boldsymbol{c}}T_{A,\boldsymbol{b}}^{-1} = T_{E,A\boldsymbol{c}}$. Conclude that $N = \{T_{E,\boldsymbol{b}} : \boldsymbol{b} \in \mathbb{R}^n\}$ is a normal subgroup of $\mathrm{Aff}(n)$.

Exercise 3.4.3. Define a multiplication on $\mathbb{Z}_n$ by $[a][b] = [ab]$. Show that this multiplication is well defined, associative, and commutative. Show that $[1]$ is an identity element for multiplication. ($[0]$ is the identity for the additive group structure.) Show that multiplication distributes over addition: $[a]([b]+[c]) = [a][b]+[a][c]$. *Remark:* Once you have shown that the multiplication is well defined, the rest follows easily from the familiar properties of $\mathbb{Z}$, which you may assume.

Exercise 3.4.4. Suppose G is finite. Verify that:

$$\#(AN) = \frac{\#(A)\,\#(N)}{\#(A \cap N)}.$$

Exercise 3.4.5. Consider the set of transformations of the complex plane: $\mathbb{C}$

$$T_{a,b;c,d}(z) = \frac{az+b}{cz+d}$$

where

$$\begin{bmatrix} a & b \\ c & d \end{bmatrix}$$

is an invertible 2-by-2 complex matrix. Show that this is a group of transformations, and is isomorphic to

$$\mathrm{GL}(2, \mathbb{C})/Z(\mathrm{GL}(2, \mathbb{C})).$$

Exercise 3.4.6. Recall that an automorphism of a group G is a group isomorphism from G to G. Denote the set of all automorphisms of G by $\mathrm{Aut}(G)$.

(a) Show that $\mathrm{Aut}(G)$ of G is also a group.
(b) Recall that for each $g \in G$, the map $c_g : G \longrightarrow G$ defined by $c_g(x) = gxg^{-1}$ is an element of $\mathrm{Aut}(G)$. Show that the map $c : g \mapsto c_g$ is a homomorphism from G to $\mathrm{Aut}(G)$.
(c) Show that the kernel of the map c is $Z(G)$.
(d) In general the map c is not surjective. The image of c is called the *group of inner automorphisms* and denoted $\mathrm{Int}(G)$. Conclude that $\mathrm{Int}(G) \cong G/Z(G)$.

Exercise 3.4.7. Let D_4 denote the group of symmetries of the square, and N the subgroup of rotations. Observe that N is normal and check that D_4/N is isomorphic to the cyclic group of order 2.

Exercise 3.4.8. Find out whether every automorphism of S_3 is inner. Note that any automorphism φ must permute the set of elements of order 2, and an automorphism φ is completely determined by what it does to order 2 elements, since all elements are products of 2-cycles. Hence, there can be at most as many automorphisms of S_3 as there are permutations of the 3 element set of 2-cycles, namely 6; that is $\#\mathrm{Aut}(S_3) \leq 6$. According to problems 3.2.11 and 3.4.6 how large is $\mathrm{Int}(S_3)$? What do you conclude?

Exercise 3.4.9. Let G be a group and let C be the subgroup generated by all elements of the form $xyx^{-1}y^{-1}$ with $x, y \in G$. C is called the *commutator subgroup* of G. Show that C is a normal subgroup and that G/C is abelian. Show that if H is a normal subgroup of G such that G/H is abelian, then $H \supseteq C$.

Exercise 3.4.10. Show that any quotient of an abelian group is abelian.

Exercise 3.4.11. Prove that if $G/Z(G)$ is cyclic, then G is abelian.

Exercise 3.4.12. Suppose $G/Z(G)$ is abelian. Must G be abelian?

3.5. Exercises on Direct Products

In this section we explore the idea of the direct product of groups by means of exercises.

Whenever one has a normal subgroup N of a group G, the group G is in some sense built up from N and the quotient group G/N, both of which are in general smaller and simpler than G itself. So a route to understanding G is to understand N and G/N, and finally to understand the way the two interact.

The simplest way in which a group can be built up from two groups is without any interaction between the two. Given any two groups A and B, we can give the Cartesian product $A \times B$ of the sets A and B a group structure, as follows. Declare the product $(a, b)(a', b')$ to be (aa', bb').

Exercise 3.5.1.

(a) This product makes $A \times B$ a group, with $e = (e_A, e_B)$ serving as the identity element, and (a^{-1}, b^{-1}) the inverse of (a, b).

(b) $\tilde{A} = A \times \{e_B\} = \{(a, e_B) : a \in A\}$ is a normal subgroup of $A \times B$, and $A \cong \tilde{A}$.

(c) Similarly, $\tilde{B} = \{e_A\} \times B$ is a normal subgroup of $A \times B$, and $B \cong \tilde{B}$.

(d) $(a, b) \mapsto a$ is a surjective homomorphism of $A \times B$ onto A with kernel $\tilde{B}$ and $(a, b) \mapsto b$ is a surjective homomorphism of $A \times B$ onto B with kernel $\tilde{A}$.

(e) $\tilde{A} \cap \tilde{B} = \{e\}$ and $xy = yx$ for $x \in \tilde{A}$ and $y \in \tilde{B}$. $\tilde{A}\tilde{B} = A \times B$.

Definition 3.5.1. $A \times B$, with this group structure, is called the *direct product* of A and B.

It is useful to be able to recognize when a group is isomorphic to a direct product of groups:

Exercise 3.5.2.

(a) Suppose M and N are normal subgroups of G, and $M \cap N = \{e\}$. Then for all $m \in M$ and $n \in N$, $mn = nm$. *Hint:* Put all the good stuff on one side of the equation: show that $mnm^{-1}n^{-1} = e$.

(b) Show that MN is a subgroup and $(m, n) \mapsto mn$ is an isomorphism of $M \times N$ onto MN.

(c) In particular, if $MN = G$, then $G \cong M \times N$.

Exercise 3.5.3.

(a) Show that $\mathbb{Z}_6 \cong \mathbb{Z}_3 \times \mathbb{Z}_2$.

(b) Show that $\mathbb{Z}_{18} \cong \mathbb{Z}_9 \times \mathbb{Z}_2$.

Exercise 3.5.4. Let G be the group of symmetries of the rectangle. Let r be the rotation or order 2 about the axis through the centroid of the faces, and let j be the rotation of order 2 about an axis passing through the centers of two opposite edges. Set $R = \{e, r\}$ and $J = \{e, j\}$. Then $G \cong R \times J \cong \mathbb{Z}_2 \times \mathbb{Z}_2$.

Sometime authors distinguish between *external* direct products and *internal* direct products. For groups A and B, the group constructed from the Cartesian product of A and B above is the *external* direct product. On the other hand if a group G has normal subgroups N and M such that $N \cap M = \{e\}$, so that G is *isomorphic* to the direct product $N \times M$, one says that G is the *internal* direct product of N and M.

Similarly, one can define the direct product of any number of groups. Given groups $A_1, A_2, \dots, A_n$, define a product on the Cartesian product set $A_1 \times A_2 \times \cdots \times A_n$ by

$$(x_1, x_2, \dots x_n)(y_1, y_2, \dots, y_n) = (x_1 y_1, x_2 y_2, \dots, x_n y_n)$$

for $x_i, y_i \in A_i$.

Exercise 3.5.5.

(a) This product makes $P = A_1 \times \cdots \times A_n$ into a group.

(b) For each i, $\tilde{A}_i = \{e\} \times \dots \{e\} \times A_i \times \{e\} \times \dots \{e\}$ is a normal subgroup of P, and $A_i \cong \tilde{A}_i$.

(c) For all i, $\tilde{A}_i \cap (\tilde{A}_1 \dots \tilde{A}_{i-1} \tilde{A}_{i+1} \dots \tilde{A}_n) = \{e\}$, and $xy = yx$ if $x \in \tilde{A}_i$ and $y \in \tilde{A}_j$, with $i \neq j$. Furthermore, $\tilde{A}_1 \tilde{A}_2 \dots \tilde{A}_n = P$.

(d) For each i, $\pi_i : (a_1, a_2, \dots, a_n) \mapsto a_i$ defines a surjective homomorphism of $A_1 \times \cdots \times A_n$ onto A_i, and

$$\ker(\pi_1) \cap \cdots \cap \ker(\pi_n) = \{e\}.$$

Moreover, for all i the restriction of π_i to $\cap_{j \neq i} \ker(\pi_j)$ maps onto A_i.

Exercise 3.5.6.

(a) Suppose $N_1, N_2, \dots, N_r$ are normal subgroups of a group G such that for all i, $N_i \cap (N_1 \dots N_{i-1} N_{i+1} \dots N_n) = \{e\}$. It follows from Exercise 3.5.2 that elements of N_i commute with elements of N_j for $i \neq j$. Show that $N_1 N_2 \dots N_r$ is a subgroup of G and

$$(n_1, n_2, \dots, n_r) \mapsto n_1 n_2 \dots n_r$$

is an isomorphism of $N_1 \times N_2 \times \cdots \times N_r$ onto $N_1 N_2 \dots N_r$. In particular, if $N_1 N_2 \dots N_r = G$, then $G \cong N_1 \times N_2 \times \cdots \times N_r$.

(b) *Caution:* When $r > 2$, it does not suffice that $N_i \cap N_j = \{e\}$ for $i \neq j$ and $N_1 N_2 \dots N_r = G$ in order for G to be isomorphic to $N_1 \times N_2 \times \cdots \times N_r$. For example, take G to be the symmetry group of the rectangular card, $G \cong \mathbb{Z}_2 \times \mathbb{Z}_2$. G has three normal subgroups of order 2; the intersection of any two is $\{e\}$ and the product of any two is G.

Exercise 3.5.7.

(a) Show that $\mathbb{Z}_{30} \cong \mathbb{Z}_5 \times \mathbb{Z}_3 \times \mathbb{Z}_2$.

(b) State and prove a result concerning a direct product decomposition of $\mathbb{Z}_n$, taking into account the decomposition of n as a product of prime numbers.

Exercise 3.5.8.

(a) Suppose G is a group and $\varphi_i : G \longrightarrow N_i$ is a homomorphism for $i = 1, 2, \dots, r$. Suppose $\ker(\varphi_1) \cap \dots \ker(\varphi_r) = \{e\}$. Show that $\varphi : x \mapsto (\varphi_1(x), \dots, \varphi_r(x))$ is an injective homomorphism *into* $N_1 \times \dots \times N_r$.

(b) φ does not have to be surjective even if $r = 2$ and both of the φ_i are surjective. For example, take $N_1 = N_1 = N$ and $G = \{(a, a) : a \in N\} \subseteq N \times N$, and $\varphi_1((a, a)) = \varphi_2((a, a)) = a$.

(c) Explore conditions for φ to be surjective.

3.6. Semi-direct Products

We now consider a slightly more complicated way in which two groups can be fit together to form a larger group.

Example 3.6.1. Consider the dihedral group D_n of order $2n$, the rotation group of the regular n-gon. D_n has a normal subgroup N of index 2 consisting of rotations about the axis through the centroid of the faces of the n-gon. N is cyclic of order n, generated by the rotation through an angle of $2\pi/n$. D_n also has a subgroup A of order 2. generated by a rotation j through an angle π about an axis through the centers of opposite edges (if n is even), or through a vertex and the opposite edge (if n is odd). One has $D_n = NA$, $N \cap A = \{e\}$, and $A \cong D_n/N$, just as in a direct product. But A is not normal, and the non-identity element j of A does not commute with N. Instead one has a commutation relation $jr = r^{-1}j$ for $r \in N$.

Example 3.6.2. Recall the affine or "$Ax + b$" group $\mathrm{Aff}(n)$ from Exercises 3.1.19 and 3.4.2 and Examples 3.4.5 and 3.4.9. It has a normal subgroup N consisting of transformations $T_{\boldsymbol{b}} : \boldsymbol{x} \mapsto \boldsymbol{x} + \boldsymbol{b}$ for $\boldsymbol{b} \in \mathbb{R}^n$. And it has the subgroup $\mathrm{GL}(n, \mathbb{R})$, which is not normal. One has $N\mathrm{GL}(n, \mathbb{R}) = \mathrm{Aff}(n)$, $N \cap \mathrm{GL}(n, \mathbb{R}) = \{E\}$, and $\mathrm{Aff}(n)/N \cong \mathrm{GL}(n, \mathbb{R})$. $\mathrm{GL}(n, \mathbb{R})$ does not commute with N; instead one has the commutation relation $AT_{\boldsymbol{b}} = T_{A\boldsymbol{b}}A$ for $A \in \mathrm{GL}(n, \mathbb{R})$ and $\boldsymbol{b} \in \mathbb{R}^n$.

Both of the last examples are instances of the following situation: A group G has a normal subgroup N and another subgroup A, which is not normal. One has $G = NA = AN$, $A \cap N = \{e\}$, and $A \cong G/N$. Since N

is normal, for each $a \in A$, the inner automorphism c_a of G restricts to an automorphism of N, and one has the commutation relation $an = c_a(n)a$ for $a \in A$ and $n \in N$.

Now, if one has groups N and A, and one has a homomorphism $\alpha : a \mapsto \alpha_a$ from A into the automorphism group $\text{Aut}(N)$ of N, one can build from this data a new group $A \underset{\alpha}{\ltimes} N$, called the semi-direct product of A and N. The semi-direct product $A \underset{\alpha}{\ltimes} N$ has the following features: it contains (isomorphic copies of) A and N as subgroups, with N normal; the intersection of these subgroups is the identity, and the product of these subgroups is $A \underset{\alpha}{\ltimes} N$; and one has the commutation relation $an = \alpha_a(n)a$ for $a \in A$ and $n \in N$.

The construction is straightforward. As a set, $A \underset{\alpha}{\ltimes} N$ is $N \times A$, but now the product is defined by $(n, a)(n', a') = (n\alpha_a(n'), aa')$.

Exercises

Exercise 3.6.1. Show that $A \underset{\alpha}{\ltimes} N$ defined as above is a group, and that it has the features described above.

Exercise 3.6.2. Show that D_n is isomorphic to a semi-direct product of $\mathbb{Z}_2$ and $\mathbb{Z}_n$.

Exercise 3.6.3. Show that the affine group $\text{Aff}(n)$ is isomorphic to a semi-direct product of $\text{GL}(n, \mathbb{R})$ and the additive group $\mathbb{R}^n$.

Exercise 3.6.4. Show that the permutation group S_n is a semi-direct product of $\mathbb{Z}_2$ and the group of even permutations A_n.

Exercise 3.6.5. Consider the set G of n-by-n matrices with entries in $\{0, \pm 1\}$ which have exactly one non-zero entry in each row and column. These are called signed permutation matrices. Show that G is a group, and that G is a semi-direct product of S_n and the group of diagonal matrices with entries in $\{\pm 1\}$. S_n acts on the group of diagonal matrices by permutation of the diagonal entries.

One final example shows that direct products and semi-direct products do not exhaust the ways in which a normal subgroup N and the quotient group G/N can be fit together to form a group G:

Exercise 3.6.6. $\mathbb{Z}_4$ has a subgroup isomorphic to $\mathbb{Z}_2$, namely the subgroup generated by [2]. The quotient $\mathbb{Z}_4/Z_2$ is also isomorphic to $\mathbb{Z}_2$. Nevertheless, $\mathbb{Z}_4$ is not a direct or semi-direct product of two copies of $\mathbb{Z}_2$.

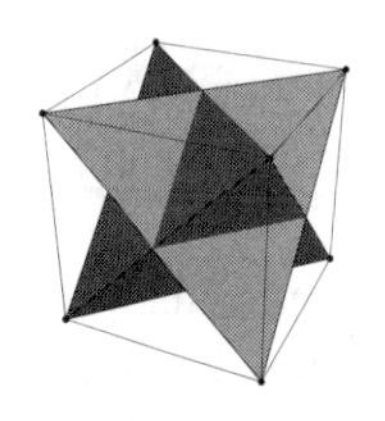

Chapter 4

Symmetries of Polyhedra

4.1. Rotations of Regular Polyhedra

There are five regular polyhedra: the tetrahedron, the cube, the octahedron, the dodecahedron (12 faces) and the icosahedron (20 faces).

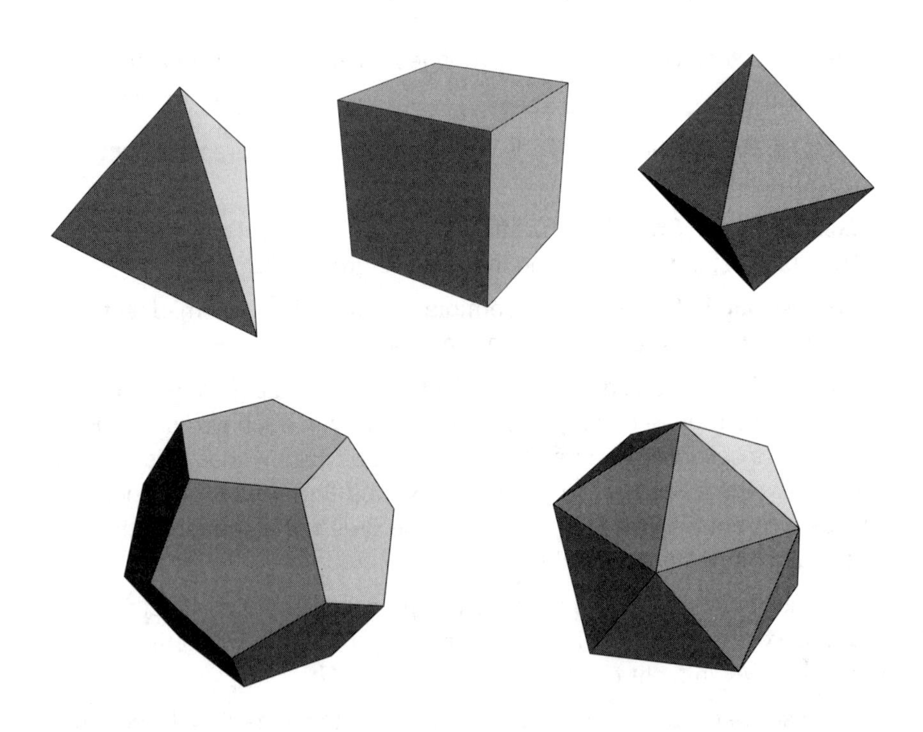

Figure 4.1.1. The Regular Polyhedra

In this section, we will work out the rotational symmetry groups of the tetrahedron, cube, and octahedron, and in the following section we will treat the dodecahedron and icosahedron. In later sections, we will work out the full symmetry groups, with reflections allowed as well as rotations.

It is pleasant to have physical models of the regular polyhedra which one can handle while studying their properties. In case you can't get any ready-made models, I have provided you with patterns for making paper models at the end of the book. (See Appendix E.) I urge you to obtain or construct models of the regular polyhedra before continuing with your reading.

Now that you have your models of the regular polyhedra, we can proceed to obtain their rotation groups. We start with the tetrahedron.

Definition 4.1.1. A line is an *n-fold axis of symmetry* for a geometric figure if the rotation by $2\pi/n$ about this line is a symmetry.

For each n-fold axis of symmetry, there are $n-1$ non-identity symmetries of the figure, namely the rotations by $2k\pi/n$ for $1 \leq k \leq n-1$.

The tetrahedron has four 3-fold axes of rotation each of which passes through a vertex and the centroid of the opposite face. These give eight non-identity group elements, each of order 3. See Figure 4.1.2.

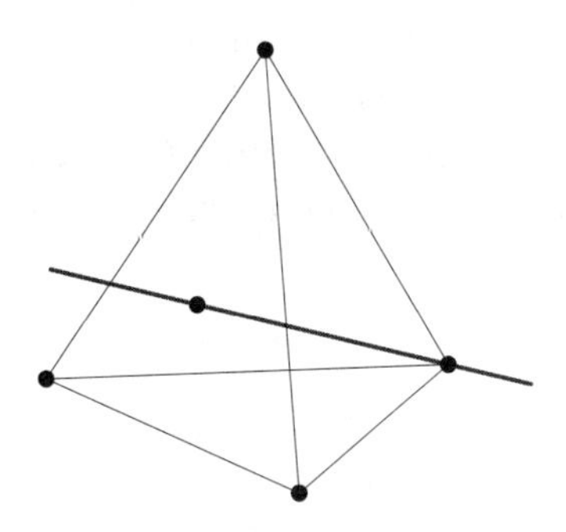

Figure 4.1.2. Three-Fold Axis of the Tetrahedron

The tetrahedron also has three 2-fold axes of symmetry, each of which passes through the centers of a pair of opposite edges. These contribute three non-identity group elements, each of order 2. See Figure 4.1.3.

Including the identity, we have 12 rotations altogether.

The rotation group acts faithfully as permutations of the four vertices; this means there is an injective homomorphism of the rotation group into S_4. Under this homomorphism the eight rotations of order three are mapped to the eight 3-cycles in S_4. The three rotations of order 2 are mapped to the

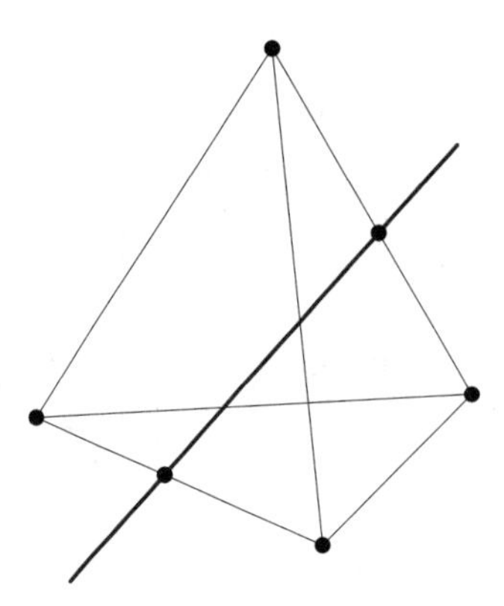

Figure 4.1.3. Two-Fold Axis of the Tetrahedron

three elements $(12)(34)$, $(13)(24)$, and $(14)(23)$. Thus the image in S_4 is precisely the group of even permutations A_4.

Proposition 4.1.2. *The rotation group of the tetrahedron is isomorphic to the group A_4 of even permutations of four objects.*

Let us also work out the matrices which implement the rotations of the tetrahedron. First we need to figure out how to write the matrix for a rotation through an angle θ about the axis determined by a unit vector $\hat{\boldsymbol{v}}$. Of course, there are two possible such rotations, which are inverses of each other; let's agree to find the one determined by the "right hand rule," as in Figure 4.1.4.

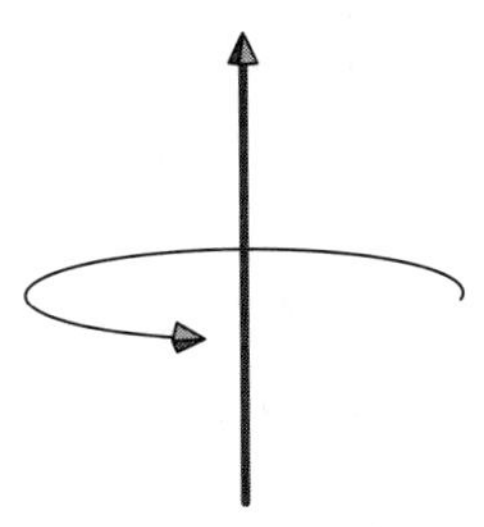

Figure 4.1.4. Right Hand Rule

If $\hat{\boldsymbol{v}}$ is the first standard co-ordinate vector

$$\hat{\boldsymbol{e}}_1 = \begin{bmatrix} 1 \\ 0 \\ 0 \end{bmatrix},$$

then the rotation matrix is

$$R_\theta = \begin{bmatrix} 1 & 0 & 0 \\ 0 & \cos\theta & \sin\theta \\ 0 & -\sin\theta & \cos\theta \end{bmatrix}.$$

To compute the matrix for the rotation about the vector $\hat{\boldsymbol{v}}$, first one needs to find additional vectors $\hat{\boldsymbol{v}}_2$ and $\hat{\boldsymbol{v}}_3$ such that $\{\hat{\boldsymbol{v}}_1 = \hat{\boldsymbol{v}}, \hat{\boldsymbol{v}}_2, \hat{\boldsymbol{v}}_3\}$ form a right handed orthonormal basis of $\mathbb{R}^3$; that is, the three vectors are of unit length, mutually orthogonal, and the determinant of the matrix $V = [\hat{\boldsymbol{v}}_1, \hat{\boldsymbol{v}}_2, \hat{\boldsymbol{v}}_3]$ with columns $\hat{\boldsymbol{v}}_i$ is 1, or equivalently, $\hat{\boldsymbol{v}}_3$ is the vector cross-product $\hat{\boldsymbol{v}}_1 \times \hat{\boldsymbol{v}}_2$. V is the matrix which rotates the standard right handed orthonormal basis $\{\hat{\boldsymbol{e}}_1, \hat{\boldsymbol{e}}_2, \hat{\boldsymbol{e}}_3\}$ onto the basis $\{\hat{\boldsymbol{v}}_1, \hat{\boldsymbol{v}}_2, \hat{\boldsymbol{v}}_3\}$. The inverse of V is the transposed matrix V^t, because the matrix entries of $V^t V$ are the inner products $\langle \hat{\boldsymbol{v}}_i | \hat{\boldsymbol{v}}_j \rangle = \delta_{ij}$. The matrix we are looking for is $V R_\theta V^t$, because the matrix first rotates the orthonormal basis$\{\hat{\boldsymbol{v}}_i\}$ onto the standard orthonormal basis$\{\hat{\boldsymbol{e}}_i\}$, then rotates through an angle θ about $\hat{\boldsymbol{e}}_1$, and then rotates the standard basis $\{\hat{\boldsymbol{e}}_i\}$ back to the basis $\{\hat{\boldsymbol{v}}_i\}$.

Consider the points:

$$\begin{bmatrix} 1 \\ 1 \\ 1 \end{bmatrix}, \quad \begin{bmatrix} 1 \\ -1 \\ -1 \end{bmatrix}, \quad \begin{bmatrix} -1 \\ -1 \\ 1 \end{bmatrix}, \quad \text{and} \quad \begin{bmatrix} -1 \\ 1 \\ -1 \end{bmatrix}.$$

They are equidistant from each other, and the sum of the four is $\boldsymbol{0}$, so the four points are the vertices of a tetrahedron whose center of mass is at the origin.

One three-fold axis of the tetrahedron passes through the origin and the point

$$\begin{bmatrix} 1 \\ 1 \\ 1 \end{bmatrix}.$$

Thus the right-handed rotation through the angle $2\pi/3$ about the unit vector

$$(1/\sqrt{3}) \begin{bmatrix} 1 \\ 1 \\ 1 \end{bmatrix}$$

is one symmetry of the tetrahedron. In Exercise 4.1.1 you are asked to compute the matrix of this rotation. The result is the permutation matrix

$$\begin{bmatrix} 0 & 0 & 1 \\ 1 & 0 & 0 \\ 0 & 1 & 0 \end{bmatrix}.$$

In Exercise 4.1.2, you are asked to show that the matrices for rotations of order 2 are the diagonal matrices with two entries of -1 and one entry of

1. Finally, you can show (Exercise 4.1.3) that these diagonal matrices and the permutation matrix

$$\begin{bmatrix} 0 & 0 & 1 \\ 1 & 0 & 0 \\ 0 & 1 & 0 \end{bmatrix}$$

generate the group of rotation matrices for the tetrahedron. Therefore the full set of rotation matrices for the tetrahedron is the set of signed permutation matrices with the property that the product of the non-zero entries is 1. That is, in each row and column there is precisely one non-zero entry, that entry is either 1 or -1, and either there are no -1's or else there are two -1's.

Now we consider the cube. The cube has four 3-fold axes through pairs of opposite vertices, giving eight rotations of order 3. See Figure 4.1.5.

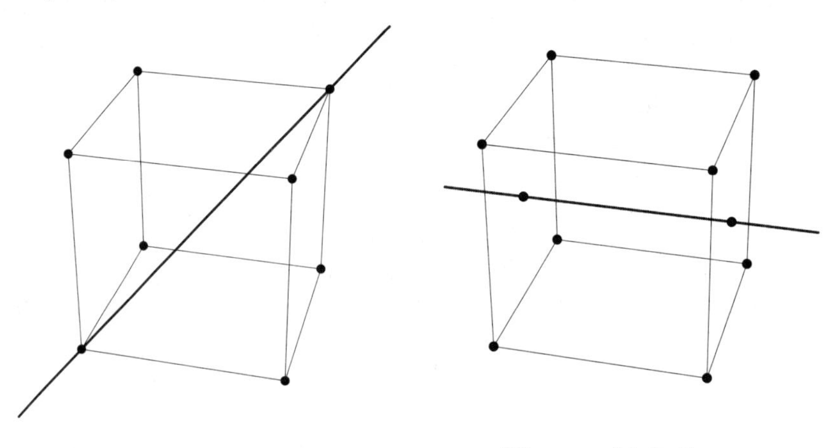

Figure 4.1.5. Three-Fold Axis of the Cube

Figure 4.1.6. Four-Fold Axis of the Cube

There are also three 4-fold axes through the centroids of opposite faces, giving nine non-identity group elements, three of order 2 and six of order 4. See Figure 4.1.6.

Finally, the cube has six 2-fold axes through centers of opposite edges, giving six order 2 elements. See Figure 4.1.7. With the identity, we have 24 rotations altogether.

Now we need to find an injective homomorphism into some permutation group induced by the action of the rotation group on some set of geometric objects associated with the cube. If we choose vertices, or edges, or faces, we get homomorphisms into S_8, S_{12}, or S_6 respectively. None of these choices look very promising, since our group has only 24 elements.

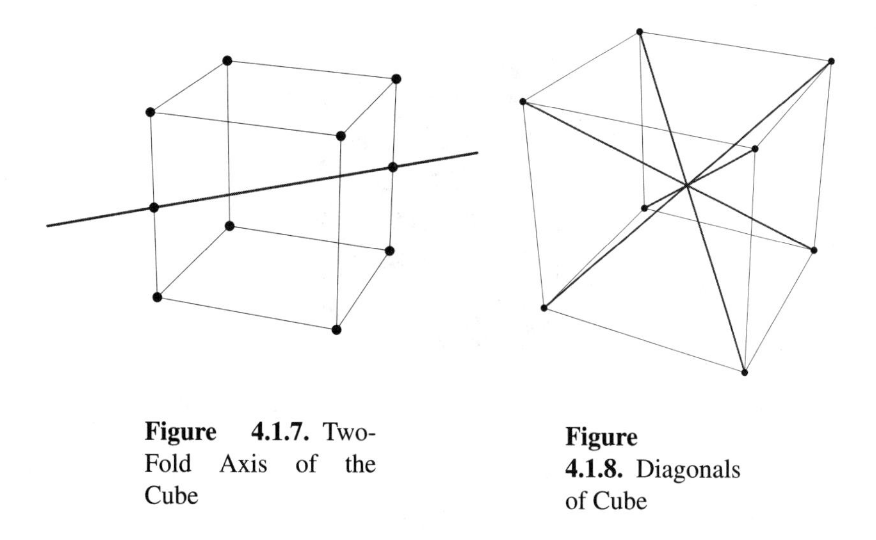

Figure 4.1.7. Two-Fold Axis of the Cube

Figure 4.1.8. Diagonals of Cube

The telling observation now is that vertices (as well as edges and faces) are really permuted in pairs. So we consider the action of the rotation group on pairs of opposite vertices, or, what amounts to the same thing, on the four diagonals of the cube. See Figure 4.1.8. This gives a homomorphism of the rotation group of the cube into S_4. Since both the rotation group and S_4 have 24 elements, to show that this is an isomorphism, it suffices to show that it is injective, that is, that no rotation leaves all four diagonals fixed. This is easy to check.

Proposition 4.1.3. *The rotation group of the cube is isomorphic to the permutation group S_4.*

The close relationship between the rotation groups of the tetrahedron and the cube suggests that there should be tetrahedra related geometrically to the cube. In fact, in choosing co-ordinates for vertices of the tetrahedron above, we have already shown how to embed a tetrahedron in the cube: Take the vertices of the cube at the points

$$\begin{bmatrix} \pm 1 \\ \pm 1 \\ \pm 1 \end{bmatrix}.$$

Then those four vertices which have the property that that the product of their coordinates is 1 are the vertices of an embedded tetrahedron. The remaining vertices (those for which the product of the co-ordinates is -1) are the vertices of a complementary tetrahedron. The even permutations of the diagonals of the cube preserve the two tetrahedra; the odd permutations interchange the two tetrahedra See Figure 4.1.9.

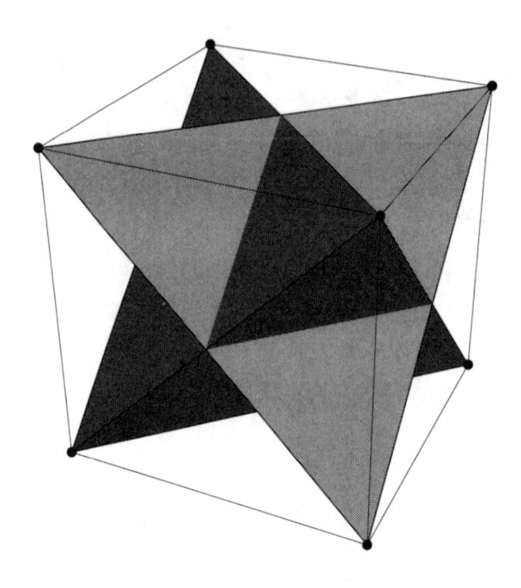

Figure 4.1.9. Cube with Inscribed Tetrahedra

This observation also lets us compute the 24 matrices for the rotations of the cube very easily. Note that the 3-fold axes for the tetrahedron are also 3-fold axes for the cube, and the 2-fold axes for the tetrahedron are 4-fold axes for the cube. In particular, the symmetries of the tetrahedron form a subgroup of the symmetries of the cube of index 2. Using these considerations, one can show that the group of rotation matrices of the cube consists of the signed permutation matrices with determinant 1. See Exercise 4.1.4.

Each convex polyhedron T has a *dual polyhedron* whose vertices are at the centroids of the faces of T; two vertices of the dual are joined by an edge if the corresponding faces of T are adjacent. *The dual polyhedron has the same symmetry group as does the original polyhedron.*

Proposition 4.1.4. *The octahedron is dual to the cube, so its group of rotations is also isomorphic to* S_4.

Exercises

Exercise 4.1.1.

(a) Given a unit vector $\hat{\boldsymbol{v}}_1$, explain how to find two further unit vectors $\hat{\boldsymbol{v}}_2$ and $\hat{\boldsymbol{v}}_3$ such that the $\{\hat{\boldsymbol{v}}_i\}$ form a right handed orthonormal basis.

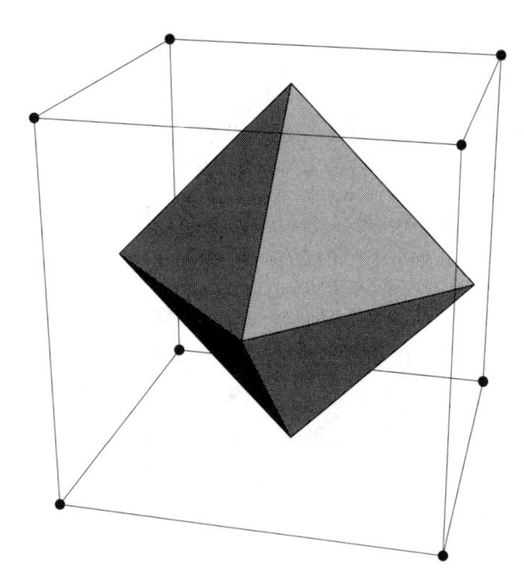

Figure 4.1.10. Cube and Octahedron

(b) Carry out the procedure for

$$\hat{\boldsymbol{v}}_1 = (1/\sqrt{3}) \begin{bmatrix} 1 \\ 1 \\ 1 \end{bmatrix}.$$

(c) Compute the matrix of rotation through $2\pi/3$ about the vector

$$\begin{bmatrix} 1 \\ 1 \\ 1 \end{bmatrix},$$

and explain why the answer has such a remarkably simple form!

Exercise 4.1.2. The midpoints of the edges of the tetrahedron are at the six points

$$\begin{bmatrix} \pm 1 \\ 0 \\ 0 \end{bmatrix}, \quad \begin{bmatrix} 0 \\ \pm 1 \\ 0 \end{bmatrix}, \quad \begin{bmatrix} 0 \\ 0 \\ \pm 1 \end{bmatrix}.$$

The matrices of the rotations by π about the axes determined by midpoints of the edges are the diagonal matrices with two entries equal to -1 and one entry equal to 1.

Exercise 4.1.3. The matrices computed in Exercises 4.1.1 and 4.1.2 generate the group of rotation matrices of the tetrahedron; the remaining matrices can be computed by matrix multiplication. Show that the rotation matrices for the tetrahedron are the matrices

$$\begin{bmatrix} \pm 1 & 0 & 0 \\ 0 & \pm 1 & 0 \\ 0 & 0 & \pm 1 \end{bmatrix}, \quad \begin{bmatrix} 0 & 0 & \pm 1 \\ \pm 1 & 0 & 0 \\ 0 & \pm 1 & 0 \end{bmatrix}, \quad \begin{bmatrix} 0 & \pm 1 & 0 \\ 0 & 0 & \pm 1 \\ \pm 1 & 0 & 0 \end{bmatrix},$$

where the product of the entries is 1. That is, there are no -1's or else two -1's.

Exercise 4.1.4. If $\mathcal{T}$ denotes the group of rotation matrices for the tetrahedron and R is any rotation matrix for the cube which is not contained in $\mathcal{T}$, then the group of rotation matrices for the cube is $\mathcal{T} \cup \mathcal{T}R$. Show that the group of rotation matrices for the cube consists of *signed permutation matrices with determinant 1,* that is, matrices with entries in $\{0, \pm 1\}$ with exactly one non-zero entry in each row and in each column, and with determinant 1.

4.2. Rotation Groups of the Dodecahedron and Icosahedron

The dodecahedron and icosahedron are dual to each other, so have the same rotational symmetry group. We need only work out the rotation group of the dodecahedron. See Figure 4.2.11.

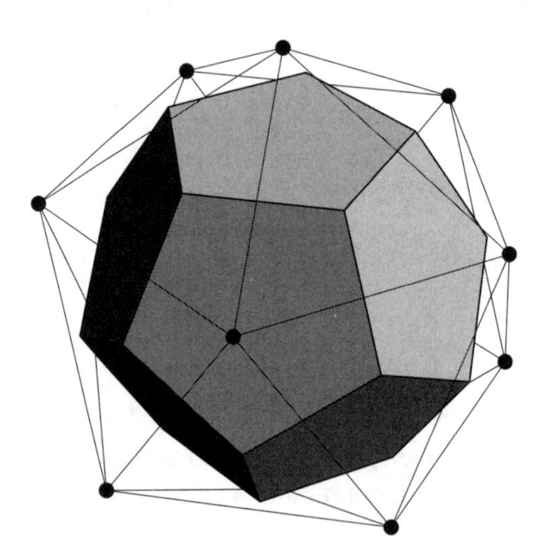

Figure 4.2.11. Dodecahedron and Icosahedron

The dodecahedron has six 5-fold axes through the centroids of opposite faces, giving 24 rotations of order 5. See Figure 4.2.12.

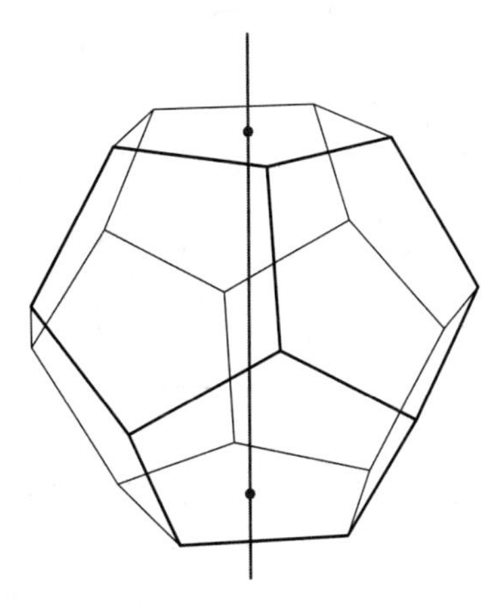

Figure 4.2.12. Five-Fold Axis of the Dodecahedron

There are ten 3-fold axes through pairs of opposite vertices, giving 20 elements of order 3. See Figure 4.2.13.

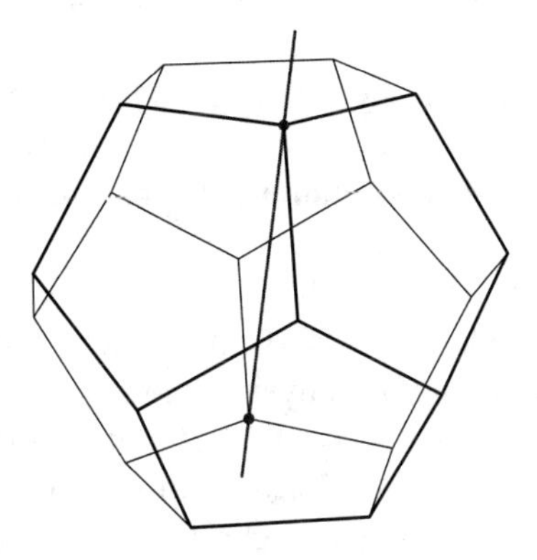

Figure 4.2.13. Three-Fold Axis of the Dodecahedron

Finally, the dodecahedron has 15 2-fold axis through centers of opposite edges, giving 15 elements of order 2. See Figure 4.2.14.

With the identity element, the dodecahedron has 60 rotational symmetries altogether. Now considering that the rotation groups of the "smaller" regular polyhedra are A_4 and S_4, and suspecting that there ought to be a lot of regularity in this subject, we might guess that the rotation group of

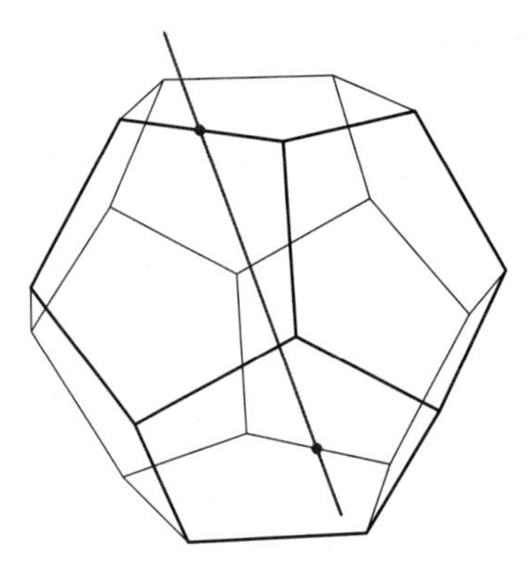

Figure 4.2.14. Two-Fold Axis of the Dodecahedron

the dodecahedron is isomorphic to the group A_5 of even permutations of five objects. So we are led to look for five geometric objects which are permuted by this rotation group. Finding the five objects is a perhaps a more subtle task than picking out the four diagonals which are permuted by the rotations of the cube. But since each face has five edges, we might suspect that each object is some equivalence class of edges which includes one edge from each face. Since there are 30 edges altogether, each such equivalence class of edges should contain six edges.

Pursuing this idea, consider any edge of the dodecahedron and its opposite edge. Notice that the plane containing these two edges bisects another pair of edges, and the plane containing *that* pair of edges bisects a third pair of edges. The resulting family of six edges contains one edge in each face of the dodecahedron. There are five such families which are permuted by the rotations of the dodecahedron. To understand these ideas, you are well advised to look closely at your physical model of the dodecahedron!

There are several other ways to pick out five objects which are permuted by the rotation group. Consider one of our families of six edges. It contains three pairs of opposite edges. Take the three lines joining the centers of the pairs of opposite edges. These three lines are mutually orthogonal; they are the axes of a cartesian coordinate system. There are five such coordinate systems which are permuted by the rotation group.

Finally, given one such coordinate system, one can locate a cube whose faces are parallel to the coordinate planes and whose edges lie on the faces of the dodecahedron! Each edge of the cube is a diagonal of a face of the dodecahedron, and exactly one of the five diagonals of each face is an edge of the cube. There are five such cubes which are permuted by the rotation group. See Figure 4.2.15.

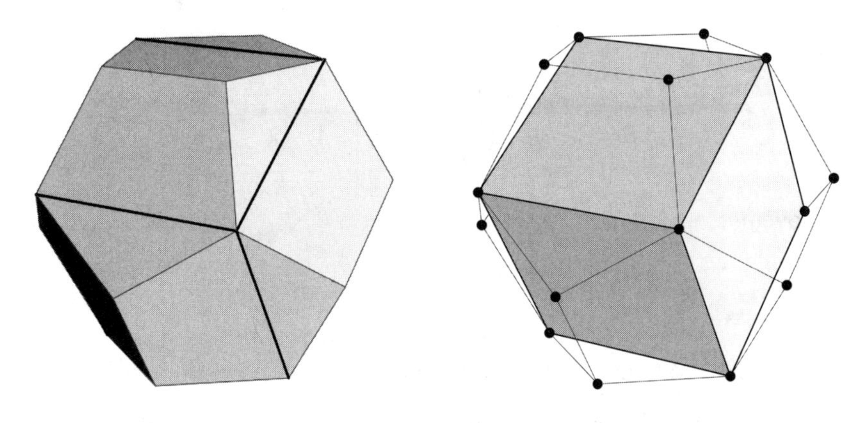

Figure 4.2.15. Cube Inscribed in the Dodecahedron

You are asked to show in Exercise 4.2.1 that the action of the rotation group on the set of five inscribed cubes is faithful; that is, the homomorphism of the rotation group into S_5 is injective.

Now, it remains to show that the image of the rotation group in S_5 is the group of even permutations A_5. One could do this by explicit computation. However, by using a previous result, we can avoid doing any computation at all! We have established earlier that for each n, A_n is the unique subgroup of S_n of index 2. (Exercise 3.2.14.) Since the image of the rotation group has 60 elements, it follows that it must be A_5.

Proposition 4.2.1. *The rotation groups of the dodecahedron and the icosahedron are isomorphic to the group of even permutations* A_5.

Exercises

Exercise 4.2.1. Show that no rotation of the dodecahedron leaves each of the five inscribed cubes fixed. Thus the action of the rotation group on the set of inscribed cubes induces an injective homomorphism of the rotation group into S_5

Exercise 4.2.2. Let

$$A = \{\begin{bmatrix} \cos 2k\pi/5 \\ \sin 2k\pi/5 \\ 1/2 \end{bmatrix} : 1 \leq k \leq 5\}$$

and

$$B = \{\begin{bmatrix} \cos(2k+1)\pi/5 \\ \sin(2k+1)\pi/5 \\ -1/2 \end{bmatrix} : 1 \leq k \leq 5\}.$$

Show that

$$\{\begin{bmatrix} 0 \\ 0 \\ \pm\sqrt{5}/2 \end{bmatrix}\} \cup A \cup B$$

is the set of vertices of an icosahedron.

Exercise 4.2.3. Each vertex of the icosahedron lies on a 5-fold axis, each midpoint of an edge on a 2-fold axis, and each centroid of a face on a 3-fold axis. Using the data of the previous exercise and the method of Exercises 4.1.1, 4.1.2, and 4.1.3, one can compute the matrices for rotations of the icosahedron. (I have only done this numerically and I don't know if the matrices have a nice closed form.)

4.3. What about Reflections?

When you thought about the nature of symmetry when you first began reading this text, you might have focused especially on reflection symmetry. (People are particularly attuned to reflection symmetry since human faces and bodies are important to us.)

A reflection in $\mathbb{R}^3$ through a plane P is the transformation which leaves the points of P fixed and sends a point $\boldsymbol{x} \notin P$ to the point on the line through $\boldsymbol{x}$ and perpendicular to P, which is equidistant from P with $\boldsymbol{x}$ and on the opposite side of P.

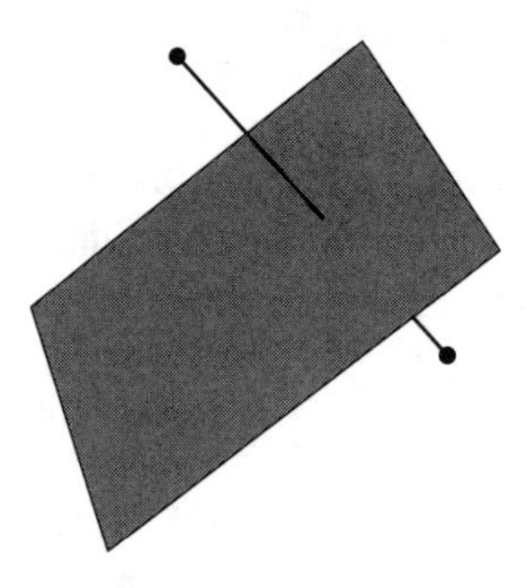

Figure 4.3.1. A Reflection

For a plane P through the origin in $\mathbb{R}^3$, the reflection through P is given by the following formula. Let $\hat{\boldsymbol{\alpha}}$ be a unit vector perpendicular to P. For any $\boldsymbol{x} \in \mathbb{R}^3$, the reflection $j_{\hat{\boldsymbol{\alpha}}}$ of $\boldsymbol{x}$ through P is given by $j_{\hat{\boldsymbol{\alpha}}}(\boldsymbol{x}) = \boldsymbol{x} - 2\langle \boldsymbol{x} | \hat{\boldsymbol{\alpha}} \rangle \hat{\boldsymbol{\alpha}}$, where $\langle \cdot | \cdot \rangle$ denotes the inner product in $\mathbb{R}^3$.

In the exercises, you are asked to verify this formula and to compute the matrix of a reflection, with respect to the standard basis of $\mathbb{R}^3$. You are also asked to find a formula for the reflection through a plane which does not pass through the origin.

A reflection which sends a geometric figure onto itself is a type of symmetry of the figure. It is not an actual motion which you could perform on the figure (while I am out of the room) but it is an ideal motion.

Let's see how we can bring reflection symmetry into our account of the symmetries of some simple geometric figures. Consider a thickened version of our rectangular card: a rectangular brick. Place the brick with its faces parallel to the coordinate planes and with its centroid at the origin of coordinates. See Figure 4.3.2.

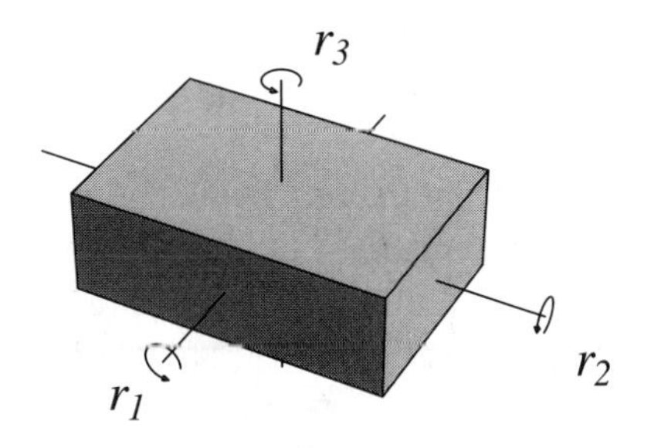

Figure 4.3.2. Rotations of a Brick

The rotational symmetries of the brick are the same as those of the rectangular card. There are four rotational symmetries: the non-motion e, and the rotations r_1, r_2, and r_3 through an angle of π about the x-, y-, and z- axes. The same matrices E, R_1, R_2, and R_3 listed in Section 1.5 implement these rotations.

In addition, the reflections in each of the coordinate planes are symmetries; write j_i for $j_{\hat{\boldsymbol{e}}_i}$, the reflection in the plane orthogonal to the unit vector $\hat{\boldsymbol{e}}_i$. (The vector $\hat{\boldsymbol{e}}_i$ has a 1 in the i-th coordinate and 0's elsewhere.) See Figure 4.3.3.

The symmetry j_i is implemented by the diagonal matrix J_i with -1 in the i-th diagonal position and 1's in the other diagonal positions. For

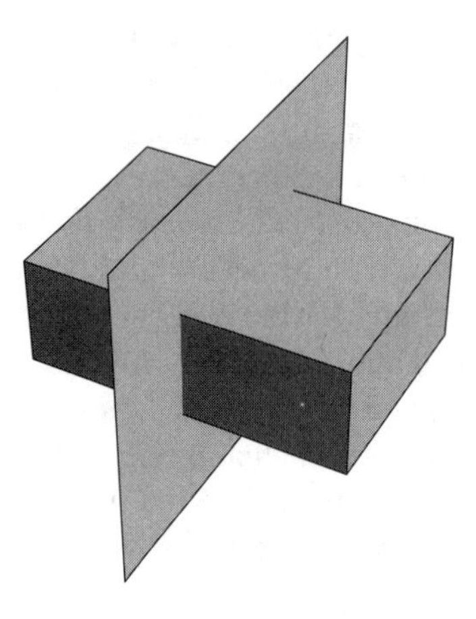

Figure 4.3.3. Reflection Symmetries

example,

$$J_2 = \begin{bmatrix} 1 & 0 & 0 \\ 0 & -1 & 0 \\ 0 & 0 & 1 \end{bmatrix}$$

There is one more symmetry which must be considered along with these, which is neither a reflection nor a rotation, but which is a product of a rotation and a reflection in several different ways. This is the inversion which sends each corner of the rectangular solid to its opposite corner. This is implemented by the matrix $-E$. Note that $-E = J_1 R_1 = J_2 R_2 = J_3 R_3$, so the inversion is equal to $j_i r_i$ for each i.

Having included the inversion as well as the three reflections, we again have a group. It is very easy to check closure under multiplication and inverse and to compute the multiplication table. The eight symmetries are represented by the eight three-by-three diagonal matrices with 1's and -1's on the diagonal; this set of matrices is clearly closed under matrix multiplication and inverse, and products of symmetries can be obtained immediately by multiplication of matrices. The product of symmetries (or of matrices) is *a priori* associative.

Now consider a thickened version of the square card: a square tile which we place with its centroid at the origin of coordinates, its square faces parallel with the x-y plane, and its other faces parallel with the other coordinate planes. This figure has the same rotational symmetries as does the square card, and these are implemented by the matrices given in Section 1.5. See Figure 4.3.4.

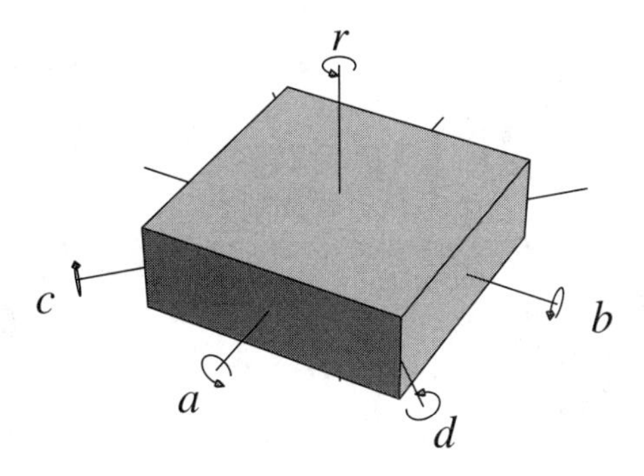

Figure 4.3.4. Rotations of the Square Tile

In addition, one can readily detect five reflection symmetries: for each of the five axes of symmetry, the plane perpendicular to the axis and passing through the origin is a plane of symmetry. See Figure 4.3.5.

Let us label the reflection through the plane perpendicular to the axis of the rotation a by j_a, and similarly for the other four rotation axes. The reflections j_a, j_b, j_c, j_d, and j_r are implemented by the matrices:

$$J_a = \begin{bmatrix} -1 & 0 & 0 \\ 0 & 1 & 0 \\ 0 & 0 & 1 \end{bmatrix} \quad J_b = \begin{bmatrix} 1 & 0 & 0 \\ 0 & -1 & 0 \\ 0 & 0 & 1 \end{bmatrix} \quad J_r = \begin{bmatrix} 1 & 0 & 0 \\ 0 & 1 & 0 \\ 0 & 0 & -1 \end{bmatrix}$$

$$J_c = \begin{bmatrix} 0 & -1 & 0 \\ -1 & 0 & 0 \\ 0 & 0 & 1 \end{bmatrix} \quad J_d = \begin{bmatrix} 0 & 1 & 0 \\ 1 & 0 & 0 \\ 0 & 0 & 1 \end{bmatrix}$$

Now, I claim there are three additional symmetries which we must include along with the five reflections and eight rotations; these symmetries are neither rotations nor reflections, but are products of a rotation and a reflection. One of these we can guess from our experience with the brick, the inversion, which is obtained, for example, as the product aj_a, and which is implemented by the matrix $-E$.

If we can't find the other two by insight, we can find them by computation: If τ_1 and τ_2 are any two symmetries, then their composition product $\tau_1\tau_2$ is also a symmetry; and if the symmetries τ_1 and τ_2 are implemented

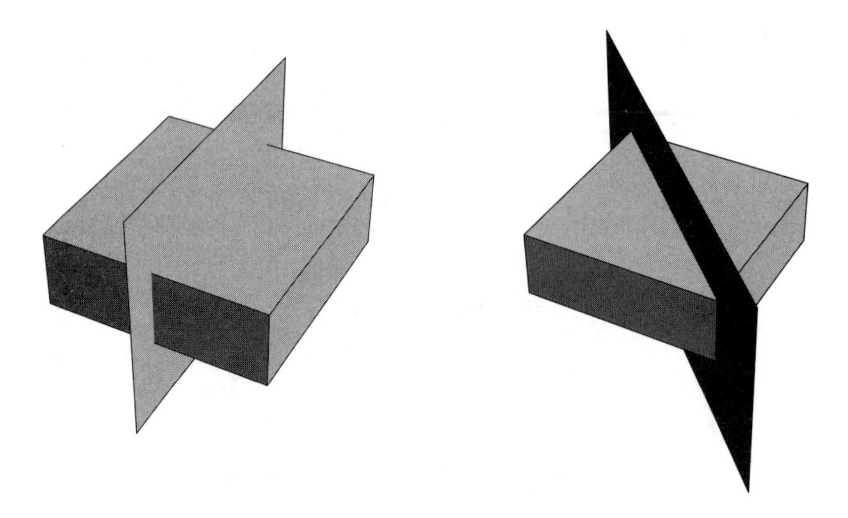

Figure 4.3.5. Reflections of the Square Tile

by matrices F_1 and F_2, then $\tau_1\tau_2$ is implemented by the matrix product F_1F_2. So we can look for other symmetries by examining products of the matrices implementing the known symmetries.

We have 14 matrices, so we can compute a lot of products before finding something new. If you are lucky, after a bit of trial and error you will discover that the combinations to try are powers of R multiplied by the reflection matrix J_r:

$$J_rR = RJ_r = \begin{bmatrix} 0 & -1 & 0 \\ 1 & 0 & 0 \\ 0 & 0 & -1 \end{bmatrix}$$

$$J_rR^3 = R^3J_r = \begin{bmatrix} 0 & 1 & 0 \\ -1 & 0 & 0 \\ 0 & 0 & -1 \end{bmatrix}$$

These are new matrices. What symmetries do they implement? Both are reflection-rotations, reflections in a plane followed by a rotation about an axis perpendicular to the plane.

(Here are all the combinations which, as one finds by experimentation, will *not* give something new: We already know that the product of two rotations of the square tile is again a rotation. The product of two reflections appears always to be a rotation. The product of a reflection and a rotation by π about the axis perpendicular to the plane of the reflection is always the inversion.)

Now we have sixteen symmetries of the square tile, eight rotations (including the non-motion), five reflections, one inversion, and two reflection-rotations. This set of sixteen symmetries is a group. It seems a bit daunting to work this out by computing 256 matrix products and recording the multiplication table, but one could do it in an hour or two, or one could get a computer to work out the multiplication table in no time at all.

However, there is a more thoughtful method to work this out which seriously reduces the necessary computation; I outline this method in the exercises.

In the next section, we will develop a more general conceptual framework in which to place this exploration, which will allow us to understand the experimental observation that for both the brick and the square tile, the total number of symmetries is twice the number of rotations. We will also see, for example, that the product of two rotations matrices is again a rotation matrix, and the product of two matrices, each of which is a reflection or a rotation-reflection, is a rotation.

Exercises

Exercise 4.3.1. Verify the formula for the reflection $J_{\hat{\boldsymbol{\alpha}}}$ through the plane perpendicular to $\hat{\boldsymbol{\alpha}}$.

Exercise 4.3.2. $J_{\hat{\boldsymbol{\alpha}}}$ is linear. Find its matrix with respect to the standard basis of $\mathbb{R}^3$. (Of course the matrix involves the coordinates of $\hat{\boldsymbol{\alpha}}$.)

Exercise 4.3.3. Consider a plane P which does not pass through the origin. Let $\hat{\boldsymbol{\alpha}}$ be a unit normal vector to P and let $\boldsymbol{x}_0$ be a point on P. Find a formula, in terms of $\hat{\boldsymbol{\alpha}}$ and $\boldsymbol{x}_0$ for the reflection of a point $\boldsymbol{x}$ through P. Such a reflection through a plane not passing through the origin is called an *affine reflection.*

Exercise 4.3.4. Here is a method to determine all the products of the symmetries of the square tile. Write J for J_r, the reflection in the x-y plane.

(a) The eight products αJ, where α runs through the set of eight rotation matrices of the square tile, are the eight non-rotation matrices. Which is which?

(b) Show that J commutes with the eight rotation matrices; that is, $J\alpha = \alpha J$ for all rotation matrices α.

(c) The information from parts (a) and (b), together with the multiplication table for the rotational symmetries, suffices to compute all products of symmetries.

(d) Verify that the sixteen symmetries form a group.

Exercise 4.3.5. Another way to work out all the products, and to understand the structure of the group of symmetries, is the following. Consider the matrix

$$S = J_c = \begin{bmatrix} 0 & 1 & 0 \\ 1 & 0 & 0 \\ 0 & 0 & 1 \end{bmatrix}.$$

(a) The set $\mathcal{D}$ of eight diagonal matrices with ± 1's on the diagonal is a subset of the matrices representing symmetries of the square tile. This set of diagonal matrices is closed under matrix multiplication.

(b) Every symmetry matrix for the square tile is either in $\mathcal{D}$ or is a product DS where $D \in \mathcal{D}$.

(c) For each $D \in \mathcal{D}$, there is a $D' \in \mathcal{D}$ (which is easy to compute) which satisfies $SD = D'S$, or equivalently, $SDS = D'$. Find the rule for determining D' from D, and use this to show how all products can be computed. Compare the results with those of the previous exercise.

4.4. Linear Isometries

The main purpose of this section is to investigate the linear isometries of three dimensional space. However, much of the work can be done without much extra effort in n-dimensional space.

Consider Euclidean n-space, $\mathbb{R}^n$ with the usual inner product $\langle \boldsymbol{x} | \boldsymbol{y} \rangle = \sum_{i=1}^n x_i y_i$, norm $||\boldsymbol{x}|| = \langle \boldsymbol{x} | \boldsymbol{x} \rangle^{1/2}$, and distance function $d(\boldsymbol{x}, \boldsymbol{y}) = ||\boldsymbol{x} - \boldsymbol{y}||$.

Definition 4.4.1. An *isometry* of $\mathbb{R}^n$ is a map $\tau : \mathbb{R}^n \longrightarrow \mathbb{R}^n$ which preserves distance, $d(\tau(\boldsymbol{a}), \tau(\boldsymbol{b})) = d(\boldsymbol{a}, \boldsymbol{b})$ for all $\boldsymbol{a}, \boldsymbol{b} \in \mathbb{R}^n$.

You are asked to show in the exercises that the set of isometries is a group, and the set of isometries τ satisfying $\tau(\boldsymbol{0}) = \boldsymbol{0}$ is a subgroup.

Lemma 4.4.2. *Let τ be an isometry of $\mathbb{R}^n$ such that $\tau(\boldsymbol{0}) = \boldsymbol{0}$. Then $\langle \tau(\boldsymbol{a}) | \tau(\boldsymbol{b}) \rangle = \langle \boldsymbol{a} | \boldsymbol{b} \rangle$ for all $\boldsymbol{a}, \boldsymbol{b} \in \mathbb{R}^n$.*

Proof. Since $\tau(\boldsymbol{0}) = \boldsymbol{0}$, τ preserves norm as well as distance, $||\tau(\boldsymbol{x})|| = ||\boldsymbol{x}||$ for all $\boldsymbol{x}$. But since $\langle \boldsymbol{a} | \boldsymbol{b} \rangle = (1/2)(||\boldsymbol{a}||^2 + ||\boldsymbol{b}||^2 - d(\boldsymbol{a}, \boldsymbol{b})^2)$, it follows that τ also preserves inner products. □

Remark 4.4.3. Recall the following property of orthonormal bases of $\mathbb{R}^n$: If $\mathbb{F} = \{\hat{f}_i\}$ is an orthonormal basis, then the expansion of a vector $\boldsymbol{x}$ with respect to $\mathbb{F}$ is $\boldsymbol{x} = \sum_i \langle \boldsymbol{x} | \hat{\boldsymbol{f}}_i \rangle \hat{\boldsymbol{f}}_i$. If $\boldsymbol{x} = \sum_i x_i \hat{\boldsymbol{f}}_i$ and $\boldsymbol{y} = \sum_i y_i \hat{\boldsymbol{f}}_i$, then $\langle \boldsymbol{x} | \boldsymbol{y} \rangle = \sum_i x_i y_i$.

Definition 4.4.4. Recall that a matrix A is said to be *orthogonal* if A^t is the inverse of A.

Exercise 4.4.3 gives another characterization of orthogonal matrices.

Lemma 4.4.5. *If A is an orthogonal matrix, then the linear map $\boldsymbol{x} \mapsto A\boldsymbol{x}$ is an isometry.*

Proof. The columns of A, say $\{\boldsymbol{v}_i\}$ are an orthonormal basis. Hence, $||A\boldsymbol{x}||^2 = ||\sum_i x_i \boldsymbol{v}_i||^2 = \sum_i x_i^2 = ||\boldsymbol{x}||^2$ by Remark 4.4.3. □

Lemma 4.4.6. *Let $\mathbb{F} = \{\hat{f}_i\}$ be an orthonormal basis. A set $\{\boldsymbol{v}_i\}$ is an orthonormal basis if and only if the matrix $A = [a_{ij}] = [\langle \boldsymbol{v}_i | \hat{\boldsymbol{f}}_j \rangle]$ is an orthogonal matrix.*

Proof. The ith row of A is the coefficient vector of $\boldsymbol{v}_i$ with respect to the orthonormal basis $\mathbb{F}$. According to Remark 4.4.3, the inner product of $\boldsymbol{v}_i$ and $\boldsymbol{v}_j$ is the same as the inner product of the ith and jth rows of A. Hence, the $\boldsymbol{v}_i$'s are an orthonormal basis if and only if the rows of A are an orthonormal basis. By Exercise 4.4.5, this is true if and only if A is orthogonal. □

Theorem 4.4.7. *Let τ be a linear map on $\mathbb{R}^n$. The following are equivalent:*

(a) *τ is an isometry.*
(b) *τ preserves inner products.*
(c) *For some orthonormal basis $\{\hat{f}_1, \ldots, \hat{f}_n\}$ of $\mathbb{R}^n$, the set $\{\tau(\hat{f}_1), \ldots, \tau(\hat{f}_n)\}$ is also orthonormal.*
(d) *For every orthonormal basis $\{\hat{f}_1, \ldots, \hat{f}_n\}$ of $\mathbb{R}^n$, the set $\{\tau(\hat{f}_1), \ldots, \tau(\hat{f}_n)\}$ is also orthonormal.*
(e) *The matrix of τ with respect to some orthonormal basis is orthogonal.*
(f) *The matrix of τ with respect to every orthonormal basis is orthogonal.*

Proof. Condition (a) implies (b) by Lemma 4.4.2. The implications (b) $\Longrightarrow$ (d) $\Longrightarrow$ (c) are trivial, and (c) implies (e) by Lemma 4.4.6. Now assume the matrix A of τ with respect to the orthonormal basis $\mathbb{F}$ is orthogonal. Let $U(\boldsymbol{x})$ be the coordinate vector of $\boldsymbol{x}$ with respect to $\mathbb{F}$; then by the remark, U is an isometry. The linear map τ is $U^{-1} \circ M_A \circ U$, where M_A means multiplication by A. By Lemma 4.4.5, M_A is an isometry, so τ is an isometry. Thus (e) $\Longrightarrow$(a). Similarly, one has (a) $\Longrightarrow$(b) $\Longrightarrow$(d) $\Longrightarrow$(f) $\Longrightarrow$(a). □

Proposition 4.4.8. *The determinant of an orthogonal matrix is* ± 1.

Proof. Let A be an orthogonal matrix. Since $\det(A) = \det(A^t)$, one has

$$1 = \det(E) = \det(A^t A) = \det(A^t)\det(A) = \det(A)^2.$$

□

Remark 4.4.9. If τ is a linear transformation of $\mathbb{R}^n$ and A and B are the matrices of τ with respect to two different bases of $\mathbb{R}^n$, then $\det(A) = \det(B)$, because A and B are related by a similarity, $A = VBV^{-1}$, where V is a change of basis matrix. Therefore one can, without ambiguity, define the determinant of τ to be the determinant of the matrix of τ with respect to any basis.

Corollary 4.4.10. *The determinant of a linear isometry is* ± 1.

Since $\det(AB) = \det(A)\det(B)$, $\det : \mathrm{O(n, \mathbb{R})} \longrightarrow \{1, -1\}$ is a group homomorphism.

Definition 4.4.11. The set of orthogonal n-by-n matrices with determinant equal to 1 is called the special orthogonal group and denoted $\mathrm{SO(n, \mathbb{R})}$.

Evidently, the special orthogonal group $\mathrm{SO(n, \mathbb{R})}$ is a normal subgroup of the orthogonal group of index 2, since it is the kernel of $\det : \mathrm{O(n, \mathbb{R})} \longrightarrow \{1, -1\}$.

We next restrict our attention to three-dimensional space and explore the role of rotations and orthogonal reflections in the group of linear isometries.

Recall from Section 4.3 that for any unit vector $\boldsymbol{\alpha}$ in $\mathbb{R}^3$, the plane P_{α} is $\{\boldsymbol{x} : \langle \boldsymbol{x} | \boldsymbol{\alpha} \rangle = 0\}$. The orthogonal reflection in P_{α} is the linear map $j_{\alpha} : \boldsymbol{x} \mapsto \boldsymbol{x} - 2\langle \boldsymbol{x} | \boldsymbol{\alpha} \rangle \boldsymbol{\alpha}$. The orthogonal reflection j_{α} fixes P_{α} pointwise and sends $\boldsymbol{\alpha}$ to $-\boldsymbol{\alpha}$. Let J_{α} denote the matrix of j_{α} with respect to the standard basis of $\mathbb{R}^3$. Call a matrix of the form J_{α} a *reflection matrix*.

Let's next sort out the role of reflections and rotations in $\mathrm{SO}(2,\mathbb{R})$. Consider an orthogonal matrix

$$\begin{bmatrix} \alpha & \gamma \\ \beta & \delta \end{bmatrix}.$$

Orthogonality implies that

$$\begin{bmatrix} \alpha \\ \beta \end{bmatrix}$$

is a unit vector and

$$\begin{bmatrix} \gamma \\ \delta \end{bmatrix} = \pm \begin{bmatrix} -\beta \\ \alpha \end{bmatrix}.$$

The vector

$$\begin{bmatrix} \alpha \\ \beta \end{bmatrix}$$

can be written as

$$\begin{bmatrix} \cos(\theta) \\ \sin(\theta) \end{bmatrix}$$

for some angle θ, so the orthogonal matrix has the form

$$\begin{bmatrix} \cos(\theta) & -\sin(\theta) \\ \sin(\theta) & \cos(\theta) \end{bmatrix} \quad \text{or} \quad \begin{bmatrix} \cos(\theta) & \sin(\theta) \\ \sin(\theta) & -\cos(\theta) \end{bmatrix}.$$

The matrix

$$R_\theta = \begin{bmatrix} \cos(\theta) & -\sin(\theta) \\ \sin(\theta) & \cos(\theta) \end{bmatrix}$$

is the matrix of the rotation through an angle θ, and has determinant equal to 1. The matrix

$$\begin{bmatrix} \cos(\theta) & \sin(\theta) \\ \sin(\theta) & -\cos(\theta) \end{bmatrix}$$

equals $R_\theta J$ where

$$J = \begin{bmatrix} 1 & 0 \\ 0 & -1 \end{bmatrix}$$

is the reflection matrix $J = J_{\hat{e}_2}$. $R_\theta J$ has determinant equal to -1.

Now consider the situation in three dimensions. Any real 3-by-3 matrix has a real eigenvalue, since the characteristic polynomial is cubic with real coefficients. A real eigenvalue of an orthogonal matrix must be ± 1 because the matrix implements an isometry.

Lemma 4.4.12. *Any element of* $\mathrm{SO}(3,\mathbb{R})$ *has* $+1$ *as an eigenvalue.*

Proof. Let $A \in SO(3, \mathbb{R})$, let τ be the linear isometry $\boldsymbol{x} \mapsto A\boldsymbol{x}$, and let $\boldsymbol{v}$ be an eigenvector with eigenvalue ± 1. If the eigenvalue is $+1$, there is nothing to do. So suppose the eigenvalue is -1. The plane $P = P_{\boldsymbol{v}}$ orthogonal to $\boldsymbol{v}$ is invariant under A, because if $\boldsymbol{x} \in P$, then $\langle \boldsymbol{v} | A\boldsymbol{x} \rangle = -\langle A\boldsymbol{v} | A\boldsymbol{x} \rangle = -\langle \boldsymbol{v} | \boldsymbol{x} \rangle = 0$. The restriction of τ to P is also orthogonal, and since $1 = \det(\tau) = (-1)(\det(\tau_{|P})$, $\tau_{|P}$ must be a reflection. But a reflection has an eigenvalue of $+1$, so in any case A has an eigenvalue of $+1$. □

Proposition 4.4.13. *An element $A \in O(3, \mathbb{R})$ has determinant 1 if and only if A implements a rotation.*

Proof. Suppose $A \in SO(3, \mathbb{R})$. Let τ denote the corresponding linear isometry $\boldsymbol{x} \mapsto A\boldsymbol{x}$. By the lemma, A has an eigenvector $\boldsymbol{v}$ with eigenvalue 1. The plane P orthogonal to $\boldsymbol{v}$ is invariant under τ, and $\det(\tau_{|P}) = \det(\tau) = 1$, so $\tau_{|P}$ is a rotation of P. Hence, τ is a rotation about the line spanned by $\boldsymbol{v}$. On the other hand, if τ is a rotation, then the matrix of τ with respect to an appropriate orthonormal basis has the form

$$\begin{bmatrix} 1 & 0 & 0 \\ 0 & \cos(\theta) & -\sin(\theta) \\ 0 & \sin(\theta) & \cos(\theta) \end{bmatrix},$$

so τ has determinant 1. □

Proposition 4.4.14. *An element of $O(3, \mathbb{R}) \setminus SO(3, \mathbb{R})$ implements either an orthogonal reflection, or a reflection-rotation, that is, the product of a reflection $j_{\boldsymbol{\alpha}}$ and a rotation about the line spanned by $\boldsymbol{\alpha}$.*

Proof. Suppose $A \in O(3, \mathbb{R}) \setminus SO(3, \mathbb{R})$. Let τ denote the corresponding linear isometry $\boldsymbol{x} \mapsto A\boldsymbol{x}$. Let $\boldsymbol{v}$ be an eigenvector of A with eigenvalue ± 1. If the eigenvalue is 1, then the restriction of τ to the plane P orthogonal to $\boldsymbol{v}$ has determinant -1, so is a reflection. Then τ itself is a reflection. If the eigenvalue is -1, then the restriction of τ to P has determinant 1, so is a rotation. In this case τ is the product of the reflection $j_{\boldsymbol{v}}$ and a rotation about the line spanned by $\boldsymbol{v}$. □

Exercises

Exercise 4.4.1.

(a) Show that j_α is isometric.

(b) Show that $\det(j_\alpha) = -1$.

(c) If τ is a linear isometry, show that $\tau j_\alpha \tau^{-1} = j_{\tau(\alpha)}$.

(d) If A is any orthogonal matrix, show that $AJ_\alpha A^{-1} = J_{A\alpha}$.

(e) Conclude that the matrix of j_α with respect to *any* orthonormal basis is a reflection matrix.

Exercise 4.4.2. Show that the set of isometries of $\mathbb{R}^n$ is a group. Show that the set of isometries τ satisfying $\tau(\boldsymbol{0}) = \boldsymbol{0}$ is a subgroup.

Exercise 4.4.3. A matrix A is orthogonal if and only if its columns are an orthonormal basis, if and only if its rows are an orthonormal basis.

Exercise 4.4.4.

(a) Show that the matrix $R_{2\theta}J = R_\theta J R_{-\theta}$ is the matrix of the reflection in the line spanned by

$$\begin{bmatrix} \cos\theta \\ \sin\theta \end{bmatrix}.$$

Write $J_\theta = R_{2\theta}J$.

(b) The reflection matrices are precisely the elements of $\mathrm{O}(2, \mathbb{R})$ with determinant equal to -1.

(c) Compute $J_{\theta_1} J_{\theta_2}$.

(d) Show that any rotation matrix R_θ is a product of two reflection matrices.

Exercise 4.4.5. Show that an element of $\mathrm{SO}(3, \mathbb{R})$ is a product of two reflection matrices. A matrix of a rotation-reflection is a product of three reflection matrices. Thus any element of $\mathrm{O}(3, \mathbb{R})$ is a product of at most three reflection matrices.

4.5. The Full Symmetry Group and Chirality

All the geometric figures in three-dimensional space which we have considered – the polygonal tiles, bricks, and the regular polyhedra – admit reflection symmetries. The rotation group of a geometric figure is the group of $g \in \mathrm{SO}(3, \mathbb{R})$ which leave the figure invariant. The full symmetry group is the group of $g \in \mathrm{O}(3, \mathbb{R})$ which leave the figure invariant. The existence of reflection symmetries means that the full symmetry group is strictly larger than the rotation group.

(You may wonder whether every symmetry of a geometric figure must be implemented by a linear isometry. Certainly it is so for any symmetry which we can readily detect of any common geometric figure. Here is the sketch of an argument – certainly there are details to be filled in, and the nature of the details may depend on how large a class of sets in $\mathbb{R}^3$ we allow as geometric figures. A geometric figure has a center of mass which is preserved under a symmetry; we might as well place the center of mass at the origin of coordinates. Now one has to argue that the symmetry, which is an isometry of the figure, can be extended to an isometry of all of $\mathbb{R}^3$. Finally, one can show that an isometry of $\mathbb{R}^3$ which fixes the origin must be linear.)

In the following discussion, I do not distinguish between linear isometries and their standard matrices.

Theorem 4.5.1. *Let S be a geometric figure with full symmetry group G and rotation group $\mathcal{R} = G \cap \mathrm{SO}(3, \mathbb{R})$. Suppose S admits a reflection symmetry J. Then $\mathcal{R}$ is an index 2 subgroup of G and $G = \mathcal{R} \cup \mathcal{R}J$.*

Proof. Suppose A is an element of $G \setminus \mathcal{R}$. Then $\det(A) = -1$ and $\det(AJ) = 1$, so $AJ \in \mathcal{R}$. Thus $A = (AJ)J \in \mathcal{R}J$. □

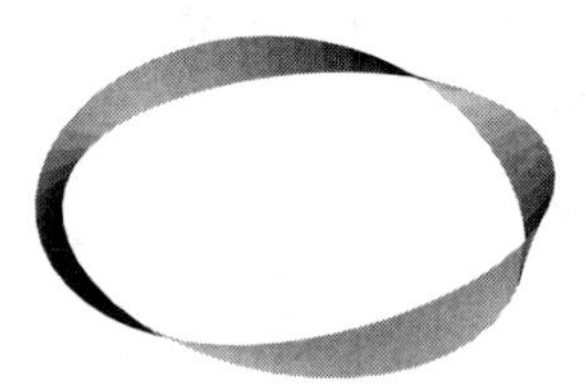

Figure 4.5.6. Twisted Band with Symmetry D_3

For example, the full symmetry group of the cube has 48 elements. What group is it? We could compute an injective homomorphism of the full symmetry group into S_6 using the action on the faces of the cube, or into S_8 using the action on the vertices. A more efficient method is given in Exercise 4.5.1; the result is that the full symmetry group is $S_4 \times \mathbb{Z}_2$.

Are there geometric figures with a non-trivial rotation group but with no reflection symmetries? Such a figure must exhibit chirality or "handedness"; it must come in two versions which are mirror images of each other. Consider, for example, a belt which is given n half twists ($n \geq 2$) and then fastened. There are two mirror image versions, with right hand and left hand twists. Either version has rotation group D_n, the rotation group of the n-gon, but no reflection symmetries. Reflections convert the right handed version into the left handed version. See Figure 4.5.6.

There exist chiral convex polyhedra with two types of regular polygonal faces, for example, the "snubcube," shown in Figure 4.5.7.

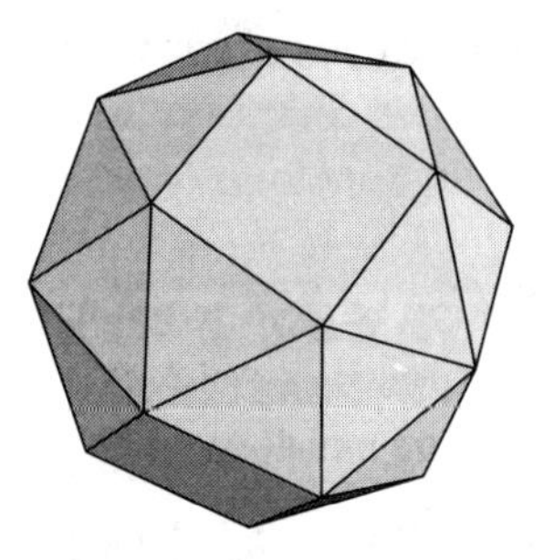

Figure 4.5.7. Snubcube

Exercises

Exercise 4.5.1. Let G denote the full symmetry group of the cube and $\mathcal{R}$ the rotation group. The inversion $i : x \mapsto -x$ with matrix $-E$ is an element of $G \setminus \mathcal{R}$, so $G = \mathcal{R} \cup \mathcal{R}i$. Observe that $i^2 = 1$, and that for any rotation r, $ir = ri$. Conclude that $G \cong S_4 \times \mathbb{Z}_2$.

Exercise 4.5.2. Show that the same trick works for the dodecahedron, and that the full symmetry group is isomorphic to $A_5 \times \mathbb{Z}_2$. Show that this group is not isomorphic to S_5.

Exercise 4.5.3. Show that the full symmetry group of the tetrahedron is S_4.

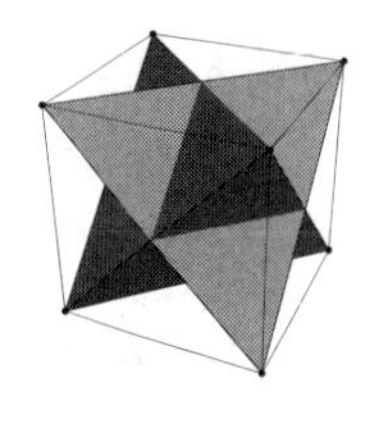

Chapter 5

Actions of Groups

5.1. Group Actions on Sets

We have observed that the symmetry group of the cube acts on various geometric sets associated with the cube, the set of vertices, the set of diagonals, the set of edges, the set of faces. In this section we look more closely at the concept of a group acting on a set.

Definition 5.1.1. An action of a group G on a set X is a homomorphism from G into $\mathrm{Sym}(X)$.

Let φ be an action of G on X. For each $g \in G$, $\varphi(g)$ is a bijection of X. The homomorphism property of φ means that for $x \in X$ and $g_1, g_2 \in G$, $\varphi(g_2g_1)(x) = \varphi(g_2)(\varphi(g_1)(x))$. Thus, if $\varphi(g_1)$ sends x to x', and $\varphi(g_2)$ sends x' to x'', then $\varphi(g_2g_1)$ sends x to x''. When it cannot cause ambiguity, it is convenient to write gx for $\varphi(g)(x)$. With this simplified notation, the homomorphism property reads like a mixed associative law: $(g_2g_1)x = g_2(g_1x)$.

Lemma 5.1.2. *Given an action of G on X, define a relation on X by* $x \sim y$ *if there exists a* $g \in G$ *such that* $gx = y$. *This relation is an equivalence relation.*

Proof. Exercise 5.1.1. □

Definition 5.1.3. Given an action of G on X, the equivalence classes of the equivalence relation associated to the action are called the *orbits* of the action. The orbit of x will be denoted $O(x)$.

Example 5.1.4. Any group G acts on itself by left multiplication. That is, for $g \in G$ and $x \in G$, gx is just the usual product of g and x in G. The homomorphism, or associative, property of the action is just the associative law of G. There is only one orbit. The action of G on itself by left multiplication is often called the *left regular action*.

Definition 5.1.5. An action of G on X is called *transitive* if there is only one orbit. That is, for any two elements $x, x' \in X$, there is a $g \in G$ such that $gx = x'$. A subgroup of $\mathrm{Sym}(X)$ is called *transitive* if it acts transitively on X.

Example 5.1.6. Let G be any group and H any subgroup. Then G acts on the set G/H of left cosets of H in G by left multiplication, $g(aH) = (ga)H$. The action is transitive.

Example 5.1.7. Any group G acts on itself by conjugation: For $g \in G$, define $c_g \in \mathrm{Aut}(G) \subseteq \mathrm{Sym}(G)$ by $c_g(x) = gxg^{-1}$. It was shown in Exercise 3.4.6 that the map $g \mapsto c_g$ is a homomorphism. The orbits of this action are called the *conjugacy classes* of G; two elements x and y are *conjugate* if there is a $g \in G$ such that $gxg^{-1} = y$. For example, it was shown in Exercise 3.1.13 that two elements of the symmetric group S_n are conjugate if, and only if, they have the same cycle structure.

Example 5.1.8. Let G be any group and X the set of subgroups of G. Then G acts on X by conjugation, $c_g(H) = gHg^{-1}$. Two subgroups in the same orbit are called *conjugate*.

You are asked in Exercise 5.1.5 to verify that this is an action.

In Section 5.3, we shall pursue the idea of classifying groups of small order, up to isomorphism. Another organizational scheme for classifying small groups is to classify those which act transitively on small sets, that is to classify transitive subgroups of S_n for small n.

Example 5.1.9. The transitive subgroups of S_3 are exactly S_3 and A_3. See exercise 5.1.9.

Example 5.1.10. The transitive subgroups of S_4 are:

- S_4,
- A_4, which is normal,
- D_4 (three conjugate copies,)
- $\mathcal{V} = \{e, (12)(34), (13)(24), (14)(23)\}$ (which is normal,) and
- $\mathbb{Z}_4$ (three conjugate copies.) See Exercise 5.1.20.

Definition 5.1.11. Let G act on X. For $x \in X$, the stabilizer of x in G is $\mathrm{Stab}(x) = \{g \in G : gx = x\}$. If it is necessary to specify the group, we will write $\mathrm{Stab}_G(x)$.

Lemma 5.1.12. *For any action of a group G on a set X and any $x \in X$,* $\text{Stab}(x)$ *is a subgroup of G.*

Proof. Exercise 5.1.6. □

Proposition 5.1.13. *Let G act on X, and let $x \in X$. Then $\psi : a\text{Stab}(x) \mapsto ax$ defines a bijection from $G/\text{Stab}(x)$ onto $O(x)$, which satisfies*

$$\psi(g(a\text{Stab}(x))) = g\psi(a\text{Stab}(x))$$

for all $g, a \in G$.

Proof. Note that $a\text{Stab}(x) = b\text{Stab}(x) \Leftrightarrow b^{-1}a \in \text{Stab}(x) \Leftrightarrow b^{-1}ax = x \Leftrightarrow bx = ax$. This calculation shows that ψ is well defined and injective. If $y \in O(x)$, then there exists $a \in G$ such that $ax = y$, so $\psi(a\text{Stab}(x)) = y$; thus ψ is surjective as well. The relation

$$\psi(g(a\text{Stab}(x))) = g\psi(a\text{Stab}(x))$$

for all $g, a \in G$ is evident from the definition of ψ. □

Corollary 5.1.14. *Suppose G is finite. Then*

$$\#O(x) = [G : \text{Stab}(x)] = \frac{\#G}{\#\text{Stab}(x)}.$$

In particular $\#O(x)$ divides $\#G$.

Proof. This follows immediately from the proposition and Lagrange's Theorem. □

Definition 5.1.15. Consider the action of a group G on its subgroups by conjugation. The stabilizer of a subgroup H is called the *normalizer* of H in G and denoted $N_G(H)$.

According to the Corollary, if G is finite, then the number of distinct subgoups xHx^{-1} for $x \in G$ is

$$[G : N_G(H)] = \frac{\#G}{\#N_G(H)}.$$

Since (clearly) $N_G(H) \supseteq H$, the number of such subgroups is no more than $[G : H]$.

Definition 5.1.16. Consider the action of a group G on itself by conjugation. The stabilizer of an element $g \in G$ is called the *centralizer* of g in G and denoted $\mathrm{Cent}(g)$, or when it is necessary to specify the group by $\mathrm{Cent}_G(x)$.

Again, according to the corollary the size of the conjugacy class of g, that is, of the orbit of g under conjugacy, is

$$[G : \mathrm{Cent}(g)] = \frac{\#G}{\#\mathrm{Cent}(g)}.$$

The *kernel* of an action of a group G on a set X is the kernel of the corresponding homomorphism $\varphi : G \to \mathrm{Sym}(X)$, i.e.

$$\{g \in G : gx = x \quad \text{for all} \quad x \in X\}.$$

According to the general theory of homomorphisms, the kernel is a normal subgroup of G. The kernel is evidently the intersection of the stablilizers of all $x \in X$. For example, the kernel of the action of G on itself by conjugation is the center of G.

Application: counting formulas. It is possible to obtain a number of well known counting formulas by means of the proposition and its corollary.

Example 5.1.17. The number of k-element subsets of a set with n elements is

$$\binom{n}{k} = \frac{n!}{k!(n-k)!}.$$

Proof. Let X be the family of k element subsets of $\{1, 2, \dots, n\}$. S_n acts transitively on X by $\pi\{a_1, a_2, \dots, a_k\} = \{\pi(a_1), \pi(a_2), \dots, \pi(a_k)\}$. (Verify!) The stabilizer of $x = \{1, 2, \dots, k\}$ is $S_k \times S_{n-k}$, the group of permutations which leaves invariant the sets $\{1, 2, \dots, k\}$ and $\{k+1, \dots, n\}$. Therefore, the number of k-element subsets is the size of the orbit of x, namely

$$\frac{\#S_n}{\#(S_k \times S_{n-k})} = \frac{n!}{k!(n-k)!}.$$

□

Example 5.1.18. The number of ordered sequences of k items chosen from a set with n elements is

$$\frac{n!}{(n-k)!}.$$

Proof. The proof is similar to that for the previous example. This time let S_n act on the set of ordered sequences of k-elements from $\{1, 2, \dots, n\}$. □

Example 5.1.19. The number of sequences of r_1 1's, r_2 2's,..., and r_k k's is

$$\frac{(r_1+r_2+\cdots+r_k)!}{r_1!r_2!\ldots r_k!}.$$

Proof. Let $n = r_1 + r_2 + \cdots + r_k$. S_n acts transitively on sequences of r_1 1's, r_2 2's, ..., and r_k k's. The stabilizer of

$$(1, \ldots, 1, 2, \ldots, 2, \ldots, k, \ldots, k)$$

with r_1 consecutive 1's, r_2 consecutive 2's, etc. is

$$S_{r_1} \times S_{r_2} \times \cdots \times S_{r_k}.$$

□

Exercises

Exercise 5.1.1. Let the group G act on a set X. Define a relation on X by $x \sim y$ if, and only if, there is a $g \in G$ such that $gx = y$. Show that this is an equivalence relation on X, and the orbit (equivalence class) of $x \in X$ is $Gx = \{gx : g \in G\}$.

Exercise 5.1.2. Verify all the assertions made in Example 5.1.4.

Exercise 5.1.3. The symmetric group S_n acts naturally on the set $\{1, 2, \ldots, n\}$. Let $\sigma \in S_n$. Show that the cycle decomposition of σ can be recovered by considering the orbits of the action of the cyclic subgroup $\langle\sigma\rangle$ on $\{1, 2, \ldots, n\}$.

Exercise 5.1.4. Verify the assertions made in Example example: action on cosets.

Exercise 5.1.5. Verify that any group G acts on the set X of its subgroups by $c_g(H) = gHg^{-1}$. Compute the example of S_3 acting by conjugation of the set X of (six) subgroups of S_3. Verify that there are four orbits, three of which consist of a single subgroup, and one of which contains three subgroups.

Exercise 5.1.6. Let G act on X, and let $x \in X$. Verify that $\mathrm{Stab}(x)$ is a subgroup of G. Verify that if x and y are in the same orbit, then the subgroups $\mathrm{Stab}(x)$ and $\mathrm{Stab}(y)$ are conjugate subgroups.

Exercise 5.1.7. Let $H = \{e, (1, 2)\} \subseteq S_3$. Find the orbit of H under conjugation by G, the stabilizer of H in G and the family of left cosets of the stabilizer in G, and verify explicitly the bijection between left cosets of the stabilizer and conjugates of H.

Exercise 5.1.8. Show that $N_G(aHa^{-1}) = aN_G(H)a^{-1}$ for $a \in G$.

Exercise 5.1.9. Show that the transitive subgroups of S_3 are exactly S_3 and A_3.

Example 5.1.20. Show that the transitive subgroups of S_4 are:

- S_4,
- A_4, which is normal,
- $D_4 = \langle(1\ 2\ 3\ 4), (1\ 2)(3\ 4)\rangle$, and two conjugate subgroups,
- $V = \{e, (12)(34), (13)(24), (14)(23)\}$, which is normal,
- $\mathbb{Z}_4 = \langle(1\ 2\ 3\ 4), (1\ 2)(3\ 4)\rangle$, and two conjugate subgroups.

Exercise 5.1.10. Suppose A is a subgroup of $N_G(H)$. Show that AH is a subgroup of $N_G(H)$, $AH = HA$, and

$$\#(AH) = \frac{\#(A)\,\#(H)}{\#(A \cap H)}.$$

Hint: H is normal in $N_G(H)$.

Exercise 5.1.11. Let A be a subgroup of $N_G(H)$. Show that there is a homomorphism $\alpha : A \longrightarrow \text{Aut}(H)$ such that $ah = \alpha(a)(h)a$ for all $a \in A$ and $h \in H$. The product in HA is determined by $(h_1a_1)(h_2a_2) = h_1\alpha(a_1)(h_2)a_1a_2$

Exercise 5.1.12. Count the number of ways to arrange four red beads, three blue beads, and two yellow beads on a straight wire.

Exercise 5.1.13. Recall that two elements of S_n are conjugate in S_n precisely if they have the same cycle structure, i.e. if when written as a product of disjoint cycles,they have the same number of cycles of each length. Let $r_1 > r_2 > ... > r_s \geq 1$ and let $m_i \in \mathbb{N}$ for $1 \leq i \leq s$. How many elements does S_n have with m_1 cycles of length r_1, m_2 cycles of length r_2, etc.? First try out some examples with small values of n.

Exercise 5.1.14. Verify that the formula

$$\pi\{a_1, a_2, \ldots, a_k\} = \{\pi(a_1), \pi(a_2), \ldots, \pi(a_k)\}$$

does indeed define an action of S_n on the set X of k-element subsets of $\{1, 2, \ldots, n\}$. (See Example 5.1.17.)

Exercise 5.1.15. Give the details of the proof of Example 5.1.18. In particular, define an action of S_n on the set of ordered sequences of k-elements from $\{1, 2, \ldots, n\}$, and verify that it is indeed an action. Show that the action is transitive. Calculate the size of the stabilizer of a particular k-element sequence.

5.2. Group Actions – Counting Orbits

How many different necklaces can we make from four red beads, three white beads and two yellow beads? Two arrangements of beads on a circular wire must be counted as the same necklace if one can be obtained from the other by sliding the beads around the wire or turning the wire over. So what we actually need to count is orbits of the action of the dihedral group D_9 (symmetries of the nonagon) on the

$$\frac{9!}{4!3!2!}$$

arrangements of the beads.

Let's consider a simpler example that we can work out by inspection:

Example 5.2.1. Consider necklaces made of two blue and two white beads. There are six arrangements of the beads at the vertices of the square, but only two orbits under the action of the dihedral group D_4, namely that with two blue beads adjacent and that with the two blue beads at opposite corners. One orbit contains four arrangements and the other two arrangements.

We see from this example that the orbits will have different sizes, so we cannot expect the answer to the problem simply to be some divisor of the number of arrangements of beads.

In order to count orbits for the action of a finite group G on a finite set X, consider the set $F = \{(g, x) \in G \times X : gx = x\}$. For $g \in G$, let $\text{Fix}(g) = \{x \in X : gx = x\}$, and let

$$\mathbf{1}_F(g, x) = \begin{cases} 1 & \text{if } (g, x) \in F \\ 0 & \text{otherwise} \end{cases}.$$

We can count F in two different ways:

$$\#F = \sum_{x \in X} \sum_{g \in G} \mathbf{1}_F(x, g) = \sum_{x \in X} \#\text{Stab}(x)$$

and

$$\#F = \sum_{g \in G} \sum_{x \in X} \mathbf{1}_F(x, g) = \sum_{g \in G} \#\mathrm{Fix}(g).$$

Dividing by $\#G$, we get:

$$\frac{1}{\#G} \sum_{g \in G} \#\mathrm{Fix}(g) = \sum_{x \in X} \frac{\#\mathrm{Stab}(x)}{\#G} = \sum_{x \in X} \frac{1}{\#O(x)}.$$

The last sum can be decomposed into a double sum:

$$\sum_{x \in X} \frac{1}{\#O(x)} = \sum_{O} \sum_{x \in O} \frac{1}{\#O},$$

where the outer sum is over distinct orbits. But

$$\sum_{O} \sum_{x \in O} \frac{1}{\#O} = \sum_{O} \frac{1}{\#O} \sum_{x \in O} 1 = \sum_{O} 1,$$

which is the number of orbits! Thus, we have the following result.

Proposition 5.2.2. (Burnside's Lemma) *Let a finite group G act on a finite set X. Then the number of orbits of the action is*

$$\frac{1}{\#G} \sum_{g \in G} \#\mathrm{Fix}(g).$$

Example 5.2.3. Let's use this result to calculate the number of necklaces which can be made from four red beads, three white beads and two yellow beads. X is the set of

$$\frac{9!}{4!3!2!} = 1260$$

arrangements of the beads, which we locate at the nine vertices of a nonagon. Let g be an element of D_9 and consider the orbits of $\langle g \rangle$ acting on vertices of the nonagon. An arrangement of the colored beads is fixed by g if, and only if, all vertices of each orbit of the action of $\langle g \rangle$ are of the same color. Every arrangement is fixed by e.

Let r be the rotation of $2\pi/9$ of the nonagon. For any k ($1 \le k \le 8$), r^k either has order 9, and $\langle r^k \rangle$ acts transitively on vertices, or r^k has order 3, and $\langle r^k \rangle$ has three orbits, each with three vertices. In either case, there are no fixed arrangements, since it is not possible to place beads of one color at all vertices of each orbit.

Now consider any rotation j of π about an axis through one vertex v of the nonagon and the center of the opposite edge. The subgroup $\{e, j\}$

has one orbit containing the one vertex v and four orbits containing two vertices. In any fixed arrangement, the vertex v must have a white bead. Of the remaining four orbits, two must be colored red, one white and one yellow; there are

$$\frac{4!}{2!1!1!} = 12$$

ways to do this. Thus, j has 360 fixed points in X. Since there are 9 such elements, there are

$$\frac{1}{\#G}\sum_{g \in G} \#\text{Fix}(g) = \frac{1}{18}(1260 + 9(12)) = 76$$

possible necklaces.

Example 5.2.4. How many different necklaces can be made with nine beads of three different colors, if any number of beads of each color can be used? Now the set X of arrangements of beads has 3^9 elements; namely each of the nine vertices of the nonagon can be occupied by a bead of any of the three colors. Likewise the number of arrangements fixed by any $g \in D_9$ is $3^{N(g)}$, where $N(g)$ is the number of orbits of $\langle g \rangle$ acting on vertices; each orbit of $\langle g \rangle$ must have beads of only one color, but any of the three colors can be used. One computes the following data:

n-fold rotation axis, $n =$	order of rotation	$N(g)$	number of such group elements
*	1	9	1
9	9	1	6
9	3	3	2
2	2	5	9

Thus, the number of necklaces is

$$\frac{1}{\#G}\sum_{g \in G} \#\text{Fix}(g) = \frac{1}{18}(3^9 + 2\,3^3 + 6\,3 + 9\,3^5) = 1219$$

Example 5.2.5. How many different ways are there to color the faces of a cube with three colors? Regard two colorings to be the same if they are related by a rotation of the cube.

It is required to count the orbits for the action of the rotation group G of the cube on the set X of 3^6 colorings of the faces of the cube. For each $g \in G$ the number of $x \in X$ which are fixed by g is $3^{N(g)}$, where $N(g)$ is the number of orbits of $\langle g \rangle$ acting on faces of the cube. One computes the following data:

n-fold rotation axis, $n =$	order of rotation	$N(g)$	number of such group elements
$*$	1	6	1
2	2	3	6
3	3	2	8
4	4	3	6
4	2	4	3

Thus, the number of colorings of the faces of the cube with three colors is:

$$\frac{1}{\#G}\sum_{g\in G}\#\mathrm{Fix}(g) = \frac{1}{24}(3^6 + 8\,3^2 + 6\,3^3 + 3\,3^4 + 6\,3^3) = 57.$$

Exercises

Exercise 5.2.1. How many necklaces can be made with six beads of three different colors?

Exercise 5.2.2. How many necklaces can be made with two red beads, two green beads and two violet beads?

Exercise 5.2.3. Count the number of ways to color the edges of a cube with four colors. Count the number of ways to color the edges of a cube with r colors; The answer is a polynomial in r.

Exercise 5.2.4. Count the number of ways to color the vertices of a cube with three colors. Count the number of ways to color the vertices of a cube with r colors.

Exercise 5.2.5. Count the number of ways to color the faces of a dodecahedron with three colors. Count the number of ways to color the faces of a dodecahedron with r colors.

5.3. Group Actions and Group Structure

In this section, we consider some applications of the idea of group actions to the study of the structure of groups.

Consider the action of a group G on itself by conjugation. Recall that he stabilizer of an element is called its *centralizer* and the orbit of an element is called its *conjugacy class*. The set of elements z whose conjugacy class consists of z alone is precisely the center of the group. If G is finite, the decomposition of G into disjoint conjugacy classes gives the equation

$$\#G = \#Z(G) + \sum_{g} \frac{\#G}{\#\text{Cent}(g)},$$

where $Z(G)$ denotes the center of G, $\text{Cent}(g)$ the centralizer of g, and the sum is over representatives of distinct conjugacy classes in $G \setminus Z(g)$. This is called the class equation.

Consider a group of order p^n, where p is a prime number and n a positive integer. Every subgroup has order a power of p by Lagrange's Theorem, so for $g \in G \setminus Z(G)$, the size of the conjugacy class of g, namely,

$$\frac{\#G}{\#\text{Cent}(g)},$$

is a positive power of p. Since p divides $\#G$ and $\#Z(G) \geq 1$, it follows that p divides $\#Z(G)$. We have proved:

Proposition 5.3.1. *If $\#G$ is a power of a prime number, then the center of G contains non-identity elements.*

We discovered quite early that any group of order 4 is either cyclic or isomorphic to $\mathbb{Z}_2 \times \mathbb{Z}_2$. We can now generalize this result to groups of order p^2 for any prime p.

Corollary 5.3.2. *Any group of order p^2, where p is a prime, is either cyclic or isomorphic to $\mathbb{Z}_p \times \mathbb{Z}_p$.*

Proof. Suppose G, of order p^2, is not cyclic. Then any non-identity element must have order p. Using the proposition, choose a non-identity element $g \in Z(G)$. Since $o(g) = p$, it is possible to choose $h \in G \setminus \langle g \rangle$. Then g and h are both of order p, and they commute.

I claim that $\langle g \rangle \cap \langle h \rangle = \{e\}$. In fact, if $\pi : G \longrightarrow G/\langle g \rangle$ is the quotient map, then $\pi(h)$ has order p in $G/\langle g \rangle$. Hence $h^a \in \langle g \rangle \Leftrightarrow \pi(h)^a = e \Leftrightarrow p|a \Leftrightarrow h^a = e$.

It follows from this that $\langle g\rangle\langle h\rangle$ contains p^2 distinct elements of G, hence $G = \langle g\rangle\langle h\rangle$. Therefore, G is abelian.

Now $\langle g\rangle$ and $\langle h\rangle$ are two normal subgroups with $\langle g\rangle \cap \langle h\rangle = \{e\}$ and $\langle g\rangle\langle h\rangle = G$. Hence $G \cong \langle g\rangle \times \langle h\rangle \cong \mathbb{Z}_p \times \mathbb{Z}_p$. □

Please look now at Exercise 5.3.1, in which you are asked to show that a group of order p^3 (p a prime) is either abelian or has center of size p.

Corollary 5.3.3. *Let G be a group of order p^n, $n > 1$. Then G has a normal subgroup $\{e\} \underset{\neq}{\subset} N \underset{\neq}{\subset} G$. Furthermore N can be chosen so that every subgroup of N is normal in G.*

Proof. If G is non-abelian, then by the proposition, $Z(G)$ has the desired properties. If G is abelian, every subgroup is normal. If g is a non-identity element, then g has order p^s for some $s \geq 1$. If $s < n$, then $\langle g\rangle$ is a proper subgroup. If $s = n$, then g^p is an element of order p^{n-1}, so $\langle g^p\rangle$ is a proper subgroup. □

Corollary 5.3.4. *Suppose $\#G = p^n$ is a power of a prime number. Then G has a sequence of subgroups*

$$\{e\} = G_0 \subseteq G_1 \subseteq G_2 \subseteq \cdots \subseteq G_n = G$$

such that the order of G_k is p^k, and G_k is normal in G for all k.

Proof. We prove this by induction on n. If $n = 1$, there is nothing to do. So suppose the result holds for all groups of order $p^{n'}$ where $n' < n$. Let N be a proper normal subgroup of G, with the property that every subgroup of N is normal in G. The order of N is p^s for some s, $1 \leq s < n$. Apply the induction hypothesis to N to obtain a sequence

$$\{e\} = G_0 \subseteq G_1 \subseteq G_2 \subseteq \cdots \subseteq G_s = N$$

with $\#G_k = p^k$. Apply the induction hypothesis again to $\bar{G} = G/N$ to obtain a sequence of subgroups

$$\{e\} = \bar{G}_0 \subseteq \bar{G}_1 \subseteq \bar{G}_2 \subseteq \cdots \subseteq \bar{G}_{n-s} = \bar{G}$$

with $\#\bar{G}_k = p^k$ and $\bar{G}_k$ normal in $\bar{G}$. Then put $G_{s+k} = \pi^{-1}\bar{G}_k$, for $1 \leq k \leq n-s$, where $\pi : G \longrightarrow G/N$ is the quotient map. Then the sequence $(G_k)_{0\leq k\leq n}$ has the desired properties. □

We now use similar techniques to investigate the existence of subgroups of order a power of a prime. The first result in this direction is Cauchy's theorem:

Theorem 5.3.5 (Cauchy's Theorem). *Suppose the prime p divides the order of a group G. Then G has an element of order p.*

The proof given here, due to McKay, is simpler and shorter than other known proofs.

Proof. Let X be the set consisting of sequences $(a_1, a_2, \dots, a_p)$ of elements of G such that $a_1 a_2 \dots a_p = e$. Note that a_1 through a_{p-1} can be chosen arbitrarily, and $a_p = (a_1 a_2 \dots a_{p-1})^{-1}$. Thus the cardinality of X is $(\#G)^{p-1}$. Recall that if $a, b \in G$ and $ab = e$, then also $ba = e$. Hence if $(a_1, a_2, \dots, a_p) \in X$, then $(a_p, a_1, a_2, \dots, a_{p-1}) \in X$ as well. Hence, the cyclic group of order p acts on X by cyclic permutations of the sequences.

Each element of X is either fixed under the action of $\mathbb{Z}_p$, or it belongs to an orbit of size p. Thus $\#X = n + kp$, where n is the number of fixed points and k is the number of orbits of size p. Note that $n \geq 1$, since $(e, e, \dots, e)$ is a fixed point of X. But p divides $\#X - kp = n$, so X has a fixed point $(a, a, \dots, a)$ with $a \neq e$. But then a has order p. □

Theorem 5.3.6 (First Sylow Theorem). *Suppose p is a prime, and p^n divides the order of a group G. Then G has a subgroup of order p^n.*

Proof. We prove this statement by induction on n, the case $n = 1$ being Cauchy's Theorem. We assume inductively that G has a subgroup H of order p^{n-1}. Then $[G : H]$ is divisible by p.

Let H act on G/H by left multiplication. We know that $[G : H]$ is equal to the number of fixed points plus the sum of the cardinalities of non-singleton orbits. The size of every non-singleton orbit divides the cardinality of H, so is a power of p. Since p divides $[G : H]$, and p divides the size of each non-singleton orbit, it follows that p also divides the number of fixed points. The number of fixed points is non-zero, since H itself is fixed.

Let's look at the condition for a coset xH to be fixed under left multiplication by H. This is so if, and only if, for each $h \in H$, $hxH = xH$. That is, for each $h \in H$, $x^{-1}hx \in H$. Thus x is in the *normalizer* of H in G, i.e., the set of $g \in G$ such that $gHg^{-1} = H$.

We conclude that the normalizer $N_G(H) \underset{\neq}{\supset} H$. More precisely, the number of fixed points for the action of H on G/H is the index $[N_G(H) : H]$, which is thus divisible by p. Of course, H is normal in $N_G(H)$, so we can consider $N_G(H)/H$, which has size divisible by p. By Cauchy's Theorem, $N_G(H)/H$ has a subgroup of order p. The inverse image of this subgroup in $N_G(H)$ is a subgroup H_1 of cardinality p^n. □

Definition 5.3.7. If p^n is the largest power of the prime p dividing the order of G, a subgroup of order p^n is called a *p-Sylow subgroup*.

The first Sylow Theorem asserts in particular the existence of a p-Sylow subgroup for each prime p.

Theorem 5.3.8. *Let G be a finite group, p a prime number, H a subgroup of G of order p^s, and P a p-Sylow subgroup of G. Then there is a $a \in G$ such that $aHa^{-1} \subseteq P$.*

Proof. Let X be the family of conjugates of P in G. According to Exercise 5.3.2, the cardinality of X is not divisible by p. Now let H act on X by conjugation. Any non-singleton orbit must have cardinality a power of p. Since $\#X$ is not divisible by p, it follows that X has a fixed point under the action of H. That is, for some $g \in G$, conjugation by elements of H fixes gPg^{-1}. Equivalently, $H \subseteq N_G(gPg^{-1}) = gN_G(P)g^{-1}$, or $g^{-1}Hg \subseteq N_G(P)$. Since $\#(g^{-1}Hg) = \#(H) = p^s$, it follows from Exercise 5.3.3 that $g^{-1}Hg \subseteq P$. □

Corollary 5.3.9 (Second Sylow Theorem). *Let P and Q be two p-Sylow subgroups of a finite group G. Then P and Q are conjugate subgroups.*

Proof. According to the theorem, there is an $a \in G$ such that $aQa^{-1} \subseteq P$. Since the two groups have the same size, it follows that $aQa^{-1} = P$ □

Theorem 5.3.10 (Third Sylow Theorem). *Let G be a finite group and let p be a prime number. The number of p-Sylow subgroups of G divides $\#G$ and is congruent to* $1 \pmod{p}$. *In other words the number of p-Sylow subgroups can be written in the form $mp + 1$.*

Proof. Let P be a p-Sylow subgroup. The family X of p-Sylow subgroups is the set of conjugates of P, according to the second Sylow theorem. Let P act on X by conjugation. If Q is a p-Sylow subgroup distinct from P, then Q is not fixed under the action of P; for if Q were fixed, then $P \subseteq N_G(Q)$, and by Exercise 5.3.3, $P \subseteq Q$. Therefore, there is exactly one fixed point, namely P for the action of P on X. All the non-singleton orbits for the action of P on X have size a power of p, so $\#(X) = mp + 1$.

On the other hand, G acts transitively on X by conjugation, so

$$\#X = \frac{\#G}{\#N_G(P)}$$

divides the order of G. □

Exercises

Exercise 5.3.1. Suppose $\#G = p^3$, where p is a prime. Show that either $\#Z(G) = p$ or G is abelian.

Exercise 5.3.2. Let P be a p-Sylow subgroup of a finite group G. Consider the set of conjugate subgroups gPg^{-1} with $g \in G$. According to Corollary 5.1.14, the number of such conjugates is the index of the normalizer of P in G, $[G : N_G(P)]$. Show that the number of conjugates is not divisible by p.

Exercise 5.3.3. Let P be a p-Sylow subgroup of a finite group G. Let H be a subgroup of $N_G(P)$ such that $\#(H) = p^s$. Show that $H \subseteq P$. *Hint:* Refer to Exercise 5.1.10 where it is shown that HP is a subgroup of $N_G(P)$ with

$$\#(HP) = \frac{\#(P)\,\#(H)}{\#(H \cap P)}.$$

Exercise 5.3.4. Let p be a prime number. Show that for each r, $1 \le r \le p-1$ there is an automorphism α_r of $\mathbb{Z}_p$ such that $\alpha_r([n]) = [rn]$ for each n. Show that any automorphism of $\mathbb{Z}_p$ is one of the α_r. Conclude that the order of the automorphism group of $\mathbb{Z}_p$ is $p-1$.

Exercise 5.3.5. Let G be a group of order pq, where p and q are prime numbers and $p > q$.

(a) According to Cauchy's Theorem or the first Sylow Theorem, G has p- and q- Sylow subgroups P and Q of order p and q respectively. Let g and h be generators of the cyclic groups P and Q. Show that $P \cap Q = \{e\}$.

(b) Show that P is normal in G.

(c) Conclude that there is a homomorphism φ from Q into $\mathrm{Aut}(P)$ such that $hg = \alpha(h)(g)h$.

(d) Show that either g and h commute, in which case G is abelian and isomorphic to $\mathbb{Z}_p \times \mathbb{Z}_q \cong \mathbb{Z}_{pq}$, or q divides $p-1$.

(e) If q divides $p-1$, show that there is a non-commutative group of order pq.

(f) Show that in a non-commutative group of order pq there are p distinct subgroups of order q.

(g) Show that any group of order 15 is cyclic.

(h) Show that there is a non-abelian group of order 21.

Exercise 5.3.6. One can show that the automorphism group of $\mathbb{Z}_p$ is actually *cyclic* of order $p-1$. We will do this in Section 6.3. Assuming this fact for now, show that there is exactly one non-abelian group of order pq, up to isomorphism, when q divides $p-1$.

Exercise 5.3.7.

(a) An abelian group is the direct product of p-Sylow subgroups for primes p dividing $\#G$.

(b) For example, an abelian group of order 28 is a direct product of $\mathbb{Z}_7$ with a group of order 4. Thus an abelian group of order 28 is either $\mathbb{Z}_7 \times \mathbb{Z}_4$ or $\mathbb{Z}_7 \times \mathbb{Z}_2 \times \mathbb{Z}_2$. Show that these two possibilities are non-isomorphic.

(c) More generally, show that an abelian group of order p^2q (p and q primes) is isomorphic to either $\mathbb{Z}_{p^2} \times \mathbb{Z}_q$ or $\mathbb{Z}_p \times \mathbb{Z}_p \times \mathbb{Z}_q$, and these two possibilities are non-isomorphic.

We have classified groups of order p, p^2, and pq completely (p and q primes). We also know that abelian groups are the direct product of Sylow subgroups, and therefore to classify abelian groups completely, it suffices to classify abelian groups of order p^n.

Exercise 5.3.8. Classify all groups of order no more than 30, except those of order 8, 12, 16, 18, 20, 24, 27, 28, and 30.

Exercise 5.3.9. Classify all abelian groups of order no more than 30, except those of order 8, 16, and 27.

The Sylow theorems provide good tools for investigating the possible structures of the remaining non-abelian groups of order no more than 30.

Exercise 5.3.10. Show that a group of order 28 has a normal subgroup N of order 7; the automorphism group of N is cyclic of order 6. A 2-Sylow subgroup of order 4 is either $\mathbb{Z}_4$ or $\mathbb{Z}_2 \times \mathbb{Z}_2$. Show that there is a unique non-trivial homomorphism of $\mathbb{Z}_4$ into $\text{Aut}(N)$, and an essentially unique automorphism of $\mathbb{Z}_2 \times \mathbb{Z}_2$ into $\text{Aut}(N)$. Conclude that there are, up to isomorphism, two non-abelian groups of order 28. One of these has to be the dihedral group D_{14}. Show that $D_{14} \cong D_7 \times \mathbb{Z}_2$. Can you describe the other non-abelian group of order 28?

Exercise 5.3.11. Do a similar analysis of the non-abelian group(s) of order 20.

Exercise 5.3.12. Try a similar analysis of the non-abelian group(s) of order 18. What new issues do you run into?

Exercise 5.3.13. Try a similar analysis of the non-abelian group(s) of order 12. What new issues do you run into?

Exercise 5.3.14. Guess what are the possible abelian groups of order 8, 16, and 27. Can you prove your conjecture?

5.4. Application: Transitive Subgroups of the Symmetric Group on Five Elements

This section can be omitted without loss of continuity. However, in the discussion of Galois groups in Chapter 9, we shall refer to the results of this section.

In this brief section, we will use the techniques of the previous sections to classify the transitive subgroups of S_5. Of course, S_5 itself and the alternating group A_5 are transitive subgroups. Also, we can readily thing of $Z_5 = \langle(12345)\rangle$ (and its conjugates), and $D_5 = \langle(12345), (12)\rangle$ (and its conjugates). (We write $\langle a, b, c, \dots\rangle$ for the subgroup generated by elements $a, b, c, \dots$.)

We know that the only subgroup of index 2 in S_5 is A_5; see Exercise 3.1.14. We will need the fact, which is proved later in the text (Section 10.3), that A_5 *is the only normal subgroup of* S_5 *other than* S_5 *and* $\{e\}$.

Lemma 5.4.1. *Let G be a subgroup of S_5 which is not equal to S_5 or A_5. Then $[S_5 : G] \geq 5$; that is, $\#G \leq 24$.*

Proof. Write $d = [S_5 : G]$. Consider the action of S_5 on the set X of left cosets of G in S_5 by left multiplication.

I claim that this action is faithful, i.e., that the corresponding homomorphism of S_5 into $\text{Sym}(X) \cong S_d$ is injective. In fact, the kernel of the homomorphism is the intersections of the stabilizers of the points of X, and, in particular, is contained in G, which is the stabilizer of $G = eG \in X$. On the other hand, the kernel is a normal subgroup of S_5. Since G does not contain S_5 or A_5, it follows from the fact mentioned just above that the kernel is $\{e\}$.

But this means that S_5 is isomorphic to a subgroup of $\text{Sym}(X) \cong S_d$, and, consequently, $5! \leq d!$, or $5 \leq d$. □

Remark 5.4.2. More, generally, A_n is the only non-trivial normal subgroup of S_n for $n \geq 5$. So the same argument shows that a subgroup of S_n which is not equal to S_n or A_n must have index at least n in S_n.

Now, suppose that G is a transitive subgroup of S_5, not equal to S_5 or A_5. We know that for any finite group acting on any set, the size of any orbit divides the order of the group. By hypothesis, G, acting on $\{1, \ldots, 5\}$, has one orbit of size 5, so 5 divides $\#G$. But, by the lemma, $\#G \leq 24$. Thus,

$$\#G = 5k, \quad \text{where} \quad 1 \leq k \leq 4.$$

By Sylow Theory, G has a normal subgroup of order 5, necessarily cyclic. Let $\sigma \in G$ be an element of order 5. Since $G \subseteq S_5$, the only possible cycle structure for σ is a 5-cycle. Without loss of generality, assume $\sigma = (1\ 2\ 3\ 4\ 5)$.

That $\langle\sigma\rangle$ is normal in G means that $G \subseteq N_{S_5}(\langle\sigma\rangle)$. What is $N_{S_5}(\langle a\rangle)$? We know that for any $\rho \in S_5$, $\rho\sigma\rho^{-1} = (\rho(1)\ \rho(2) \ldots \rho(5))$. For ρ to normalize $\langle\sigma\rangle$, it is necessary and sufficient that $(\rho(1)\ \rho(2) \ldots \rho(5))$ be a power of $(1\ 2\ 3\ 4\ 5)$. We can readily find one permutation ρ which will serve, namely $\rho = (2\ 3\ 5\ 4)$, which satisfies $\rho\sigma\rho^{-1} = \sigma^2$.

Observe that A_5 does *not* normalize $\langle\sigma\rangle$, so the lemma, applied to $N_{S_5}(\langle\sigma\rangle)$ gives that the cardinality of this group is no more than 20. On the other hand, $\langle\sigma, \rho\rangle$ is a subgroup of $N_{S_5}(\langle\sigma\rangle)$ isomorphic to $\mathbb{Z}_4 \ltimes \mathbb{Z}_5$ and has cardinality 20. Therefore $N_{S_5}(\langle\sigma\rangle) = \langle\sigma, \rho\rangle$.

Now, we have $\langle\sigma\rangle \subseteq G \subseteq \langle\sigma, \rho\rangle$. The possibilities for G are $\langle\sigma\rangle \cong \mathbb{Z}_5$, $\langle\sigma, \rho^2\rangle \cong D_5$ and, finally, $\langle\sigma, \rho\rangle \cong \mathbb{Z}_4 \ltimes \mathbb{Z}_5$. The normalizer of each of these groups is $\langle\sigma, \rho\rangle$; hence the number of conjugates of each of the groups is 5.

To summarize:

Proposition 5.4.3. *The transitive subgroups of S_5 are:*

(a) S_5,
(b) A_5,
(c) $\langle(1\ 2\ 3\ 4\ 5), (2\ 3\ 5\ 4)\rangle \cong \mathbb{Z}_4 \ltimes \mathbb{Z}_5$ *(and its five conjugates),*
(d) $\langle(1\ 2\ 3\ 4\ 5), (2\ 5)(3\ 4)\rangle \cong D_5$ *(and its five conjugates), and*
(e) $\langle(1\ 2\ 3\ 4\ 5)\rangle \cong \mathbb{Z}_5$ *(and its five conjugates).*

Remark 5.4.4. There are 16 conjugacy classes of transitive subgroups of S_6, and seven of S_7. See J. D. Dixon and B. Mortimer, *Permutation Groups*, Springer Verlag, 1996, pp. 58-64. Transitive subgroups of S_n at least for $n \leq 11$ have been classified. Consult Dixon and Mortimer for further details.

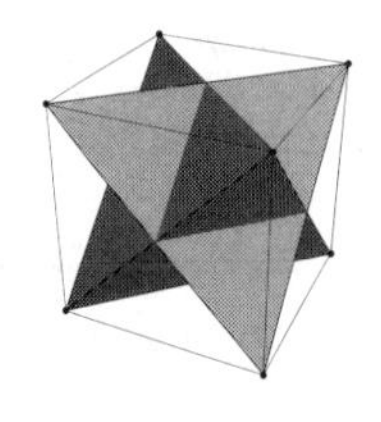

Chapter 6

Finite Abelian Groups

6.1. Some Number Theory

In this section, we will extend the theory of divisibility in the integers a little beyond what is probably familiar to you. We will see a pleasant interaction between group theory and number theory which enriches both topics.

Recall from Section 2.3 that a pair of non-zero integers m and n are said to be *relatively prime* if their greatest common divisor is 1. In this case, there exist integers a and b such that $1 = am + bn$.

For $n \in \mathbb{N}$, the number of $k \in \mathbb{N}$ such that $k \leq n$ and k is relatively prime to n is denoted $\varphi(n)$; the function $n \mapsto \varphi(n)$ is called the *Euler φ function.* Here are some tables of φ.

n	1	2	3	4	5	6	7	8	9	10	11	12	13	14	15
$\varphi(n)$	1	1	2	2	4	2	6	4	6	4	10	4	12	6	8

n	16	17	18	19	20	21	22	23	24	25	26	27	28	29	30
$\varphi(n)$	8	16	6	18	8	12	10	22	8	20	12	18	12	28	8

n	31	32	33	34	35	36	37	38	39	40	41	42	43	44	45
$\varphi(n)$	30	16	20	16	24	12	36	18	24	16	40	12	42	20	24

Examine the values of $\varphi(n)$ for n a prime number. What do you notice? Can you find any other patterns in the data? What about $\varphi(n)$ when n is a power of a prime? What about the parity of $\varphi(n)$? Empirical discovery of patterns is a tradition in number theory, and empirical investigations are made more feasible by the availability of computers.

Proposition 6.1.1. *If p is prime, then* g.c.d.$(k, p) = 1$ *for* $1 \leq k < p$. *Hence* $\varphi(p) = p - 1$.

Proof. 1 is the only divisor of p which is less than p. □

Fix a natural number n. For $k \in \mathbb{Z}$, let $[k]$ denote the image of k in $\mathbb{Z}_n$. We have seen that $\mathbb{Z}_n$ has an associative multiplication defined by $[a][b] = [ab]$, and $[1]$ is a multiplicative identity. When does a congruence class $[k]$ have a multiplicative inverse?

Proposition 6.1.2. *A congruence class* $[k]$ *has a multiplicative inverse in* $\mathbb{Z}_n$ *if, and only if,* $\text{g.c.d.}(k, n) = 1$.

Proof. $[k][a] = 1 \Leftrightarrow ka - 1 \in n\mathbb{Z} \Leftrightarrow 1 = ka + nb$ for some $b \in \mathbb{Z} \Leftrightarrow 1 = \text{g.c.d.}(k, n)$ by Exercise 2.3.6. □

Corollary 6.1.3. *Let* $\Phi(n)$ *be the set of* $[k]$ *such that* $1 \le k < n$ *and* $\text{g.c.d.}(k, n) = 1$. $\Phi(n)$ *is a commutative group (under multiplication) of order* $\varphi(n)$.

Corollary 6.1.4. *If* $\text{g.c.d.}(k, n) = 1$ *then* $k^{\varphi(n)} \equiv 1 \pmod{n}$.

Proof. Since the order of $\Phi(n)$ is $\varphi(n)$, $[1] = [k]^{\varphi(n)} = [k^{\varphi(n)}]$. □

Corollary 6.1.5. *If* p *is prime, then* $k^{p-1} \equiv 1 \pmod{p}$ *for all integers* k *not divisible by* p.

For example, 7 divides $k^6 - 1$ for $1 \le k \le 6$.

Please look now at Exercise 6.1.4, in which you are asked to determine the structure of $\Phi(n)$ for $n \le 15$.

Computational experiments suggest the following result.

Theorem 6.1.6. *When* p *is prime,* $\Phi(p)$ *is cyclic of order* $p - 1$.

The proof requires developing some ideas about divisibility of polynomials, to which we devote Section 7.2

Exercises

Exercise 6.1.1. Write a computer program to tabulate the Euler φ function for $n \le 100$. Examine your results.

Exercise 6.1.2. Show more generally that $\varphi(p^n) = p^{n-1}(p - 1) = p^n(1 - 1/p)$. In particular, $\varphi(2^n) = 2^{n-1}$.

Exercise 6.1.3. Another theorem immediately suggested by the data is that $\varphi(n)$ is even if $n \geq 3$. In particular, $\text{g.c.d.}(\varphi(n), \varphi(m)) \geq 2$ if $n, m \geq 3$. Prove this.

Exercise 6.1.4. Find the structure of the group $\Phi(n)$ for $n \leq 15$. There is only one abelian group of order $1, 2, 6$, or 10, so everything is clear except for $n = 5, 8, 10, 12, 13$, and 15. For these values of n, it is clearly possible to find the structure of $\Phi(n)$ by computation. Concentrate first on the groups $\Phi(n)$ of order 4; are they all cyclic? (No, they are not.) What about the groups $\Phi(p)$ when p is prime?

Exercise 6.1.5. This exercise concerns the order of elements in a cyclic group. Fix a natural number n.

(a) If s divides n, then the order of $[s]$ in $\mathbb{Z}_n$ is n/s.

(b) If $\text{g.c.d.}(s, n) = 1$, then the order of $[s]$ in $\mathbb{Z}_n$ is n. *Hint:* On the one hand, $o([s])$ divides n. On the other hand, if $[ks] = [0]$, then n divides ks. But $1 = \alpha s + \beta n$ for some integers α and β, so $k = \alpha ks + \beta kn$. Conclude that n divides k. In particular, n divides $o([s])$.

(c) Let a be an element of a group with $o(a) = n$. If s divides n, then $o(a^s) = n/s$. If $\text{g.c.d.}(s, n) = 1$, then $o(a^s) = n$, so $\langle a \rangle = \langle a^s \rangle$.

(d) For any s, let $\alpha = \text{g.c.d.}(s, n)$. Then $s = (s/\alpha)\alpha$, where (s/α) is relatively prime to n, and α divides n. Thus $[s/\alpha]$ is a generator of $\mathbb{Z}_n$ by part (c), and the order of $[s]$ in $\mathbb{Z}_n$ is $o[s] = o[\alpha(s/\alpha)] = n/\alpha$.

Exercise 6.1.6.

(a) If $n = ab$, where a and b are relatively prime, then $\mathbb{Z}_n \cong \mathbb{Z}_a \times \mathbb{Z}_b$. *Hint:* Map $\mathbb{Z}_n$ to $\mathbb{Z}_a \times \mathbb{Z}_b$ by $[x]_n \mapsto ([x]_a, [x]_b)$. Here $[x]_r$ denotes the equivalence class $\pmod{r}$. Show that this map is well-defined and an isomorphism.

(b) If $n = p_1^{r_1} p_2^{r_2} \cdots p_k^{r_k}$ is the prime decomposition of n, then $\mathbb{Z}_n \cong \mathbb{Z}_{p_1^{r_1}} \times \mathbb{Z}_{p_2^{r_2}} \times \cdots \times \mathbb{Z}_{p_k^{r_k}}$. *Hint:* Use part (a) and induction.

(c) Define a multiplication on $\mathbb{Z}_a \times \mathbb{Z}_b$ by

$$([x], [y])([x'], [y']) = ([xx'], [yy']).$$

Observe that the group isomorphisms in parts (a) and (b) also respect multiplicative structures.

Exercise 6.1.7. Let $n = ab$ be a product of relatively prime natural numbers a, b. According to Exercise 6.1.6, the group isomorphism $\mathbb{Z}_n \cong \mathbb{Z}_a \times \mathbb{Z}_b$ also respects the multiplicative structure of $\mathbb{Z}_n$ and of $\mathbb{Z}_a \times \mathbb{Z}_b$.

(a) Show that this implies that $\Phi(n) \cong \Phi(a) \times \Phi(b)$.

(b) If $n = p_1^{\alpha_1} \cdots p_k^{\alpha_k}$ is the prime decomposition of n, then

$$\Phi(n) \cong \Phi(p_1^{\alpha_1}) \times \cdots \times \Phi(p_k^{\alpha_k}).$$

(c) Conclude that also $\varphi(ab) = \varphi(a)\varphi(b)$ if a, b are relatively prime. Consequently, if $n = p_1^{\alpha_1} \cdots p_k^{\alpha_k}$, then $\varphi(n) = \prod_i \varphi(p_i^{\alpha_i}) = \prod_i p_i^{\alpha_i}(1 - 1/p_i) = n \prod_i (1 - 1/p_i)$.

Remark 6.1.7. By the previous exercise, to find the structure of $\Phi(n)$, it suffices to find the structure of $\Phi(p^\alpha)$. It is a fact that $\Phi(p^\alpha)$ is cyclic if p is an odd prime and that $\Phi(2^\alpha)$ is not cyclic if $\alpha \geq 3$. The proof of both facts can be found in George Andrews, *Number Theory*, Dover Publications, New York, 1994 (original edition: W.B. Saunders, 1971), exercises for Section 7-2. The problem of finding the structure of $\Phi(2^\alpha)$ for $\alpha \geq 2$ could be approached by computational experiments.

Exercise 6.1.8. Assuming the results mentioned in the previous remark, how that $\Phi(n)$ is cyclic precisely when $n = 1, 2$, or 4, or when $n = p^\alpha$ or $n = 2p^\alpha$ for an odd prime p

6.2. Finite Abelian Groups

In this section, we will obtain a definitive structure theorem and classification of finite abelian groups. All the groups in this section will be abelian, and, following a common convention, we will use additive notation for the group operation.

The exposition here is independent of the material on the Sylow Theory in Section 5.3. The following lemma is the most subtle item in this section, and the proof relies heavily on the result of Exercise 6.1.5.

Lemma 6.2.1. *Let G be a finite abelian group. Let $a \in G$ be an element of maximum order m. Let $\pi : G \longrightarrow G/\langle a \rangle$ be the quotient map. Let $\bar{b} \in G/\langle a \rangle$ be an element of order r. Then there is an element $b \in G$ such that $\pi(b) = \bar{b}$, and $o(b) = r$.*

Proof. Let b_1 be any pre-image of $\bar{b}$ in G. Then $o(b_1)\bar{b} = \pi(o(b_1)b_1) = \pi(0) = 0$, so r divides the order of b_1. On the other hand, $o(b_1) \leq m$ by hypothesis. Since $0 = r\bar{b} = \pi(rb_1)$, we have $rb_1 \in \langle a \rangle$, $rb_1 = na$ for some n, $0 \leq n \leq m - 1$. Write $n = qr + s$ with $0 \leq s < r$. Then $rb_1 = qra + sa$, or $r(b_1 - qa) = sa$.

Set $b = b_1 - qa$, so $\pi(b) = \pi(b_1) = \bar{b}$. Suppose that $s > 0$. Then $o(sa) = m/\alpha$, where $\alpha = \text{g.c.d.}(s, m)$, by Exercise 6.1.5. Since $o(b)$ is divisible by r, $o(b) = ro(rb) = ro(sa) = r(m/\alpha)$, again using Exercise 6.1.5. Then since $o(b) \leq m$, we have $rm/\alpha \leq m$, or $r \leq \alpha \leq s$. This is a contradiction, so we must have $s = 0$, which means $rb = 0$. It follows that $o(b) = r$. □

Proposition 6.2.2. *Let G be a finite abelian group. Then G is a direct product of cyclic groups.*

Proof. We prove this by induction on the order of the abelian group, there being nothing to prove if $\#(G) = 1$. So assume $\#(G) > 1$ and that the assertion holds for all finite abelian groups of size strictly less than $\#(G)$. Let a_1 be an element of G of maximum order $r_1 > 1$ and put $A_1 = \langle a_1 \rangle$.

Applying the induction hypothesis to G/A_1, suppose that G/A_1 is a direct product of subgroups $\bar{A}_2, \bar{A}_3, \dots, \bar{A}_k$, where $\bar{A}_i = \langle \bar{a}_i \rangle$ is cyclic of order r_i for $2 \le i \le k$. Let $\pi : G \longrightarrow G/A_1$ be the quotient map. Using Lemma 6.2.1, let a_i be a pre-image of $\bar{a}_i$ of order r_i for each i, and let $A_i = \langle a_i \rangle$. For each $g \in G$, there exist n_i for $2 \le i \le k$ such that $0 \le n_i < r_i$ and $\pi(g) = \sum_{i \ge 2} n_i \bar{a}_i$. Thus, $g - \sum_{i \ge 2} n_i a_i \in A_1 = \langle a_i \rangle$, so there is an n_1, $0 \le n_1 < r_1$ such that $g = \sum_{i \ge 1} n_i a_i$. On the other hand, if $0 \le n_i < r_i$ for all i and $\sum_{i \ge 1} n_i a_i = 0$, then applying π gives $\sum_{i \ge 2} n_i \bar{a}_i = 0$, so $n_i = 0$ for $i \ge 2$, hence also $n_1 = 0$. Thus, G is the direct product of the subgroups A_i, $1 \le i \le k$. □

We can now obtain a complete structure theorem for abelian groups whose order is a power of a prime.

Theorem 6.2.3.

(a) *Let G be an abelian group of order p^n, where p is a prime. There exist natural numbers $n_1 \ge n_2 \ge \dots \ge n_s$ such that $\sum_i n_i = n$, and $G \cong \mathbb{Z}_{p^{n_1}} \times \dots \times \mathbb{Z}_{p^{n_s}}$.*

(b) *The sequence of exponents in part (b) is unique. That is, if $m_1 \ge m_2 \ge \dots \ge m_r$, $\sum_j m_j = n$, and $G \cong \mathbb{Z}_{p^{m_1}} \times \dots \times \mathbb{Z}_{p^{m_r}}$, then $s = r$ and $n_i = m_i$ for all i.*

Proof. It follows from Lemma 6.2.1 and Exercise 6.1.6 that a finite abelian group G is a direct product of cyclic groups, each of which has order a power of a prime. The only primes which can occur are primes dividing the order of G, by Lagrange's Theorem. This observation already suffices to prove part (a).

We prove the uniqueness statement (b) by induction on n, the case $n = 1$ being trivial. So suppose the uniqueness statement holds for all abelian groups of order $p^{n'}$ where $n' < n$. Consider the homomorphism $\varphi(x) = x^p$ of G into itself. Suppose

$$(n_1, \dots, n_s) = (n_1, \dots, n_{s'}, 1, 1, \dots, 1)$$

and

$$(m_1, \ldots, m_r) = (m_1, \ldots, m_{r'}, 1, 1, \ldots, 1),$$

where $n_{s'} > 1$ and $n_{r'} > 1$. Then the isomorphism

$$G \cong \mathbb{Z}_{p^{n_1}} \times \cdots \times \mathbb{Z}_{p^{n_s}}$$

gives

$$\varphi(G) \cong \mathbb{Z}_{p^{n_1 - 1}} \times \cdots \times \mathbb{Z}_{p^{n_{s'}-1}},$$

and the isomorphism

$$G \cong \mathbb{Z}_{p^{m_1}} \times \cdots \times \mathbb{Z}_{p^{m_r}}$$

gives

$$\varphi(G) \cong \mathbb{Z}_{p^{m_1 - 1}} \times \cdots \times \mathbb{Z}_{p^{m_{r'}-1}}.$$

The first isomorphism yields

$$\#\varphi(G) = \sum_{i=1}^{s'} (n_i - 1) = \sum_{i=1}^{s} (n_i - 1) = n - s,$$

and similarly the second isomorphism yields $\#\varphi(G) = n - r$. In particular, $r = s$. Now applying the induction hypothesis to $\varphi(G)$ gives also $s' = r'$ and $n_i - 1 = m_i - 1$ for $1 \le i \le s'$. This implies $n_i = m_i$ for $1 \le i \le s$. □

The final step is to obtain a decomposition of a finite abelian group into a direct product of groups, each of which has order a power of a prime. This, together with the previous theorem, gives a complete structure theorem for finite abelian groups.

Theorem 6.2.4. *Let G be an abelian group of order n and let $n = p_1^{r_1} p_2^{r_2} \ldots p_k^{r_k}$ be the prime decomposition of n. Let*

$$G[p_i] = \{g \in G : o(g) \text{ is a power of } p_i\}.$$

Then $G[p_i]$ is a subgroup of order $p_i^{r_i}$ and $G \cong G[p_1] \times \cdots \times G[p_k]$.

Proof. Now suppose G has order $n = p_1^{r_1} p_2^{r_2} \ldots p_k^{r_k}$, where the p_i are distinct primes. Then gathering together the cyclic factors whose order is a power of p_i, for each i, one can write $G \cong A_1 \times \cdots \times A_k$, where A_i is a subgroup of G, and a direct product of cyclic groups, each of order a power of p_i. It follows that $\#G = \prod \#A_i$, and $\#A_i$ is a power of p_i. By the uniqueness of the prime decomposition of n, we must have $\#A_i = p_i^{r_i}$. Now consider an element of G, $g = a_1 + a_2 + \cdots + a_k$, where $a_i \in A_i$. For each i, $o(a_i)$ is a power of p_i, and the order of g is $\prod o(a_i)$. It follows that $A_i = \{g \in G : o(g) \text{ is a power of } p_i\} = G[p_i]$. □

Exercises

Exercise 6.2.1. Find all abelian groups of order ≤ 50, up to isomorphism.

Exercise 6.2.2. How many abelian groups are there of order 128, up to isomorphism?

Exercise 6.2.3. Show that $\mathbb{Z}_a \times \mathbb{Z}_b$ is not cyclic if $\text{g.c.d.}(a, b) \geq 2$.

Exercise 6.2.4. If G is a finite abelian group and m is the maximum of orders of elements of G, show that the order of any element of G divides m.

Exercise 6.2.5. If G is a finite abelian group, show that there exist natural numbers $m_1, m_2, \dots, m_r$ such that m_{i+1} divides m_i for all i and $G \cong \mathbb{Z}_{m_1} \times \cdots \times \mathbb{Z}_{m_r}$. Show that the numbers m_i are unique. *Hint:* You need to work out how the m_i are related to the exponents in the structure theorem; the uniqueness follows because the m_i and the exponents determine one another.

Exercise 6.2.6. Determine the decomposition $G \cong \mathbb{Z}_{m_1} \times \cdots \times \mathbb{Z}_{m_r}$ given in the last exercise for finite abelian groups G or order ≤ 50.

Exercise 6.2.7. If G is an abelian group and a and b are elements or order k and l respectively, such that $\langle a \rangle \cap \langle b \rangle = \{0\}$, then

$$o(a + b) = (kl)/\text{g.c.d.}(k, l).$$

6.3. Symmetries of Groups

A mathematical object is a set with some structure. A bijection of the set which preserves the structure is undetectable insofar as that structure is concerned. For example, a rotational symmetry of the cube moves the individual points of the cube around but preserves the structure of the cube. A structure preserving bijection of any sort of object can be regarded as a symmetry of the object, and the set of symmetries always constitutes a group.

If, for example, the structure under consideration is a group, then a structure preserving bijection is a group automorphism. In this section, we will work out a few examples of automorphism groups of groups.

Recall that the inner automorphisms of a group are those of the form $c_g(x) = gxg^{-1}$ for some g in the group. The map $g \mapsto c_g$ is a homomorphism of G into $\text{Aut}(G)$ with image the subgroup $\text{Int}(G)$ of inner automorphisms. The kernel of this homomorphism is the center of the group and therefore $\text{Int}(G) \cong G/Z(G)$. Observe that the group of inner automorphisms of an abelian group is trivial, since $G = Z(G)$.

Proposition 6.3.1. $\text{Int}(G)$ *is a normal subgroup of* $\text{Aut}(G)$.

Proof. Compute that for any automorphism α of G, $\alpha c_g \alpha^{-1} = c_{\alpha(g)}$. □

Remark 6.3.2. The symmetric group S_n has trivial center, so $\text{Int}(S_n) \cong S_n/Z(S_n) \cong S_n$. We showed that every automorphism of S_3 is inner, so $\text{Aut}(S_3) \cong S_3$. An interesting question is whether this is also true for S_n for all $n \geq 3$. The rather unexpected answer that it is true except when $n = 6$. ($S_6 \cong \text{Int}(S_6)$ is an index 2 subgroup of $\text{Aut}(S_6)$.) See W. R. Scott, *Group Theory*, Dover Publications, 1987, pp. 309-314 (original edition, Prentice-Hall, 1964).

Exercises

Exercise 6.3.1. Show that any homomorphism of $\mathbb{Z}$ into itself has the form $x \mapsto nx$ for some n. Show that a homomorphism is injective unless $n = 0$, and it is an automorphism if, and only if, $n = \pm 1$. Show that the automorphism group of $\mathbb{Z}$ is isomorphic to $\mathbb{Z}_2$.

Exercise 6.3.2. Show that a homomorphism of the additive group $\mathbb{Z}^2$ into itself is determined by a 2-by-2 matrix of integers. Show that the homomorphism is injective if, and only if, the determinant of the matrix is non-zero, and bijective if, and only if, the determinant of the matrix is ± 1. Conclude that the group of automorphisms of $\mathbb{Z}^2$ is isomorphic to the group of 2-by-2 matrices with integer coefficients and determinant equal to ± 1.

Exercise 6.3.3. Show that any homomorphism of the additive group $\mathbb{Q}$ into itself has the form $x \mapsto rx$ for some $r \in \mathbb{Q}$. Show that a homomorphism is an automorphism unless $r = 0$. Conclude that the automorphism group of $\mathbb{Q}$ is isomorphic to $\mathbb{Q}^*$, namely, the multiplicative group of non-zero rational numbers.

Exercise 6.3.4. Show that any group homomorphism of the additive group $\mathbb{Q}^2$ is determined by rational 2-by-2 matrix. Show that any group homomorphism is actually a linear map, and the group of automorphisms is the same as the group of invertible linear maps.

Exercise 6.3.5. Think about whether the results of the last two exercises hold if $\mathbb{Q}$ is replaced by $\mathbb{R}$. What issue arises?

Exercise 6.3.6. We can show that $\text{Aut}(\mathbb{Z}_2 \times \mathbb{Z}_2) \cong S_3$ in two ways. One way is to show that any automorphism is determined by a 2-by-2 matrix with entries in $\mathbb{Z}_2$, that there are six such matrices, and that they form a group isomorphic to S_3. Another way is to recall that $\mathbb{Z}_2 \times \mathbb{Z}_2$ can be described as a group with four elements e, a, b, c, with each non-identity element of order 2 and the product of any two non-identity elements equal to the third. Show that any permutation of $\{a, b, c\}$ determines an automorphism and, conversely, any automorphism is given by a permutation of $\{a, b, c\}$.

Exercise 6.3.7. Fix a natural number n.

(a) For any integer r, define $\psi_r : \mathbb{Z}_n \mapsto \mathbb{Z}_n$ by $\psi_r([k]) = [rk]$. Show that ψ_r is a well defined homomorphism of $\mathbb{Z}_n$.

(b) Show that $\psi_r = \psi_s \Leftrightarrow [r] = [s]$.

(c) Show that ψ_r is an automorphism of $\mathbb{Z}_n$ if, and only if, r is relatively prime to n. *Hint:* Recall that r is relatively prime to n if, and only if, there is an s such that $[rs] = [1]$.

(d) Show that $[r] \mapsto \psi_r$ is an isomorphism from $\Phi(n)$ to $\text{Aut}(\mathbb{Z}_n)$. See Section 6.1 for a discussion of $\Phi(n)$.

(e) Conclude from Theorem 6.1.6 that if p is a prime, then $\text{Aut}(\mathbb{Z}_p) \cong \mathbb{Z}_{p-1}$.

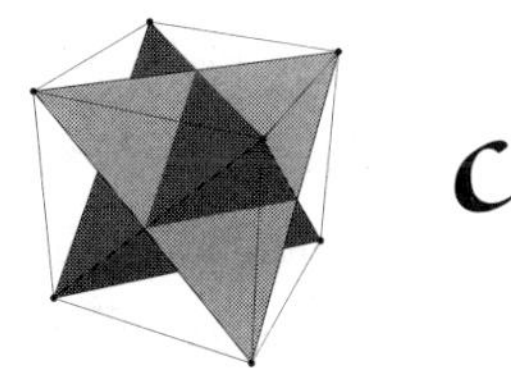

Chapter 7

Rings

7.1. First Look at Rings and Fields

In this section, we will take a first look at *rings* and *fields*, give some examples of these structures, and comment on their role in mathematics.

You are familiar with a number of structures which have two operations, addition and multiplication; think, for example, of the integers, or the real numbers, or the set of 3-by-3 or n-by-n matrices with real entries, or the set of polynomials with real coefficients. The two operations on each of these sets satisfy a number of familiar rules of arithmetic. The following definition is nothing more or less than a list of the common laws of arithmetic satisfied in these examples:

Definition 7.1.1. A *ring* is a set R with two operations: addition, denoted here by $+$, and multiplication, denoted by juxtaposition, satisfying the following requirements:

1. Under addition, R is an abelian group.
2. Multiplication is associative.
3. Multiplication distributes over addition: $a(b+c) = ab + ac$, and $(b+c)a = ba + ca$ for all $a, b, c \in R$.

Multiplication is *not* required to be commutative in a ring, in general. If multiplication *is* commutative, the ring is called a *commutative ring*.

You are asked to supply details for some of the following examples in the exercises. Additional examples also appear in the exercises.

Example 7.1.2. The set of integers $\mathbb{Z}$ is a commutative ring.

Example 7.1.3. The set of n-by-n matrices with real entries is a noncommutative ring, with the two operations entry-by-entry addition and matrix multiplication

Example 7.1.4. The set of real valued functions on a set X is a ring with the pointwise operations: $(f+g)(x) = f(x)+g(x)$, $fg(x) = f(x)g(x)$.

Example 7.1.5. The set $\mathbb{R}[x]$ of polynomials with real coefficients in one variable is a ring, with the usual addition and multiplication of polynomials. Likewise the set $\mathbb{R}[X, Y]$ of polynomials in two variables is a ring.

There are many, many variations on these examples: matrices with complex entries, with integer entries, with polynomial entries; functions with complex values, or integer values; continuous functions or differentiable functions; polynomials with any number of variables and with complex, or integer coefficients.

Many rings have an identity element for multiplication, often denoted by 1.

Example 7.1.6. The identity matrix (diagonal matrix with diagonal entries equal to 1) is the multiplicative identity in the n-by-n matrices. The constant polynomial 1 is the multiplicative identity in the ring $\mathbb{R}[x]$. The constant function 1 is the multiplicative identity in the ring of real valued functions on a set X. Examples of rings without multiplicative identity are given in the exercises.

In a ring with a multiplicative identity, non-zero elements may or may not have multiplicative inverses; an element with a multiplicative inverse is called a *unit*.

Example 7.1.7. Some non-zero square matrices have multiplicative inverses, and some do not. The condition for a square matrix to be invertible is that the rows (or columns) are linearly independent, or equivalently that the determinant is non-zero.

Example 7.1.8. In the ring of integers, the only units are ± 1. In the real numbers, every non-zero element is a unit.

Example 7.1.9. In the ring $\mathbb{R}[x]$ the units are non-zero constant polynomials.

Example 7.1.10. In the ring of continuous real valued functions defined on an interval $[a, b]$, the units are the functions which never take the value zero.

Definition 7.1.11. A non-empty subset S of a ring R is called a *subring* if S is a ring *with the two ring operations inherited from* R.

For S to be a subring of R, it is necessary and sufficient that

1. For all elements x and y of S, the sum and product $x + y$ and xy are elements of S.
2. For all $x \in S$, the additive opposite $-x$ is an element of H.

Example 7.1.12. $\mathbb{Z}$ is a subring of $\mathbb{Q}$. $\mathbb{N}$ is not a subring of $\mathbb{Z}$, although $\mathbb{N}$ is closed under both addition and multiplication; the problem is that N is not a subgroup of $\mathbb{Z}$ under addition.

Example 7.1.13. The set of 3-by-3 matrices with rational entries is a subring of the ring of 3-by-3 matrices with real entries. The set of upper triangular 3-by-3 matrices is a subring of the ring of 3-by-3 matrices with real entries.

Example 7.1.14. The set of continuous functions from $\mathbb{R}$ to $\mathbb{R}$ is a subring of the ring of all functions from $\mathbb{R}$ to $\mathbb{R}$. The set of rational valued functions on a set X is a subring of the ring of real valued functions on X.

A number of further examples of subrings are given in the exercises.

Example 7.1.15. If R and S are two rings, we already have the structure of an abelian group on the Cartesian product $R \times S$. Defining a multiplication by $(r, s)(r', s') = (rr', ss')$ makes $R \times S$ into a ring. One can do the same with the Cartesian product of several rings. An alternative notation for $R \times S$ is $R \oplus S$, and often $R \oplus S$ is referred to as the *direct sum* of R and S.

Definition 7.1.16. Two rings R and S are *isomorphic* if there is a bijection between them which preserves both the additive and multiplicative structures. That is, there is a bijection $\varphi : R \to S$ satisfying $\varphi(a + b) = \varphi(a) + \varphi(b)$ and $\varphi(ab) = \varphi(a)\varphi(b)$ for all $a, b \in R$.

Example 7.1.17. According to Exercise 6.1.6, if a and b are relatively prime, then $\mathbb{Z}_{ab}$ and $\mathbb{Z}_a \oplus \mathbb{Z}_b$ are isomorphic rings.

Definition 7.1.18. A *field* is a commutative ring with multiplicative identity element 1 in which *every non-zero element is a unit.*

Example 7.1.19. $\mathbb{R}$, $\mathbb{Q}$, and $\mathbb{C}$ are fields. $\mathbb{Z}$ is not a field. $\mathbb{R}[x]$ is not a field.

Example 7.1.20. If p is a prime, then $\mathbb{Z}_p$ is a field. This follows at once from Propositions 6.1.1 and 6.1.2.

Exercises

Exercise 7.1.1. Show that if a ring R has a multiplicative identity, then the multiplicative identity is unique. Show that if an element $r \in R$ has a left multiplicative inverse r' and a right multiplicative inverse r'', then $r' = r''$.

Exercise 7.1.2. Show that the only units in the ring of integers are ± 1.

Exercise 7.1.3. Show that the set of integers modulo n, $\mathbb{Z}_n$ is a commutative ring with multiplicative identity $[1]$. Show that the units are $\{[k] : \text{g.c.d.}(k, n) = 1\}$.

Exercise 7.1.4. Show that the set of polynomials with real coefficients, denoted $\mathbb{R}[x]$, is a commutative ring with the usual addition and multiplication of polynomials. Show that the constant polynomial 1 is the multiplicative identity, and the only units are the constant polynomials.

Exercise 7.1.5. A *Laurent polynomial* is a "polynomial" in which negative as well as positive powers of the variable x are allowed, for example, $p(x) = 7x^{-3} + 4x^{-2} + 4 + 2x$. Show that the set of Laurent polynomials with real coefficients forms a ring with identity. This ring is denoted by $\mathbb{R}[x, x^{-1}]$. What are the units?

Exercise 7.1.6. Show that the set of polynomials with real coefficients in three variables, $\mathbb{R}[x, y, z]$ is a ring with identity. What are the units?

Exercise 7.1.7.

(a) Let X be any set. The set of functions from X to $\mathbb{R}$ is a ring. What is the multiplicative identity? What are the units?

(b) Let $S \subseteq X$. The set of functions from X to $\mathbb{R}$ whose restriction to S is zero is a ring. Does this ring have a multiplicative identity element?

Exercise 7.1.8. Show that the set of *continuous* functions from $\mathbb{R}$ into $\mathbb{R}$ is a ring. What is the multiplicative identity? What are the units?

Exercise 7.1.9. Show that the set of continuous functions $f : \mathbb{R} \to \mathbb{R}$ such that $\lim_{x \to \pm\infty} f(x) = 0$ is a ring *without multiplicative identity.*

Exercise 7.1.10. Consider the set of infinite-by-infinite matrices (a_{ij}), where $i, j \in \mathbb{N}$, which have only finitely many non-zero entries. (For each such matrix, there is a natural number n such that $a_{ij} = 0$ if $i \geq n$ or $j \geq n$.) Show that the set of such matrices is a ring without identity element.

Exercise 7.1.11. Show that (a) the set of upper triangular matrices and (b) the set of upper triangular matrices with zero entries on the diagonal are both subrings of the ring of all n-by-n matrices with real coefficients. The second example is a ring without multiplicative identity.

Exercise 7.1.12. Show that the set of matrices with integer entries is a subring of the ring of all n-by-n matrices with real entries. Show that the set of matrices with entries in $\mathbb{N}$ is closed under addition and multiplication but is not a subring.

Exercise 7.1.13. Show that the set of symmetric polynomials in 3 variable is a subring of the ring of all polynomials in 3 variables. Recall that a polynomial is symmetric if it remains unchanged when the variables are permuted, $p(x, y, z) = p(y, x, z)$, etc.

Exercise 7.1.14. Experiment with variations on the examples and exercises above by changing the domain of coefficients of polynomials, values of functions, and entries of matrices. For example, polynomials with coefficients in the natural numbers, complex valued functions, matrices with complex entries. What is allowed and what is not allowed for producing rings?

Exercise 7.1.15. Suppose $\varphi : R \to S$ is a ring isomorphism. Show that R has a multiplicative identity if, and only if, S has a multiplicative identity. Show that R is commutative if, and only if, S is commutative.

Exercise 7.1.16. Consider the *group* $\mathbb{Z}_2$ written as $\{e, \xi\}$, where $\xi^2 = e$. Consider formal sums $ae + b\xi$, with $a, b \in \mathbb{R}$. Define a sum by $(ae + b\xi) + (a'e + b'\xi) = (a + a')e + (b + b')\xi$ and a product by $(ae + b\xi)(a'e + b'\xi) = (aa' + bb')e + (ab' + a'b)\xi$. Show that this structure is a ring. This example is denoted by $\mathbb{R}\mathbb{Z}_2$ and is called the *group ring* of the group $\mathbb{Z}_2$.

Exercise 7.1.17. The map $a + b\xi \mapsto (a + b, a - b)$ is an isomorphism from the group ring $\mathbb{R}\mathbb{Z}_2$ to the ring $\mathbb{R} \times \mathbb{R}$.

Exercise 7.1.18. A trigonometric polynomial is a finite linear combination of the functions $t \mapsto e^{int}$, where n is an integer, for example, $f(t) = 3e^{-i2t} + 4e^{it} + i\sqrt{3}e^{i7t}$. Show that the set of trigonometric polynomials is a subring of the ring of continuous complex valued functions on $\mathbb{R}$. Show that the ring of trigonometric polynomials is isomorphic to the ring of Laurent polynomials with complex coefficients.

Exercise 7.1.19. Show that the set $\mathbb{R}(x)$ of rational functions $p(x)/q(x)$, where $p(x), q(x) \in \mathbb{R}[x]$ and $q(x) \neq 0$, is a field. Note the use of parentheses to distinguish this ring $\mathbb{R}(x)$ of rational functions from the ring $\mathbb{R}[x]$ of polynomials.

Exercise 7.1.20. Show that the set of real numbers of the form $a + b\sqrt{2}$, where a and b are rational is a field. Show that $a + b\sqrt{2} \mapsto a - b\sqrt{2}$ is an isomorphism of this field onto itself.

Exercise 7.1.21. Show that the units of the ring $\mathbb{Z}_m$ are the elements $[k]$ such that k and m are relatively prime. Use the fact that k and m are relatively prime if, and only if, there exist $r, s \in Z$ such that $rk + sm = 1$.

Exercise 7.1.22. Let R be a ring. Show that the set of 2-by-2 matrices with entries in R is a ring.

Exercise 7.1.23. Let R be a ring and X a set. Show that the set $\text{Fun}(X, R)$ of functions on X with values in R is a ring. Show that R is isomorphic to the subring of constant functions on X. Show that $\text{Fun}(X, R)$ is commutative if, and only if, R is commutative. Suppose that R has an identity; show that $\text{Fun}(X, R)$ has an identity and describe the units of $\text{Fun}(X, R)$.

7.2. Divisibility of Polynomials

In this section, we take a close look at the ring of polynomials over a field, and develop a theory of divisibility which exactly parallels the theory of divisibility for the integers.

Let K denote any field. One can consider polynomials with coefficients in K, which are expressions of the form $a_n x^n + a_{n-1}x^{n-1} + \cdots + a_0$, where the a_i are elements in K. One can define addition and multiplication of polynomials exactly as when the field K is one of the familiar fields $\mathbb{R}$ or $\mathbb{Q}$. Addition is determined by the rule $ax^n + bx^n = (a + b)x^n$, together with the requirements of commutativity and associativity of addition. Multiplication is determined by the rule $ax^r bx^s = abx^{r+s}$, together with the requirement of distributivity of multiplication over addition. The set of polynomials over K then becomes a commutative ring with identity element 1; this ring is denoted $K[x]$.

Definition 7.2.1. The *degree* of a polynomial $\sum_k a_k x^k$ is the largest k such that $a_k \neq 0$. (The degree of a constant polynomial c is zero, unless $c = 0$. By convention, the degree of the constant polynomial 0 is $-\infty$.) The degree of $p \in K[x]$ is denoted $\deg(p)$. If p is a non-zero polynomial, the *leading coefficient* of p is the coefficient of $x^{\deg(p)}$. A polynomial is said to be *monic* if its leading coefficient is 1.

Example 7.2.2. The degree of $p = (\pi/2)x^7 + ix^4 - \sqrt{2}x^3$ is 7; the leading coefficient is $\pi/2$; $(2/\pi)p$ is a monic polynomial.

Recall the process of long division of polynomials from school mathematics. This works in $K[x]$ for any field K.

Proposition 7.2.3. *Let K be any field, and let p and d be elements of $K[x]$. Then there exist polynomials q and r in $K[x]$ such that $p = dq + r$ and* $\deg(r) < \deg(d)$

Let's recall how to do this with an example. Let $p = 7x^5 + 3x^2 + x + 2$ and $d = 5x^3 + 2x^2 + 3x + 4$. Then the first contribution to q is $\frac{7}{5}x^2$, and

$$p - \frac{7}{5}x^2 d = -\frac{14}{5}x^4 - \frac{21}{5}x^3 - \frac{13}{5}x^2 + x + 2$$

The next contribution to q is $-\frac{14}{25}x$, and

$$p - (\frac{7}{5}x^2 - \frac{14}{25}x)\, d = -\frac{77}{25}x^3 - \frac{23}{25}x^2 + \frac{81}{25}x + 2.$$

The next contribution to q is $-\frac{77}{125}$, and

$$p - (\frac{7}{5}x^2 - \frac{14}{25}x - \frac{77}{125})\, d = \frac{39}{125}x^2 + \frac{636}{125}x + \frac{558}{125}.$$

Thus,

$$q = \frac{7}{5}x^2 - \frac{14}{25}x - \frac{77}{125}$$

and

$$r = \frac{39}{125}x^2 + \frac{636}{125}x + \frac{558}{125}.$$

This example illustrates all the features of the proof of Proposition 7.2.3. Of course, since the procedure is an algorithm in which one basic step is repeated until a remainder with small degree is obtained, the formalized proof is by induction on the degree of p.

Proof of Proposition 7.2.3. If $\deg(p) < \deg(d)$, then put $d = 0$ and $r = p$. So assume now that $\deg(p) \geq \deg(d)$, and that the result is true when p is replaced by any polynomial of lower degree. Write $p = a_n x^n + a_{n-1}x^{d-1} + \cdots + a_0$ and $d = b_s x^s + b_{s-1}x^{s-1} + \cdots + b_0$, where $n = \deg(p)$ and $s = \deg(d)$, and $s \leq n$. Put $p' = p - (a_n/b_s)x^{n-s}d$. Note that $\deg(p') < \deg(p)$. Therefore, by the induction hypothesis, there exist polynomials q' and r with $\deg(r) < \deg(d)$ such that $p' = q'd + r$. Putting $q = q' + (a_n/b_s)x^{n-s}$, we have $p = qd + r$. □

We can now draw some remarkable conclusions from Proposition 7.2.3: The polynomial ring $K[x]$ has a divisibility theory exactly analogous to that of the integers $\mathbb{Z}$. Say that a polynomial d *divides* a polynomial p if there is a polynomial q such that $qd = p$. Also say that a polynomial $p \in K[x]$ of degree at least 1 is *irreducible over* K if it has no divisors in $K[x]$ of degree ≥ 1, aside from constant multiples of itself. For example, $x^2 + 1$ is irreducible over $\mathbb{R}$ but not over $\mathbb{C}$, since $x^2 + 1 = (x - i)(x + i)$.

Proposition 7.2.4. *Every polynomial in* $K[x]$ *can be factored as a product of irreducible polynomials.*

Proof. This follows immediately from the definition of irreducibility and induction on the degree of the polynomial. Compare the proof of Proposition C.1 in Appendix C. □

Definition 7.2.5. An *ideal* I in a commutative ring R is an additive subgroup of R such that whenever $r \in R$ and $a \in I$, then $ra \in I$.

Definition 7.2.6. A polynomial $f \in K[x]$ is a *greatest common divisor* of non-zero polynomials $p, q \in K[x]$ if

1. f divides p and q in $K[x]$ and
2. whenever $g \in K[x]$ divides p and q, then g also divides f.

We are about to show that two non-zero polynomials over a field always have a greatest common divisor. Notice that a greatest common divisor is unique up to multiplication by a unit of $K[x]$, that is, by a non-zero element of K. There is a unique greatest common divisor which is *monic*, i.e., whose leading coefficient is 1. When we need to refer to *the* greatest common divisor, we will mean the one which is monic. We denote the greatest common divisor of p and q by $\text{g.c.d.}(p, q)$.

Theorem 7.2.7. *Any two non-zero polynomials* $f, g \in K[x]$ *have a greatest common divisor, which is an element of the ideal* $I(f, g)$.

Proof. One can simply mimic the proof of Proposition 2.3.4, with Proposition 7.2.3 replacing 2.3.1. Namely (assuming $\deg(f) \leq \deg(g)$), one does repeated division with remainder, each time obtaining a remainder of smaller degree. The process must terminate with a zero remainder after at most

$\deg(f)$ steps:

$$
\begin{aligned}
g &= q_1 f + f_1 \\
f &= q_2 f_1 + f_2 \\
&\dots \\
f_{k-2} &= q_k f_{k-1} + f_k \\
&\dots \\
f_{r-1} &= q_{r+1} f_r.
\end{aligned}
$$

Here $\deg f > \deg(f_1) > \deg(f_2) > \dots$. By the argument of Proposition 2.3.4, the final non-zero remainder f_r is an element of $I(f, g)$ and is a greatest common divisor of f and g. □

Definition 7.2.8. Two polynomials $f, g \in K[x]$ are *relatively prime* if $\text{g.c.d.}(f, g) = 1$.

Theorem 7.2.9. *The factorization of a polynomial in $K[x]$ into irreducible factors is unique. That is, the irreducible factors appearing are unique up to multiplication by non-zero elements in K.*

Proof. Mimic the proof of Theorem 2.3.9. □

Proposition 7.2.10. *Let $p \in K[x]$ and $a \in K$. Then $p(a) = 0$ if, and only if, $x - a$ is an irreducible factor of p.*

Proof. Suppose that $p(a) = 0$, and write $p(x) = q(x)(x - a) + r$, where the remainder r is a constant. Substituting a for x gives $0 = r$, so $p = q(x - a)$. Conversely, if $x - a$ divides p, it is clear that $p(a) = 0$. □

Definition 7.2.11. Say an element $\alpha \in K$ is a *root* of a polynomial $p \in K[x]$ if $p(\alpha) = 0$. Say the *multiplicity of the root* α is k if $x - \alpha$ appears exactly k times in the irreducible factorization of p.

Corollary 7.2.12. *A polynomial $p \in K[x]$ of degree n has at most n roots in K, counting with multiplicities. That is, the sum of multiplicities of all roots is at most n.*

Proof. If $p = (x - \alpha_1)^{m_1}(x - \alpha_2)^{m_2} \cdots (x - \alpha_k)^{m_k} q_1 \cdots q_s$, where the q_i are irreducible, then evidently $m_1 + m_2 + \cdots + m_k \le \deg(p)$. □

After this excursion into the divisibility theory for polynomials over a field, we are finally in a position to prove Theorem 6.1.6.

Proof of Theorem 6.1.6. Consider the field $K = \mathbb{Z}_p$ and the multiplicative group $\Phi(p)$ of non-zero elements. Suppose $m \le p-1$ is the maximum of the orders of elements of $\Phi(p)$. According to Exercise 6.2.4, for all $a \in \Phi(p)$, $a^m = [1]$. Thus, the polynomial $x^m - 1$ has $p-1$ distinct solutions in K, so by the previous proposition, $p - 1 \le m$. Therefore, $m = p-1$, and there is an element of $\Phi(p)$ of order $p-1$. □

Exercises

Exercise 7.2.1.

(a) Let R be a commutative ring and $a \in R$. Show that $I(a) = \{ra : r \in R\}$ is an ideal in R.

(b) Let R be a commutative ring and $a, b \in R$. Show that $I(a, b) = \{ra + sb : r, s \in R\}$ is an ideal in R.

Exercise 7.2.2. Let h be an element of $I(f, g)$ of least degree. Then h is a greatest common divisor of f and g. *Hint:* Apply division with remainder.

Exercise 7.2.3. Two polynomials $f, g \in K[x]$ are relatively prime if, and only if, $1 \in I(f, g)$.

Exercise 7.2.4. If $p \in K[x]$ is irreducible and $f \in K[x]$, then either p divides f, or p and f are relatively prime.

Exercise 7.2.5. Let $p \in K[x]$ be irreducible.

(a) If p divides a product fg of elements of $K[x]$, then p divides f or p divides g.

(b) If p divides a product $f_1 f_2 \dots f_r$ of several elements of $K[x]$, then p divides one of the f_i.

Hint: Mimic the arguments of 2.3.7 and 2.3.8.

Exercise 7.2.6. Suppose F is a field with finitely many elements. Then the set F^* of non-zero elements is an abelian group. Use the same argument to show that F^* is a cyclic group.

Exercise 7.2.7. Let I be any ideal in $K[x]$. Show that there is an element $p \in K[x]$ such that $I = pK[x] = \{pq : q \in K[x]\}$. *Hint:* Let p be an element of I of least degree. Since I is an ideal, $pK[x] \subseteq I$. Show the opposite inclusion by using division with remainder. Compare Exercise 2.4.3.

Exercise 7.2.8. For each of the following pairs of polynomials f, g, find the greatest common divisor and find polynomials r, s such that $rf + sg = \text{g.c.d.}(f, g)$.

(a) $x^3 - 3x + 3,\ x^2 - 4$.
(b) $-4 + 6x - 4x^2 + x^3,\ x^2 - 4$.

Exercise 7.2.9. Write a computer program to compute the greatest common divisor of two polynomials f, g with real coefficients. Make your program find polynomials r, s such that $rf + sg = \text{g.c.d.}(f, g)$.

Exercise 7.2.10. Explore the idea of the greatest common divisor of *several* non-zero polynomials, $f_1, f_2, \dots, f_k \in K[x]$.

(a) Make a reasonable definition of $\text{g.c.d}(f_1, f_2, \dots, f_k)$.
(b) Let $I = I(f_1, f_2, \dots, f_k) =$

$$\{m_1 f_1 + m_2 f_2 + \cdots + m_k f_k : m_1, \dots, m_k \in \mathbb{Z}\}.$$

Show that I is an ideal in $K[x]$.
(c) Show that $f = \text{g.c.d}(f_1, f_2, \dots, f_k)$ is an element of I of smallest degree and that $I = fK[x]$.
(d) Develop an algorithm to compute $\text{g.c.d}(f_1, f_2, \dots, f_k)$.
(e) Develop a computer program to compute the greatest common divisor of any finite collection of non-zero polynomials with real coefficients.

Exercise 7.2.11.

(a) Suppose that $p(x) = a_n x^n + \cdots + a_1 x + a_0 \in Z[x]$. Suppose that $r/s \in \mathbb{Q}$ is a root of p, where r and s are relatively prime integers. Show that s divides a_n and r divides a_0. *Hint:* Start with the equation $p(r/s) = 0$, multiply by s^n, and note, for example, that all the terms except $a_n r^n$ are divisible by s.
(b) Conclude that any rational root of a monic polynomial in $\mathbb{Z}[x]$ is an integer.
(c) Since $x^2 - 2$ has no integer root, it has no rational root. Therefore, $\sqrt{2}$ is irrational.

Exercise 7.2.12.

(a) Show that a quadratic or cubic polynomial $f(x)$ in $K[x]$ is irreducible if, and only if, $f(x)$ has no root in K.

(b) A monic quadratic or cubic polynomial $f(x) \in \mathbb{Z}[x]$ is irreducible if, and only if, it has no integer root.

(c) $x^3 - 3x + 1$ is irreducible in $\mathbb{Q}[x]$.

7.3. Basic Results on Rings

In this section, we return to the study of rings in general and develop some basic principles.

It should not be a big surprise that certain concepts which were fundamental to our study of groups are also important for the study of rings. One could expect these concepts to play a role for any reasonable algebraic structure. We have already discussed the idea of a subring, which is analogous to the idea of a subgroup. The next concept from group theory which we might expect to play a fundamental role in ring theory is the notion of a homomorphism.

Definition 7.3.1. A *homomorphism* $\varphi : R \to S$ of rings is a map satisfying $\varphi(x + y) = \varphi(x) + \varphi(y)$, and $\varphi(xy) = \varphi(x)\varphi(y)$ for all $x, y \in R$.

In particular, a ring homomorphism is a homomorphism for the abelian group structure of R and S, so we know, for example, that $\varphi(-x) = -\varphi(x)$ and $\varphi(0) = 0$. Even if R and S both have an identity element 1, it is not automatic that $\varphi(1) = 1$. If we want to specify that this is so, we will call the homomorphism a *unital* homomorphism.

Example 7.3.2. The map $\varphi : \mathbb{Z} \to \mathbb{Z}/n\mathbb{Z}$ defined by $\varphi(a) = [a] = a + n\mathbb{Z}$ is a unital ring homomorphism. In fact, it follows from the definition of the operations in $\mathbb{Z}/$ that $\varphi(a + b) = [a + b] = [a] + [b] = \varphi(a) + \varphi(b)$, and similarly $\varphi(ab) = [ab] = [a][b] = \varphi(a)\varphi(b)$ for integers a and b. (The actual work here was in verifying that the operations in $\mathbb{Z}/n\mathbb{Z}$ were well defined.)

Example 7.3.3. Consider the ring $K[x]$ of polynomials over a field K. For any $a \in K$,the map $p \mapsto p(a)$ is a unital ring homomorphism from $K[x]$ to K.

Example 7.3.4. Consider the ring $C(\mathbb{R})$ of continuous real valued functions on $\mathbb{R}$. Let S be any subset of R, for example, $S = [0, 1]$. The map $f \mapsto f_{|S}$ which associates to each function its restriction to S is a unital ring homomorphism from $C(\mathbb{R})$ to $C(S)$. Likewise for any $t \in \mathbb{R}$ the map $f \mapsto f(t)$ is a unital ring homomorphism from $C(\mathbb{R})$ to $\mathbb{R}$.

Numerous further examples of ring homomorphisms are given in the exercises.

Again extrapolating from our experience with group theory, we would expect the kernel of a ring homomorphism to be a special sort of subring. Of course, the kernel of a ring homomorphism $\varphi : R \to S$ is $\{x \in R : \varphi(x) = 0\}$. Observe that a ring homomorphism is injective if, and only if, its kernel is $\{0\}$. The following definition captures the special properties of the kernel of a homomorphism.

Definition 7.3.5. An *ideal* I in a ring R is a subring of R satisfying $xr, rx \in I$ for all $x \in I$ and $r \in R$. A *left ideal* I of R is a subring of R such that $rx \in I$ whenever $r \in R$ and $x \in I$. A *right ideal* is defined similarly. Note that for commutative rings, all of these notions coincide.

The verification that the kernel of a ring homomorphism is an ideal is left as an exercise.

Proposition 7.3.6. *If* $\varphi : R \to S$ *is a ring homomorphism, then* $\ker(\varphi)$ *is an ideal of of* R.

Proof. Exercise. □

The most direct way to provide examples of ideals is the following:

Example 7.3.7. If R is any ring and $x \in R$ then $Rx = \{yx : y \in R\}$ is a left ideal. Similarly xR is a right ideal and RxR is a (two-sided) ideal. Verification of these statements is left as an exercise.

Definition 7.3.8. Ideals of this type are called *principal* ideals. More explicitly, xR is called the principal right ideal generated by x in R and RxR the principal ideal generated by x in R.

You are asked to show in Exercise 7.3.10 that every ideal in the ring of integers is principal. Similarly, every ideal in the ring $K[x]$ of polynomials in one variable over a field is principal, according to Exercise 7.2.7.

In Exercise 7.3.11, you are asked to show that the ring M of n-by-n matrices with real entries has no ideals other than (0) and M itself.

Definition 7.3.9. A ring R is said to be *simple* if it has no ideals other than (0) and R.

Thus the result of Exercise 7.3.11 is that matrix rings over $\mathbb{R}$ are simple. Of course, the result holds equally well for matrix rings over any field.

In Section 3.4, it was shown that given a group G and a normal subgroup N, one can construct a quotient group G/N and a natural homomorphism from G onto G/N. The program of Section 3.4 can not be carried out more or less verbatim with rings and ideals in place of groups and normal subgroups:

For a ring R and an ideal I, we can form the quotient *group* R/I, whose elements are cosets $a + I$ of I in R. The additive group operation in R/I is $(a + I) + (b + I) = (a + b) + I$. Now attempt to define a multiplication in R/I in the obvious way: $(a + I)(b + I) = (ab + I)$. One has to check that this this is well defined. But this follows from the closure of I under multiplication by elements of R; namely, if $a + I = a' + I$ and $b + I = b' + I$, then $(ab - a'b') = a(b - b') + (a - a')b' \in aI + Ib \subseteq I$. Thus, $ab + I = a'b' + I$, and the multiplication in R/I is well-defined.

Theorem 7.3.10. *If I is an ideal in a ring R, then R/I has the structure of a ring, and the quotient map $a \mapsto a + I$ is a surjective ring homomorphism from R to R/I with kernel equal to I. If R has a multiplicative identity, then so does R/I, and the quotient map is unital.*

Proof. Once one has checked that the multiplication in R/I is well defined, it is straightforward to check that it is associative, that the distributive law holds, and that the quotient map is a ring homomorphism. If 1 is the multiplicative identity in R, then $1 + I$ is the multiplicative identity in R/I. □

Example 7.3.11. The ring $\mathbb{Z}/n\mathbb{Z}$ is the quotient of the ring $\mathbb{Z}$ by the principal ideal $n\mathbb{Z}$. The homomorphisms $a \mapsto [a] = a + n\mathbb{Z}$ is the quotient homomorphism.

Example 7.3.12. Any ideal in $\mathbb{R}[x]$ is of the form $(f) = f\mathbb{R}[x]$ for some polynomial f according to Exercise 7.2.7. For any $g(x) \in \mathbb{R}[x]$, there exist polynomials q, r such that $g(x) = q(x)f(x) + r(x)$, and $\deg(r) < deg(f)$. Thus $g(x) + (f) = r(x) + (f)$. In other words, $\mathbb{R}[x]/(f) = \{r(x) + (f) : \deg(r) < \deg(f)\}$. The multiplication in $\mathbb{R}[x]/(f)$ is as follows: given polynomials $r(x)$ and $s(x)$ each of degree less than the degree of f,the product $(r(x) + (f))(s(x) + (f)) = ((r(x)s(x) + (f)) = (a(x) + (f))$, where $a(x)$ is the remainder upon division of $r(x)s(x)$ by $f(x)$.

Let's look at the particular example $f(x) = x^2 + 1$. Then $\mathbb{R}[x]/(f)$ consists of cosets $a + bx + (f)$ represented by linear polynomials. Furthermore, one has the computational rule

$$x^2 + (f) = x^2 + 1 - 1 + (f) = -1 + (f).$$

Thus

$$(a + bx + (f))(a' + b'x + (f)) = (aa' - bb') + (ab' + a'b)x + (f).$$

Theorem 7.3.13 (Homomorphism Theorem for Rings). *Let $\varphi : R \longrightarrow S$ be a surjective homomorphism of rings with kernel I. Let $\pi : R \longrightarrow R/I$ be the quotient homomorphism. There is a ring isomorphism $\tilde{\varphi} : R/I \longrightarrow S$ satisfying $\tilde{\varphi} \circ \pi = \varphi$. (See the diagram below.)*

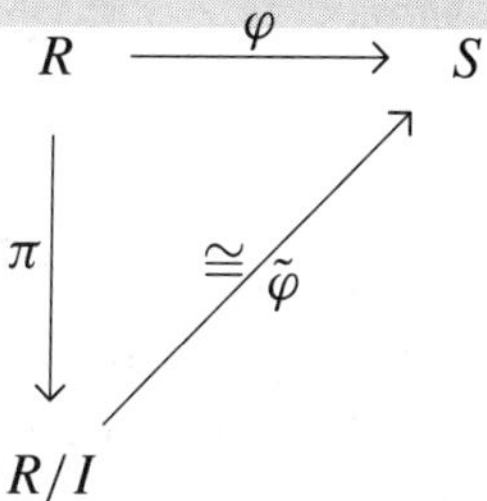

Proof. The Homomorphism Theorem for groups (Theorem 3.4.6) gives us an isomorphism of abelian groups $\tilde{\varphi}$ satisfying $\tilde{\varphi} \circ \pi = \varphi$. We have only to verify that $\tilde{\varphi}$ also respects multiplication. But this follows at once from the definition of the product on R/I: $\tilde{\varphi}(a+I)(b+I) = \tilde{\varphi}(ab+I) = \varphi(ab) = \varphi(a)\varphi(b) = \tilde{\varphi}(a+I)\tilde{\varphi}(b+I)$. □

Example 7.3.14. Define a homomorphism $\varphi : \mathbb{R}[x] \to \mathbb{C}$ by evaluation of polynomials at $i \in \mathbb{C}$, $\varphi(g(x)) = g(i)$. For example, $\varphi(x^3 - 1) = i^3 - 1 = -i - 1$. This homomorphism is surjective because $\varphi(a + bx) = a + bi$. The kernel of φ consists of all polynomials g such that $g(i) = 0$. The kernel contains at least the ideal $(x^2 + 1) = (x^2 + 1)\mathbb{R}[x]$ because $i^2 + 1 = 0$. On the other hand, if $g \in \ker(\varphi)$, write $g(x) = (x^2 + 1)q(x) + (a + bx)$; evaluating at i, one gets $0 = a + bi$, which is possible only if $a = b = 0$. Thus g is a multiple of $x^2 + 1$. That is $\ker(\varphi) = (x^2 + 1)$. By the Homomorphism Theorem for rings, $\mathbb{R}[x]/(x^2 + 1) \cong \mathbb{C}$ as rings. In particular, since $\mathbb{C}$ is a field, $\mathbb{R}[x]/(x^2 + 1)$ is a field. Note that we have already calculated explicitly in Example 7.3.12 that multiplication in $\mathbb{R}[x]/(x^2 + 1)$ satisfies the same rule as multiplication in $\mathbb{C}$.

Proposition 7.3.15. *Let $\varphi : R \longrightarrow S$ be a ring homomorphism of R* onto *S, and let I denote its kernel.*

(a) *If B is a subring of S, then $\varphi^{-1}(B)$ is a subring of R containing I.*
(b) *$B \mapsto \varphi^{-1}(B)$ is a bijection between the subrings of S and the subrings of R containing I.*
(c) *$J \subseteq S$ is an ideal of S if, and only if, $\varphi^{-1}(J)$ is an ideal of R.*

Proof. By Proposition 3.4.12, $B \mapsto \varphi^{-1}(B)$ is a bijection between the subgroups of S and the subgroups of R containing I. It remains to show that this bijection carries subrings to subrings and ideals to ideals. We leave this as an exercise. □

Proposition 7.3.16. *Let $\varphi : R \longrightarrow S$ be a surjective ring homomorphism. Let $\bar{I}$ be an ideal of S and let $I = \varphi^{-1}(\bar{I})$. Then $x + I \mapsto \varphi(x) + \bar{I}$ is an isomorphism of R/I onto $S/\bar{I}$.*

Proof. Exercise. □

Proposition 7.3.17. *Let $\varphi : R \longrightarrow S$ be a surjective homomorphism of rings with kernel I. Let A be a subring of R. Then $\varphi^{-1}(\varphi(A)) = A + I = \{a + r : a \in A \text{ and } r \in I\}$ is a subring of R containing I. Furthermore, $(A + I)/I \cong \varphi(A) \cong A/(A \cap I)$.*

Proof. Exercise. □

Definition 7.3.18. An ideal M in a ring R is called *proper* if $M \neq R$. An proper ideal M in a ring R is called *maximal* if there are no ideals strictly between M and R, that is, no ideals which properly contain M and are properly contained in R.

Recall that a ring is called simple if it has no ideals other that the trivial ideal (0) and the whole ring.

Proposition 7.3.19. *An ideal M in R is maximal if, and only if, R/M is simple.*

Proof. Exercise. □

Proposition 7.3.20. *A commutative ring R with 1 is a field if, and only if, R is simple.*

Proof. Suppose R is simple and $x \in R$ is a non-zero element. The ideal Rx is non-zero since $x = 1x \in Rx$; because R is simple, $R = Rx$. Hence there is a $y \in R$ such that $1 = yx$. Conversely, suppose R is a field and M is a non-zero ideal. Since M contains a non-zero element x, it also contains $r = rx^{-1}x$ for any $r \in R$; that is, $M = R$. □

Corollary 7.3.21. *If M is a proper ideal in a commutative ring R with 1, then R/M is a field if, and only if, M is maximal.*

Proof. This follows from 7.3.19 and 7.3.20. □

If R is a ring and S is a subset, then there is a smallest subring of R which contains S, called the *subring generated by* S. (Compare the discussion of the subgroup generated by a subset at the end of Section 2.4.) The constructive view of the subring generated by S is that it consists of all sums of products of elements of S. Another view is that it is the intersection of the family of all subrings of R which contain S. If the subring generated by S is all of R one says that S generates R (as a ring).

Likewise, there is a smallest ideal of R which contains S, called the *ideal generated by* S. Assuming that R has an identity 1, the ideal generated by S consists of all sums of elements of the form $r_1 s r_2$, where $r_i \in R$ and $s \in S$. The ideal generated by S is in general larger than the subring generated by S.

Exercises

Exercise 7.3.1. Show that $A \mapsto \begin{bmatrix} A & 0 \\ 0 & A \end{bmatrix}$, and $A \mapsto \begin{bmatrix} A & 0 \\ 0 & 0 \end{bmatrix}$ are homomorphisms of the ring of 2-by-2 matrices into the ring of 4-by-4 matrices. The former is unital, but the latter is not.

Exercise 7.3.2. Define a map φ from the ring $\mathbb{R}[x]$ of polynomials with real coefficients into the ring M of 3-by-3 matrices by

$$\varphi(\sum a_i x^i) = \begin{bmatrix} a_0 & a_1 & a_2 \\ 0 & a_0 & a_1 \\ 0 & 0 & a_0 \end{bmatrix}.$$

Show that φ is a unital ring homomorphism. What is the kernel of this homomorphism?

Exercise 7.3.3. If $\varphi : R \to S$ is a ring homomorphism and R has a unit element 1, show that $e = \varphi(1)$ satisfies $e^2 = e$ and $ex = xe = exe$ for all $x \in \varphi(R)$.

Exercise 7.3.4. If $\varphi : R \to S$ is a ring homomorphism, then $\varphi(R)$ is a subring of S.

Exercise 7.3.5. If $\varphi : R \to S$ and $\psi : S \to T$ are ring homomorphisms, then the composition $\psi \circ \varphi$ is a ring homomorphism.

Exercise 7.3.6. If $\varphi : R \to S$ is a ring homomorphism, then $\ker(\varphi)$ is an ideal of of R.

Exercise 7.3.7. Let S be a subset of a set X. Let R be the ring of real valued functions on X, and let I be the set of real valued functions on X whose restriction to S is zero. Show that I is an ideal in R.

Exercise 7.3.8. Let R be the ring of 3-by-3 upper triangular matrices and I be the set of upper triangular matrices which are zero on the diagonal. Show that I is an ideal in R.

Exercise 7.3.9. If R is any ring and $x \in R$ then $Rx = \{yx : y \in R\}$ is a left ideal. Similarly, xR is a right ideal and RxR is a (two-sided) ideal.

Exercise 7.3.10.

(a) Using division with remainder, show that every ideal in the ring $\mathbb{Z}$ is principal. *Hint:* Let I be an ideal. Let d be the smallest element in $I \cap \mathbb{N}$. Show that every element of I is divisible by d, and conclude that $I = \mathbb{Z}d$.

(b) Using division with remainder, show that every ideal in the ring $K[x]$ is principal, where K is a field. *Hint:* Let I be an ideal. Let f be an element in I of least degree. Show that every element of I is divisible by f, and conclude that $I = K[x]f$.

Exercise 7.3.11. Show that the ring M of n-by-n matrices over $\mathbb{R}$ has no ideals other than 0 and M. Conclude that any ring homomorphism $\varphi : M \to S$ is either identically zero or is injective. *Hint:* To begin with, work in the 2-by-2 or 3-by-3 case; when you have these cases, you will understand the general case as well. Let I be a non-zero ideal, and let $x \in I$ be a non-zero element. Introduce the matrix units E_{ij}, which are matrices with a 1 in the (i, j) position and zeros elsewhere. Observe that the set of E_{ij} is a basis for the linear space of matrices. Show that $E_{ij}E_{kl} = \delta_{jk}E_{il}$. Note that the identity E matrix satisfies $E = \sum_{i=1}^{n} E_{ii}$, and write $x = ExE = \sum_{i,j} E_{ii}xE_{jj}$. Conclude that $y = E_{ii}xE_{jj} \neq 0$ for some pair (i, j). Now, since y is a matrix, it is possible to write $y = \sum_{r,s} y_{rs}E_{r,s}$. Conclude that $y = y_{i,j}E_{i,j}$, and $y_{i,j} \neq 0$, and that therefore $E_{ij} \in I$. Now use the multiplication rules for the matrix units to conclude that $E_{rs} \in I$ for all (r, s), and hence $I = M$.

Exercise 7.3.12. A non-zero homomorphism of a simple ring is injective. In particular, a non-zero homomorphism of a field is injective.

Exercise 7.3.13. Complete the proof of Proposition 7.3.15.

Exercise 7.3.14. Prove Proposition 7.3.16, following the pattern of the proof of Proposition 3.4.17.

Exercise 7.3.15. Prove Proposition 7.3.17, following the pattern of the proof of Proposition 3.4.18.

Exercise 7.3.16. Prove that an ideal M in R is maximal if, and only if, R/M is simple.

Exercise 7.3.17.

(a) We know that every ideal in $\mathbb{Z}$ is principal, that is, of the form $n\mathbb{Z}$, by Exercise 2.4.3. Show that $n\mathbb{Z}$ is maximal if, and only if, $\pm n$ is a prime.

(b) Likewise every ideal in $K[x]$ is principal when K is a field by Exercise 7.2.7. Show that an ideal $(f) = fK[x]$ is maximal if, and only if, f is irreducible.

(c) Conclude that $\mathbb{Z}_n = \mathbb{Z}/n\mathbb{Z}$ is a field if, and only if, $\pm n$ is prime, and that $K[x]/(f)$ is a field if, and only if, (f) is irreducible.

Exercise 7.3.18.

(a) Let R be a ring, P and Q ideals in R, and suppose that $P \cap Q = (0)$, and $P + Q = R$. Show that the map $x \mapsto (x + P, x + Q)$ is an isomorphism of R onto $R/P \oplus R/Q$. *Hint:* Injectivity is clear. For surjectivity, show that for each $a, b \in R$, there exist $x \in R$, $p \in P$, and $q \in Q$ such that $x + p = a$, and $x + q = b$.

(b) More generally, if $P + Q = R$, show that $R/(P \cap Q) \cong R/P \oplus R/Q$.

Exercise 7.3.19.

(a) Show that integers m and n are relatively prime if, and only if, $m\mathbb{Z} + n\mathbb{Z} = \mathbb{Z}$ if, and only if, $m\mathbb{Z} \cap n\mathbb{Z} = mn\mathbb{Z}$. Conclude that if m and n are relatively prime, then $\mathbb{Z}_{mn} \cong \mathbb{Z}_m \oplus \mathbb{Z}_n$ as rings. (This statement (or the surjectivity of the map $x \mapsto ([x]_m, [x]_n)$, or the ring theoretic generalization above) is often called the *Chinese Remainder Theorem.*

(b) State and prove a generalization of this result for the ring of polynomials $K[x]$ over a field K.

Exercise 7.3.20. Work out the rule of computation in the ring $\mathbb{R}[x]/(f)$, where $f(x) = x^2 - 1$. Note that the quotient ring consists of elements $a + bx + (f)$. Compare Example 7.3.12.

Exercise 7.3.21. Work out the rule of computation in the ring $\mathbb{R}[x]/(f)$, where $f(x) = x^3 - 1$.Note that the quotient ring consists of elements $a + bx + cx^2 + (f)$. Compare Example 7.3.12.

Exercise 7.3.22. For any ring R, one can define a ring $R[x]$ of polynomials with coefficients in R (in the evident way). If J is an ideal of R, show that

$J[x]$ is an ideal in $R[x]$ and furthermore $R[x]/J[x] \cong (R/J)[x]$. *Hint:* Find a natural homomorphism from $R[x]$ onto $(R/J)[x]$ with kernel $J[x]$.

Exercise 7.3.23. For any ring R, and any natural number n, one can define the matrix ring $\mathrm{Mat_n(R)}$ consisting of n-by-n matrices with entries in R. If J is an ideal of R, show that $\mathrm{Mat_n(J)}$ is an ideal in $\mathrm{Mat_n(R)}$ and furthermore $\mathrm{Mat_n(R)/Mat_n(J)} \cong \mathrm{Mat_n(R/J)}$. *Hint:* Find a natural homomorphism from $Mat_n(R)$ onto $\mathrm{Mat_n(R/J)}$ with kernel $\mathrm{Mat_n(J)}$.

7.4. Integral Domains

Definition 7.4.1. An integral domain is a commutative ring with identity element 1 in which the product of any two non-zero elements is non-zero.

You may think at first that the product of non-zero elements in a ring is always non-zero, but you already know of examples where this is not the case. Let R be the ring of real-valued functions on a finite set X and let A be a proper subset of X. Let f be the characteristic function of A, that is, the function satisfying $f(a) = 1$ if $a \in A$ and $f(x) = 0$ if $x \in X \setminus A$. Then f and $1 - f$ are non-zero elements of R whose product is zero.

For another example, let x be the 4-by-4 matrix with zeros on and below the diagonal and 1's above the diagonal. Compute that $x^3 \neq 0$ but $x^4 = 0$. Let R be the ring generated by the identity matrix 1 and x, that is, the set matrices which are sums of powers of x. This is a commutative ring with identity in which the product of the non-zero elements x and x^3 is zero. This example suggests the following concept:

Definition 7.4.2. An element n in a ring is *nilpotent* if $n^k = 0$ for some k.

You already know several important examples of integral domains: The ring of integers $\mathbb{Z}$ is an integral domain, any field is an integral domain, and the ring $K[x]$ of polynomials over a field K is an integral domain; you are asked to show this in an exercise.

There are two common constructions of fields from integral domains. One construction is treated in Corollary 7.3.21 and Exercise 7.3.17. Namely, if R is a commutative ring with 1 and M is a maximal ideal, then R/M is a field.

Another construction is that of the *field of fractions* of an integral domain. This construction is known to you from the formation of the rational numbers as fractions of integers. Given an integral domain R, we wish to construct a field from symbols of the form a/b, where $a, b \in R$ and $b \neq 0$.

You know that in the rational numbers, $2/3 = 8/12$; the rule is $a/b = a'/b'$ if $ab' = a'b$. It seems prudent to adopt the same rule for any integral domain R.

Lemma 7.4.3. *Let R be an integral domain. Let S be the set of symbols of the form a/b, where $a, b \in R$ and $b \neq 0$. Show that the relation $a/b \sim a'/b'$ if $ab' = a'b$ is an equivalence relation on S.*

Proof. Exercise 7.4.4 □

Let us denote the quotient of S by the equivalence relation $\sim$ by $Q(R)$. For $a/b \in S$, write $[a/b]$ for the equivalence class of a/b, an element of $Q(R)$.

Now, we attempt to define operations on $Q(R)$, following the guiding example of the rational numbers. The sum $[a/b] + [c/d]$ will have to be defined as $[(ad + bc)/bd]$ and the product $[a/b][c/d]$ as $[ac/bd]$. *As always, when one defines operations on equivalence classes in terms of representatives of those classes, the next thing which has to be done is to check that the operations are well-defined.* This is to be done in Exercise 7.4.5.

It is now straightforward, if slightly tedious, to check that $Q(R)$ is a field, and that R can be considered as a subring of $Q(R)$; the map $a \mapsto [a/1]$ is an injective ring homomorphism of R into $Q(R)$. See Exercises 7.4.6 and 7.4.7.

Thus we have the result:

Proposition 7.4.4. *If R is an integral domain, then $Q(R)$ is a field containing R as a subring.*

Example 7.4.5. $Q(K[x])$ is the field of rational functions in one variable. This field is denoted $K(x)$.

Example 7.4.6. $Q(K[x_1, \dots, x_n])$ is the field of rational functions in n variables. This field is denoted $K(x_1, \dots, x_n)$.

Example 7.4.7. If R is the ring of symmetric polynomials in n variables, then $Q(R)$ is the ring of symmetric rational functions in n variables.

Definition 7.4.8. If the subring C of an integral domain R generated by the identity is isomorphic to Z, the integral domain is said to have *characteristic 0*. If the subring C is isomorphic to $\mathbb{Z}_p$ for a prime p, then R is said to have characteristic p.

Exercises

Exercise 7.4.1. Show that if K is a field, then the ring of polynomials $K[x]$ is an integral domain. More generally show that if R is an integral domain, then the ring of polynomials $R[x]$ with coefficients in R is an integral domain.

Exercise 7.4.2. Generalize the results of the previous problem to rings of polynomials in several variables.

Exercise 7.4.3.

(a) Show that any subring of an integral domain is an integral domain.

(b) The *Gaussian integers* are the complex numbers whose real and imaginary parts are integers. Show that the set of Gaussian integers is an integral domain.

(c) Show that the ring of *symmetric* polynomials in n variables is an integral domain.

Exercise 7.4.4. Prove Lemma 7.4.3.

Exercise 7.4.5. Show that addition and multiplication on $Q(R)$ is well defined. This amounts to the following: Suppose $a/b \sim a'/b'$ and $c/d \sim c'/d'$ and show that $(ad+bc)/bd \sim (a'd'+c'b')/b'd'$ and $ac/bd \sim a'c'/b'd'$.

Exercise 7.4.6.

(a) Show that $Q(R)$ is a ring with identity element $[1/1]$ and $0 = [0/1]$.

(b) Show that $[a/b] = 0$ if, and only if, $a = 0$.

(c) Show that if $[a/b] \neq 0$, then $[b/a]$ is the multiplicative inverse of $[a/b]$. Thus $Q(R)$ is a field.

Exercise 7.4.7.

(a) Show that $a \mapsto [a/1]$ is an injective unital ring homomorphism of R into $Q(R)$. In this sense $Q(R)$ is a field *containing* R.

(b) If F is a field such that $F \supseteq R$, show that there is an injective unital homomorphism $\varphi : Q(R) \to F$ such that $\varphi([a/1]) = a$ for $a \in R$.

Exercise 7.4.8. If R is the ring of Gaussian integers, show that $Q(R)$ is isomorphic to the subfield of $\mathbb{C}$ consisting of complex numbers with rational real and imaginary parts.

Exercise 7.4.9. In an integral domain R, consider the additive cyclic subgroup C generated by the identity. Show that C is actually a subring, and is the image of $\mathbb{Z}$ under a ring homomorphism. Since C must be an integral domain, conclude that $C \cong Z$ or $C \cong \mathbb{Z}_p$ for some prime p.

7.5. Euclidean Domains and Unique Factorization

We have seen two examples of integral domains with a good theory of divisibility, the integers $\mathbb{Z}$ and the ring of polynomials $K[x]$ over a field K. For both of these rings R,

1. Every ideal is principal. That is, if M is an ideal, then there is an element $f \in R$ such that $M = fR$.
2. Any two non-zero elements f and g have a greatest common divisor which is contained in the ideal $I(f, g) = \{af + bg : a, b \in R\}$.
3. Every non-zero element has an essentially unique factorization into irreducible factors.

The key to these results is that both rings have a "Euclidean" function $d : R \setminus \{0\} \to \mathbb{N}$ with the property that $d(fg) \geq \max\{d(f), d(g)\}$ and for each $f, g \in R \setminus \{0\}$ there exist $q, r \in R$ such that $f = qg + r$ and $r = 0$ or $d(r) < d(g)$.

For the integers, the Euclidean function is $d(n) = |n|$. For the polynomials, the function is $d(f) = \deg(f)$.

Definition 7.5.1. Call an integral domain R a *Euclidean domain* if it admits a function d with the properties above.

Let us consider one more example of a Euclidean domain, in order to justify having made the definition.

Example 7.5.2. Let R be the ring of Gaussian integers, namely the complex numbers with integer real and imaginary parts. For the "degree" map d take $d(z) = |z|^2$. We have $d(zw) = d(z)d(w)$. The trick to establishing the Euclidean property is to work temporarily in the field of fractions, the set of complex numbers with rational coefficients. Let z, w be non-zero elements in R. Since z/w is a complex number with rational coefficients, there is an element q of R such that $|\Re(q - z/w)| \leq 1/2$ and $|\Im(q - z/w)| \leq 1/2$. Write $z = qw + r$. Since z and $qw \in R$, it follows that $r \in R$. But $r = (z - qw) = (z/w - q)w$, so $d(r) = |z/w - q|^2 d(w) \leq (1/2)d(w)$. This completes the proof that the ring of Gaussian integers is Euclidean.

Let us introduce several definitions related to divisibility before we continue the discussion:

Definition 7.5.3. A non-zero element of a commutative ring with identity is *irreducible* or *prime* if its only divisors are itself and units. Two or more non-zero elements are said to be *relatively prime* if they have no common divisors other than units.

Now, given any Euclidean domain R, one can follow the proofs which we have given for the integers and the polynomials over a field to establish the following results:

Theorem 7.5.4. *Let R be a Euclidean domain.*

(a) *Two non-zero elements f and $g \in R$ have a greatest common divisor which is contained in the ideal $I(f, g)$. The greatest common divisor is unique up to multiplication by a unit.*
(b) *Two elements f and $g \in R$ are relatively prime if, and only if, $1 \in I(f, g) = Rf + Rg$.*
(c) *Every ideal in R is principal.*
(d) *If an irreducible p divides the product of two non-zero elements, then it divides one or the other of them.*
(e) *Every non-zero element has a factorization into irreducibles, which is unique (up to units and up to order of the factors).*

Proof. Exercise. □

Now let's give an example of a garden variety integral domain which is *not* a Euclidean domain.

Example 7.5.5. The ring of polynomials in two variables over a field K is not a Euclidean domain. How does one show such a thing? Probably *not* directly by showing that no function on $K[x, y]$ has the properties of the degree function on $K[x]$. Rather one has to show that some property shared by all Euclidean domains fails. One can show for example that there exist relatively prime elements such that 1 is not in the ideal generated by the two elements, or that there exist ideals which are not principal. See Exercises 7.5.3 and 7.5.4.

Nevertheless, one can show that the ring $R = K[x, y]$ does have unique factorization.

Definition 7.5.6. An integral domain is a *unique factorization domain* if every non-zero element has a factorization into irreducibles which is unique up to order and multiplication by units.

Lemma 7.5.7. *In a unique factorization domain, any two non-zero elements have a greatest common divisor which is unique up to multiplication by units.*

Proof. Exercise 7.5.5. □

Theorem 7.5.8. *If R is a unique factorization domain, then $R[x]$ is a unique factorization domain.*

It follows from this result and induction on the number of variables that polynomial rings $K[x_1, \cdots, x_n]$ over a field K have unique factorization; see Exercise 7.5.6.

Let R be a unique factorization domain and let F denote the field of fractions of R. The key to showing that $R[x]$ is a unique factorization domain is to compare factorizations in $R[x]$ with factorizations in the Euclidean domain $F[x]$.

Call an element of $R[x]$ *primitive* if its coefficients are relatively prime. Any element $g(x) \in R[x]$ can be written as $g(x) = df(x)$, where $d \in R$ is a greatest common divisor of the coefficients of $g(x)$ and $f(x)$ is primitive. This decomposition is unique up to units of R. Furthermore, any element $\varphi(x) \in F[x]$ can be written as $\varphi(x) = (1/b)g(x)$, where b is a non-zero element of R and $g(x) \in R[x]$. For example, just take b to be the product of the denominators of the coefficients of $\varphi(x)$ (in $F[x]$). Combining these two observations, one can write:

$$\varphi(x) = (d/b)f(x), \tag{7.5.1}$$

where $f(x)$ is primitive in $R[x]$. This decomposition is unique up to units in R.

Lemma 7.5.9. (Gauss' Lemma.) *Let R be a unique factorization domain with field of fractions F.*

(a) *The product of two primitive elements of $R[x]$ is primitive.*

(b) *Suppose $f(x) \in R[x]$. Then $f(x)$ has a factorization $f(x) = \varphi(x)\psi(x)$ in $F[x]$ with* $\deg(\varphi), \deg(\psi) \geq 1$ *if, and only if, $f(x)$ has such a factorization in $R[x]$.*

Proof. Suppose that $f(x) = \sum a_i x^i$ and $g(x) = \sum b_j x^j$ are primitive in $R[x]$. Suppose p is a prime in R. There is a first index r such that p does not divide a_r and a first index s such that p does not divide b_s. The coefficient of x^{r+s} in $fg(x)$ is $a_r b_s + \sum_{i<r} a_i b_{r+s-i} + \sum_{j<s} a_{r+s-j} b_j$. By assumption, all the summands are divisible by p, except for $a_r b_s$, which is not. So the

coefficient of x^{r+s} in $fg(x)$ is not divisible by p. It follows that $fg(x)$ is also primitive. This proves part (a).

Suppose that $f(x)$ has the factorization $f(x) = \varphi(x)\psi(x)$ in $F[x]$ with $\deg(\varphi), \deg(\psi) \geq 1$. Write $f(x) = e\tilde{f}(x)$, $\varphi(x) = (a/b)\tilde{\varphi}(x)$ and $\psi(x) = (c/d)\tilde{\psi}(x)$, where $\tilde{f}(x)$, $\tilde{\varphi}(x)$, and $\tilde{\psi}(x)$ are primitive in $R[x]$. Then $f(x) = e\tilde{f}(x) = (ac/bd)\tilde{\varphi}(x)\tilde{\psi}(x)$. By part (a), the product $\tilde{\varphi}(x)\tilde{\psi}(x)$ is primitive in $R[x]$. By the uniqueness of such decompositions, it is necessary that $eu = (ac/bd)$, where u is a unit in R, so $f(x)$ factors as $f(x) = ue\tilde{\varphi}(x)\tilde{\psi}(x)$ in $R[x]$. □

Corollary 7.5.10. *If a polynomial in $\mathbb{Z}[x]$ has a non-trivial factorization in $\mathbb{Q}[x]$, then it has a non-trivial factorization in $\mathbb{Z}[x]$.*

Proof of Theorem 7.5.8. Let $g(x)$ be an element of $R[x]$. First, $g(x)$ can be written as $af(x)$, where $f(x)$ is primitive and $a \in R$; furthermore, this decomposition is unique up to units in R. The element a has a unique factorization in R, by assumption, so it remains to show that $f(x)$ has a unique factorization into irreducibles in $R[x]$. But using the factorization of $f(x)$ in $F[x]$ and the technique of the lemma, one can write

$$f(x) = p_1(x)p_2(x)\cdots p_s(x),$$

where the $p_i(x)$ are primitive elements of $R[x]$ and irreducible in $F[x]$. The uniqueness of this factorization follows from the uniqueness of prime factorization in $F[x]$ together with the uniqueness of the factorization (7.5.1). □

Exercises

Exercise 7.5.1. Prove Theorem 7.5.4; Verify that the proofs given for the ring of integers and the ring of polynomials over a field go through essentially without change for any Euclidean domain.

Exercise 7.5.2. Suppose a and b are relatively prime elements in a Euclidean domain R. Show that $aR \cap bR = abR$ and $aR + bR = R$. Show that $R/abR \cong R/aR \oplus R/bR$. This is a generalization of the Chinese Remainder Theorem, Exercise 7.3.19.

Exercise 7.5.3. Consider the ring of polynomials in two variables over any field of your choice, say $R = \mathbb{C}[x, y]$.

(a) The elements x and y are relatively prime.

(b) It is not possible to write $1 = p(x, y)x + q(x, y)y$, with $p, q \in R$.
(c) Conclude that R is not a Euclidean domain.

Exercise 7.5.4. Another property of all Euclidean domains is that every ideal is principal. Show that the ring $R = \mathbb{C}[x, y]$ does not share this property, as the ideal $Rx + Ry$ is not principal.

Exercise 7.5.5. . Prove Lemma 7.5.7.

Exercise 7.5.6.

(a) Let R be a commutative ring with identity 1. Show that the polynomial rings $R[x_1, \dots, x_{n-1}, x_n]$ and $(R[x_1, \dots, x_{n-1}])[x_n]$ can be identified.
(b) Assuming Theorem 7.5.8, show by induction that if K is a field, then $K[x_1, \dots, x_{n-1}, x_n]$ is a unique factorization domain for all n.

Exercise 7.5.7. Use Gauss' lemma or the idea of its proof to show that if a polynomial $a_n x^n + \cdots + a_1 x + a_0 \in \mathbb{Z}[x]$ has a rational root r/s, where r and s are relatively prime, then s divides a_n and r divides a_0. In particular, if the polynomial is monic, then its only rational roots are integers.

7.6. Irreducibility criteria

In this section, we will consider some elementary techniques for determining whether a polynomial is irreducible.

We restrict ourselves to the problem of determining whether a polynomial in $\mathbb{Z}[x]$ is irreducible. Recall that an integer polynomial factors over the integers if, and only if, it factors over the rational numbers, according to Lemma 7.5.9 and Corollary 7.5.10.

A basic technique in testing for irreducibility is to reduce the polynomial modulo a prime. The natural homomorphism of $\mathbb{Z}$ onto $\mathbb{Z}_p$, where p is a prime, extends to a homomorphism of $\mathbb{Z}[x]$ onto $\mathbb{Z}_p[x]$, $\sum a_i x^i \mapsto \sum [a_i] x^i$.

Proposition 7.6.1. *Consider a polynomial $f(x) = \sum a_i x^i \in \mathbb{Z}[x]$ of degree n. Suppose that the leading coefficient a_n is not divisible by the prime p. Consider the polynomial $\tilde{f}(x) \in \mathbb{Z}_p[x]$ obtained by reducing all coefficients of f modulo p. If $\tilde{f}$ is irreducible in $\mathbb{Z}_p[x]$, then f is irreducible in $\mathbb{Q}[x]$.*

Proof. This follows at once, because the map $f \mapsto \tilde{f}$ is a homomorphism. If f had a factorization $f = gh$ in $\mathbb{Z}[x]$, then $\tilde{f}$ would also factor as $\tilde{f} = \tilde{g}\tilde{h}$. □

Efficient algorithms are known for factorization in $\mathbb{Z}_p[x]$. The common computer algebra packages such as *Mathematica* and *Maple* have these algorithms built in. The *Mathematica* command **Factor[f, Modulus → p]** can be used for reducing a polynomial f modulo a prime p and factoring the reduction. Unfortunately, the condition of the proposition is merely a sufficient condition. It is quite possible (but rare) for a polynomial to be irreducible over $\mathbb{Q}$ but nevertheless for its reductions modulo every prime to be reducible.

Example 7.6.2. Let $f(x) = 83 + 82\,x - 99\,x^2 - 87\,x^3 - 17\,x^4$. The reduction of f modulo 3 is $\{2 + x + x^4\}$, which is irreducible over $\mathbb{Z}_3$. Hence, f is irreducible over $\mathbb{Q}$.

Example 7.6.3. Let $f(x) = -91 - 63\,x - 73\,x^2 + 22\,x^3 + 50\,x^4$. The reduction of f modulo 17 is $16\left(6 + 12\,x + 5\,x^2 + 12\,x^3 + x^4\right)$, which is irreducible over $\mathbb{Z}_{17}$. Therefore, f is irreducible over $\mathbb{Q}$.

A related sufficient condition for irreducibility is Eisenstein's criterion:

Proposition 7.6.4 (Eisenstein's Criterion). *Let*

$$f(x) = x^n + a_{n-1}x^{n-1} + \cdots + a_1 x + a_0.$$

Suppose p is a prime which divides all the coefficients a_i and such that p^2 does not divide a_0. Then $f(x)$ is irreducible over $\mathbb{Q}$.

Proof. If f has a non-trivial factorization over $\mathbb{Q}$, then it also has a non-trivial factorization over $\mathbb{Z}$, with both factors monic polynomials. Write $f(x) = a(x)b(x)$, where $a(x) = \sum_{i=0}^{r} \alpha_i x^i$ and $b(x) = \sum_{j=0}^{s} \beta_i x^j$. Since $a_0 = \alpha_0\beta_0$, exactly one of α_0 and β_0 is divisible by p; suppose without loss of generality that p divides β_0, and p does not divide α_0. Considering the equations

$$\begin{aligned} a_1 &= \beta_1\alpha_0 + \beta_0\alpha_1 \\ &\cdots \\ a_{s-1} &= \beta_{s-1}\alpha_0 + \cdots + \beta_0\alpha_{s-1} \\ a_s &= \alpha_0 + \beta_{s-1}\alpha_1 + \cdots + \beta_0\alpha_s, \end{aligned}$$

one obtains by induction that β_j is divisible by p for all j $(0 \le j \le s-1)$. Finally the last equation yields that α_0 is divisible by p, a contradiction. □

Example 7.6.5. $x^3 + 14x + 7$ is irreducible by the Eisenstein criterion.

Example 7.6.6. Sometimes the Eisenstein criterion can be applied after a linear change of variables. For example, for the so-called cyclotomic polynomial

$$f(x) = x^{p-1} + x^{p-2} + \cdots + x^2 + x + 1,$$

where p is a prime, one has

$$f(x+1) = \sum_{s=0}^{p-1} \binom{p}{s+1} x^s.$$

This is irreducible by Eisenstein's criterion, so f is irreducible as well. You are asked to provide the details for this example in Exercise 7.6.3.

There is a simple criterion for a polynomial in $\mathbb{Z}[x]$ to have (or not to have) a linear factor, the so called *rational root test.*

Proposition 7.6.7 (Rational root test). *Let $f(x) = a_n x^n + a_{n-1} x^n - 1 + \cdots a_1 x + a_0 \in \mathbb{Z}[x]$. If r/s is a rational root of f, where r and s are relatively prime, then s divides a_n and r divides a_0.*

Proof. Exercise 7.6.1 □

A quadratic or cubic polynomial is irreducible if, and only if, it has no linear factors, so the rational root test is a definitive test for irreducibility for such polynomials. The rational root test can sometimes be used as an adjunct to prove irreducibility of higher degree polynomials: If an integer polynomial of degree n has no rational root, but for some prime p its reduction mod p has irreducible factors of degrees 1 and $n-1$, then then f is irreducible (exercise).

Exercises

Exercise 7.6.1. Prove the rational root test.

Exercise 7.6.2. Show that if a polynomial $f(x) \in \mathbb{Z}[x]$ of degree n has no rational root, but for some prime p its reduction mod p has irreducible factors of degrees 1 and $n-1$, then then f is irreducible.

Exercise 7.6.3. Provide the details for Example 7.6.6.

Exercise 7.6.4. Show that for each natural number n

$$(x-1)(x-2)(x-3)\cdots(x-n) - 1$$

is irreducible over the rationals.

Exercise 7.6.5. Show that for each natural number $n \neq 4$

$$(x-1)(x-2)(x-3)\cdots(x-n)+1$$

is irreducible over the rationals.

Exercise 7.6.6. Determine whether the following polynomials are irreducible over the rationals. You may wish to do computer computations of factorizations modulo primes.

(a) $8 - 60\,x - 54\,x^2 + 89\,x^3 - 55\,x^4$
(b) $42 - 55\,x - 66\,x^2 + 44\,x^3$
(c) $42 - 55\,x - 66\,x^2 + 44\,x^3 + x^4$
(d) $5 + 49\,x + 15\,x^2 - 27\,x^3$
(e) $-96 + 53\,x - 26\,x^2 + 21\,x^3 - 75\,x^4$

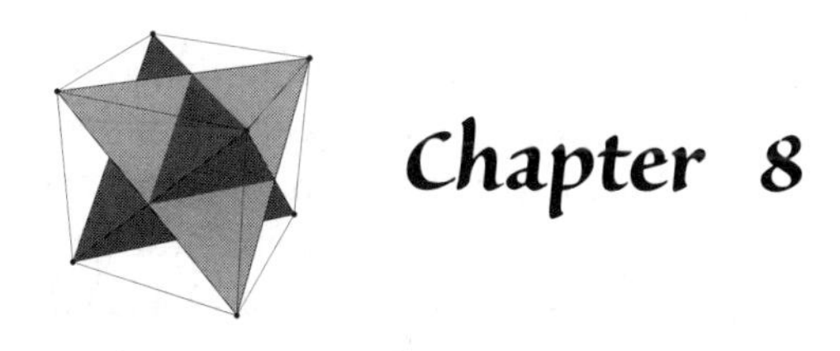

Chapter 8

Field Extensions – First Look

8.1. A Brief History

The most traditional concern of algebra is the solution of polynomial equations and related matters such as computation with radicals. Methods of solving linear and quadratic equations were already known to the ancients, and Arabic scholars preserved and augmented this knowledge during the middle ages.

You have learned the quadratic formula for the roots of a quadratic equation in school, but more fundamental is the algorithm of completing the square, which justifies the formula. Similar procedures for equations of the third and fourth degree were discovered by several mathematicians in 16th century Italy, who also introduced complex numbers (with some misgivings). And there the matter stood, more or less, for another 250 years.

At the end of the 18th century, no general method or formula, of the sort which had worked for equations of lower degree, was known for equations of degree 5, nor was it known that no such method was possible. The sort of method sought was one of "solution by radicals," that is, by algebraic operations and by introduction of n-th roots.

In 1798 C. F. Gauss showed that every polynomial equation with real or complex coefficients has a complete set of solutions in the complex numbers. Gauss published several proofs of this theorem, known as the Fundamental Theorem of Algebra. The easiest proofs known involve some complex analysis, and you can find a proof in any text on that subject.

A number of mathematicians worked on the problem of solution of polynomial equations during the period 1770-1820, among them J-L. Lagrange, A. Cauchy, and P. Ruffini. Their insights included a certain appreciation for the role of symmetry of the roots. Finally N. H. Abel, in 1824-1829, succeeded in showing that the general fifth degree equation could not be solved by radicals.

It was E. Galois, however, who, in the years 1829-1832, provided the most satisfactory solution to the problem of solution of equations by radicals and, in doing so, radically changed the nature of algebra. Galois associated with a polynomial $p(x)$ over a field K a canonical smallest field L containing K in which the polynomial has a complete set of roots *and, moreover, a canonical group of symmetries of L which acts on the roots of* $p(x)$. Galois' brilliant idea was to study the polynomial equation $p(x) = 0$ by means of this symmetry group. In particular, he showed that solvability of the equation by radicals corresponded to a certain property of the group, which also became known as *solvability*. The *Galois group* associated to a polynomial equation of degree n is always a subgroup of the permutation group S_n. It turns out that subgroups of S_n for $n \leq 4$ are solvable, but S_5 is not solvable. In Galois' theory, the non-solvability of S_5 implies the impossibility of an analogue of the quadratic formula for equations of degree 5.

Neither Abel nor Galois had much time to enjoy his success. Abel died at the age of 26 in 1829, and Galois died in a duel in 1832 at the age of 20. Galois' memoir was first published by Liouville in 1846, 14 years after Galois' death. Galois had submitted two manuscripts on his theory in 1829 to the Académie des Sciences de Paris, which apparently were lost by Cauchy. A second version of his manuscript was submitted in 1830, but Fourier, who received it, died before reading it, and that manuscript was also lost. A third version was submitted to the Academy in 1831, and referred to Poisson, who reported that he was unable to understand it and recommended revisions. But Galois was killed before he could prepare another manuscript.

In this chapter, we will take a first look at polynomial equations, field extensions and symmetry. Chapters 9 and 10 contain a more systematic treatment of Galois theory.

8.2. Solving the Cubic Equation

Consider a cubic polynomial equation,

$$x^3 + ax^2 + bx + c = 0,$$

where the coefficients lie in some field K, which for simplicity we assume to be contained in the field of complex numbers $\mathbb{C}$. Necessarily, K contains the rational field $\mathbb{Q}$.

If $\alpha_1, \alpha_2, \alpha_3$ are the roots of the equation in $\mathbb{C}$, then

$$x^3 + ax^2 + bx + c = (x - \alpha_1)(x - \alpha_2)(x - \alpha_3),$$

from which it follows that

$$\begin{aligned} \alpha_1 + \alpha_2 + \alpha_3 &= -a \\ a_1 a_2 + a_1 a_3 + a_2 a_3 &= b \\ a_1 a_2 a_3 &= -c. \end{aligned}$$

It is simpler to deal with a polynomial with zero quadratic term, and this can be accomplished by a linear change of variables $y = x + a/3$. One computes that the equation is transformed into

$$y^3 + (-\frac{a^2}{3} + b)\, y + (\frac{2a^3}{27} - \frac{ab}{3} + c)$$

Changing notation, we can suppose without loss of generality that we have at the outset a polynomial equation without quadratic term

$$f(x) = x^3 + px + q = 0,$$

with roots $\alpha_1, \alpha_2, \alpha_3$ satisfying

$$\begin{aligned} \alpha_1 + \alpha_3 + \alpha_3 &= 0 \\ a_1 a_2 + a_1 a_3 + a_2 a_3 &= p \\ a_1 a_2 a_3 &= -q. \end{aligned} \tag{8.2.1}$$

If one experiments with changes of variables in the hope of somehow simplifying the equation, one might eventually come upon the idea of expressing the variable x as a difference of two variables $x = v - u$. The result is:

$$(v^3 - u^3) + q - 3uv(v - u) + p(v - u) = 0.$$

A sufficient condition for a solution is:

$$\begin{aligned} (v^3 - u^3) + q &= 0 \\ 3uv &= p. \end{aligned}$$

Now, using the second equation to eliminate u from the first gives a quadratic equation for v^3:

$$v^3 - \frac{p^3}{27v^3} + q = 0.$$

The solutions to this are:

$$v^3 = -\frac{q}{2} \pm \sqrt{\frac{q^2}{4} + \frac{p^3}{27}}.$$

It turns out that we get the same solutions to our original equation regardless of the choice of the square root. Let ω denote the primitive third root of unity $\omega = e^{2\pi i/3}$, and let A denote one cube root of

$$-\frac{q}{2} + \sqrt{\frac{q^2}{4} + \frac{p^3}{27}}.$$

Then the solutions for v are A, ωA, and $\omega^2 A$, and the solutions for $x = v - \dfrac{p}{3v}$ are:

$$\alpha_1 = A - \frac{p}{3A}, \; a_2 = \omega A - \omega^2 \frac{p}{3A}, \; \alpha_3 = \omega^2 A - \omega \frac{p}{3A}.$$

These are generally known as *Cardano's formulas*, but Cardano credits Scipione del Ferro and N. Tartaglia and for their discovery (prior to 1535).

What we will be concerned with here is the structure of the *field extension* $K \subseteq K(\alpha_1, \alpha_2, \alpha_3)$ and the symmetry of the roots.

Here is some general terminology and notation: If $F \subseteq L$ are fields, one says that F is a *subfield* of L or that L is a *field extension* of F. If $F \subseteq L$ is a field extension and $S \subseteq L$ is any subset, then $F(S)$ denotes the smallest subfield of L which contains F and S. If $F \subseteq L$ is a field extension and $g(x) \in F[x]$ has a complete set of roots in L, i.e., $g(x)$ factors into linear factors in $L[x]$, then the smallest subfield of L containing F and the roots of $g(x)$ in L is called a *splitting field* of $g(x)$ over F.

Returning to our more particular situation, $K(\alpha_1, \alpha_2, \alpha_3)$ is the splitting field (in $\mathbb{C}$) of the cubic polynomial $f(x) \in K[x]$.

One thing which is noticeable is that the roots of $f(x)$ are obtained by rational operations and by extraction of cube and square roots (in $\mathbb{C}$); in fact, A is obtained by first taking a square root and then taking a cube root. And ω also involves a square root, namely $\omega = 1/2 + \sqrt{-3}/2$. So it would seem that it might be necessary to obtain $K(\alpha_1, \alpha_2, \alpha_3)$ by three stages, $K \subseteq K_1 \subseteq K_2 \subseteq K_3 = K(\alpha_1, \alpha_2, \alpha_3)$, where at each stage one enlarges the field by adjoining a new cube or square root. In fact, we will see that at most two stages are necessary.

An element which is important for understanding the splitting field is

$$\delta = (\alpha_1 - \alpha_2)(\alpha_1 - \alpha_3)(\alpha_2 - \alpha_3) = \det \begin{bmatrix} 1 & 1 & 1 \\ \alpha_1 & \alpha_2 & \alpha_3 \\ \alpha_1^2 & \alpha_2^2 & \alpha_3^2 \end{bmatrix}.$$

The square δ^2 of this element is invariant under permutations of the α_i; a general result, which we will discuss later, says that δ^2 is, therefore, a polynomial expression in the coefficients p, q of the polynomial and, in particular, $\delta^2 \in K$. We can compute δ^2 explicitly in terms of p and q; the result is $\delta^2 = -4p^3 - 27q^2$. You are asked to verify this in Exercise 8.2.7. The element δ^2 is called the *discriminant* of the polynomial f.

Now, it turns out that the nature of the field extension $K \subseteq K(\alpha_1, \alpha_2, \alpha_3)$ depends on whether δ is in the ground field K or not. Before discussing this further, it will be convenient to introduce some remarks of a general nature about algebraic elements of a field extension. We do this in the following section, and complete the discussion of the cubic equation in Section 8.4.

Exercises

Exercise 8.2.1. Show that a cubic polynomial in $K[x]$ either has a root in K or is irreducible over K.

Exercise 8.2.2. Verify the reduction of the monic cubic polynomial to one with no quadratic term.

Exercise 8.2.3. How can you deal with a cubic polynomial which is not monic?

Exercise 8.2.4. Verify in detail the derivation of Cardano's formulas.

Exercise 8.2.5. Consider the polynomial $p(x) = x^3 + 2x^2 + 2x - 3$. Show that p is irreducible over $\mathbb{Q}$. *Hint*: Show that if p has a rational root, then it must have an integer root, and the integer root must be a divisor of 3. Carry out the reduction of p to a polynomial f without quadratic term. Use the method described above to find the roots of f. Use this information to find the roots of p.

Exercise 8.2.6. Repeat the previous exercise with various cubic polynomials of your choice.

Exercise 8.2.7. Let V denote the Vandermonde matrix

$$\begin{bmatrix} 1 & 1 & 1 \\ \alpha_1 & \alpha_2 & \alpha_3 \\ \alpha_1^2 & \alpha_2^2 & \alpha_3^2 \end{bmatrix}.$$

(a) Show that $\delta^2 = \det(VV^t) = \det \begin{bmatrix} 3 & 0 & \sum \alpha_i^2 \\ 0 & \sum \alpha_i^2 & \sum \alpha_i^3 \\ \sum \alpha_i^2 & \sum \alpha_i^3 & \sum \alpha_i^4 \end{bmatrix}$.

(b) Use equations (8.2.1) as well as the fact that the α_i are roots of $f(x) = 0$ to compute that $\sum \alpha_i^2 = -2p$, $\sum_i a_i^3 = -3q$, and $\sum_i a_i^4 = 2p^2$.

(c) Compute that $\delta^2 = -4p^3 - 27q^2$.

Exercise 8.2.8.

(a) Show that $x^3 - 2$ is irreducible in $\mathbb{Q}[x]$. Compute its roots and also δ^2 and δ.

(b) $\mathbb{Q}(\sqrt[3]{2}) \subseteq \mathbb{R}$, so is not equal to the splitting field E of $x^3 - 2$. The splitting field is $\mathbb{Q}(\sqrt[3]{2}, \omega)$, where $\omega = e^{2\pi i/3}$. Show that ω satisfies a quadratic polynomial over $\mathbb{Q}$.

(c) Show that $x^3 - 3x + 1$ is irreducible in $\mathbb{Q}[x]$. Compute its roots, and also δ^2 and δ. *Hint:* The quantity A is a root of unity, and the roots are twice the real part of certain roots of unity.

8.3. Adjoining Algebraic Elements to a Field

The fields in this section are general, not necessarily subfields of the complex numbers.

A field extension L of a field K is, in particular, a *vector space* over K. We should examine this statement carefully. We have used quite a bit of linear algebra in this text but, for the most part, we have only dealt with linear operators or matrices acting on standard real or complex vector spaces $\mathbb{R}^n$ or $\mathbb{C}^n$. It might be helpful to review the definition of a vector space over an arbitrary field.

Definition 8.3.1. A *vector space* V over a field K is a abelian group with a product $K \times V \to V$, $(s, v) \mapsto sv$ satisfying

1. $1v = v$ for all $v \in V$.
2. $(st)v = s(tv)$ for all $s, t \in K$, $v \in V$.
3. $s(v + w) = sv + sw$ for all $s \in K$ and $v, w \in V$.
4. $(s + t)v = sv + tv$ for all $s, t \in K$ and $v \in V$.

Compare this definition with that contained in your linear algebra text; notice that we could state the definition more concisely by referring to the notion of an abelian group. You are asked to check in the exercises that if $K \subseteq L$ are fields, then L is a vector space over K.

Let us recall several important definitions from linear algebra. Let V be a vector space over a field K. A subset $S \subseteq V$ is said to be *linearly independent* if, and only if, for all $n \in \mathbb{N}$, $\alpha_1, \dots, \alpha_n \in K$, and *distinct* $v_1, \dots, v_n \in S$, if

$$\alpha_1 v_1 + \cdots + \alpha_n v_n = 0,$$

then $\alpha_1 = \alpha_2 = \cdots = \alpha_n = 0$. A subset $S \subseteq V$ is said to *span* V if, and only if, for each $v \in V$, there exist $n \in \mathbb{N}$, $\alpha_1, \dots, \alpha_n \in K$, and $v_1, \dots, v_n \in S$ such that

$$v = \alpha_1 v_1 + \cdots + \alpha_n v_n.$$

A *basis* of V is a subset which is both linearly independent and spanning. It is a theorem that any two bases have the same cardinality, called the *dimension* of V.

The following conditions are equivalent for a vector space V:

- V has a finite spanning set.

- V has a finite basis.
- Every linearly independent subset of V is finite.

The vector space V is said to be *finite dimensional* if it has a finite basis. Otherwise, V is said to be *infinite dimensional.*

The following conditions are equivalent for a vector space V:

- V is infinite dimensional.
- V has an infinite linearly independent subset.
- For every $n \in \mathbb{N}$, V has a linearly independent subset with n elements.

Let us now return to the consideration of a field extension $K \subseteq L$. Since L is a K-vector space, in particular, L has a dimension over K, possibly infinite. The dimension of L as a K-vector space is denoted $\dim_K(L)$ (or, sometimes, $[L : K]$.) Dimensions of field extensions have the following multiplicative property:

Proposition 8.3.2. *If $K \subseteq L \subseteq M$ are fields, then:*

$$\dim_K(M) = \dim_K(L)\dim_L(M).$$

Proof. Suppose that $\{\lambda_1, \dots, \lambda_r\}$ is a subset of L which is linearly independent over K, and that $\{\mu_1, \dots, \mu_s\}$ is a subset of M which is linearly independent over L. I claim that $\{\lambda_i\mu_j : 1 \le i \le r, 1 \le j \le s\}$ is linearly independent over K. In fact, if $0 = \sum_{i,j} k_{ij}\lambda_i\mu_j = \sum_j(\sum_i k_{ij}\lambda_i)\mu_j$, with $k_{ij} \in K$, then linear independence of $\{\mu_j\}$ over L implies that $\sum_i k_{ij}\lambda_i = 0$ for all j, and then linear independence of $\{\lambda_i\}$ over K implies that $k_{ij} = 0$ for all i, j, which proves the claim.

In particular, if either $\dim_K(L)$ or $\dim_L(M)$ is infinite, then there are arbitrarily large subsets of M which are linearly independent over K, so $\dim_K(M)$ is also infinite.

Suppose now that $\dim_K(L)$ and $\dim_L(M)$ are finite, that $\{\lambda_1, \dots, \lambda_r\}$ is a basis of L over K, and that $\{\mu_1, \dots, \mu_s\}$ is a basis of M over L. The fact that $\{\mu_j\}$ spans M over L and that $\{\lambda_i\}$ spans L over K implies that the set of products $\{\lambda_i\mu_j\}$ spans M over K (exercise). Hence, $\{\lambda_i\mu_j\}$ is a basis of M over K. □

Definition 8.3.3. A field extension $K \subseteq L$ is called *finite* if L is a finite dimensional vector space over K.

Now, consider a field extension $K \subseteq L$. For any element $\alpha \in L$ there is a ring homomorphism from $K[x]$ into L given by $\varphi_\alpha(f(x)) = f(\alpha)$. That is,

$$\varphi_\alpha(k_0 + k_1x + \cdots + k_nx^n) = k_0 + k_1\alpha + \cdots + k_n\alpha^n \tag{8.3.1}$$

Definition 8.3.4. An element α in a field extension L of K is said to be *algebraic over* K if there is some polynomial $p(x) \in K[x]$ such that $p(\alpha) = 0$; equivalently, there are elements $k_0, k_1, \dots k_n \in K$ such that $k_0 + k_1\alpha + \cdots + k_n\alpha^n = 0$. Any element which is not algebraic is called *transcendental.* The field extension is called *algebraic* if every element of L is algebraic over K.

In particular, the set of complex numbers which are algebraic over $\mathbb{Q}$ are called *algebraic numbers* and those which are transcendental over $\mathbb{Q}$ are called *transcendental numbers*. It is not difficult to see that there are (only) countably many algebraic numbers, and that, therefore, there are uncountably many transcendental numbers. Here is the argument: The rational numbers are countable, so for each natural number n, there are only countably many distinct polynomials with rational coefficients with degree no more than n. Consequently, there are only countably many polynomials with rational coefficients altogether, and each of these has only finitely many roots in the complex numbers. Therefore, the set of algebraic numbers is a countable union of finite sets, so countable. On the other hand, the complex numbers are uncountable, so there are uncountably many transcendental numbers.

You are asked to prove the following result in the exercises:

Proposition 8.3.5. *if $K \subseteq L$ is a finite field extension, then L is algebraic over K, and moreover there are finitely many elements $a_1, \dots, a_n \in L$ such that $L = K(a_1, \dots, a_n)$.*

Proof. Exercise. □

The set I_α of polynomials $p(x) \in K[x]$ satisfying $p(\alpha) = 0$ is the kernel of the homomorphism $\varphi_\alpha : K[x] \to L$, so is an ideal in $K[x]$. We have $I_\alpha = f(x)K[x]$, where $f(x)$ is an element of minimal degree in I_α (Exercise 7.2.7). The polynomial $f(x)$ is necessarily irreducible; if it factored as $f(x) = f_1(x)f_2(x)$, where $\deg(f) > \deg(f_i) > 0$, then $0 = f(\alpha) = f_1(\alpha)f_2(\alpha)$. But then one of the f_i would have to be in I_α, while $\deg(f_i) < \deg(f)$, a contradiction. The generator $f(x)$ of I_α is unique up to multiplication by a non-zero element of K, so there is a unique monic polynomial $f(x)$ such that $I_\alpha = f(x)K[x]$, called the *minimal polynomial for α over K*.

We have proved the following proposition:

Proposition 8.3.6. *If $K \subseteq L$ are fields and $\alpha \in L$ is algebraic over K, then there is a unique monic irreducible polynomial $f(x) \in K[x]$ such that the set of polynomials $p(x) \in K[x]$ satisfying $p(\alpha) = 0$ is $f(x)K[x]$.*

Fix the algebraic element $\alpha \in L$, and consider the set of elements of L of the form

$$k_0 + k_1\alpha + \cdots + k_n\alpha^n, \tag{8.3.2}$$

with $n \in \mathbb{N}$ and $k_0, k_1, \ldots, k_n \in K$. By equation (8.3.1), this is the range of the homomorphism φ_α, so, by the Homomorphism Theorem for rings, it is isomorphic to $K[x]/I_\alpha = K[x]/(f(x))$, where we have written $(f(x))$ for $f(x)K[x]$. But since $f(x)$ is irreducible, the ideal $(f(x))$ is maximal, and, therefore, the quotient ring $K[x]/(f(x))$ is a field (Exercise 7.3.17).

Proposition 8.3.7. *Suppose $K \subseteq L$ are fields, $\alpha \in L$ is algebraic over K, and $f(x) \in K[x]$ is the minimal polynomial for α over K.*

(a) *$K(\alpha)$, the subfield of L generated by K and α, is isomorphic to the quotient field $K[x]/(f(x))$.*

(b) *$K(\alpha)$ is the set of elements of the form*

$$k_0 + k_1\alpha + \cdots + k_{d-1}\alpha^{d-1},$$

where d is the degree of f.

(c) $\dim_K(K(\alpha)) = \deg(f)$.

Proof. It is shown above that the ring of polynomials in α with coefficients in K is a field, isomorphic to $K[x]/(f(x))$. Therefore, $K(\alpha) = K[\alpha] \cong K[x]/(f(x))$. This shows part (a). For any $p \in K[x]$, write $p = qf + r$, where $r = 0$ or $\deg(r) < \deg(f)$. Then $p(\alpha) = r(\alpha)$, since $f(\alpha) = 0$. This means that $\{1, \alpha, \ldots, \alpha^{d-1}\}$ spans $K(\alpha)$ over K. But this set is also linearly independent over K, because α is not a solution to any equation of degree less than d. This shows parts (b) and (c). □

Example 8.3.8. Consider $f(x) = x^2 - 2 \in \mathbb{Q}(x)$, which is irreducible by Eisenstein's criterion (Proposition 7.6.4). The element $\sqrt{2} \in \mathbb{R}$ is a root of $f(x)$. (The existence of this root in $\mathbb{R}$ is a fact of *analysis*.) The field $\mathbb{Q}(\sqrt{2}) \cong \mathbb{Q}[x]/(x^2 - 2)$ consists of elements of the form $a + b\sqrt{2}$, with $a, b \in \mathbb{Q}$. The rule for addition in $\mathbb{Q}(\sqrt{2})$ is

$$(a + b\sqrt{2}) + (a' + b'\sqrt{2}) = (a + a') + (b + b')\sqrt{2},$$

and the rule for multiplication is

$$(a+b\sqrt{2})(a'+b'\sqrt{2}) = (aa'+2bb') + (ab'+ba')\sqrt{2}.$$

The inverse of $a+b\sqrt{2}$ is:

$$(a+b\sqrt{2})^{-1} = \frac{a}{a^2-2b^2} - \frac{b}{a^2-2b^2}\sqrt{2}.$$

Example 8.3.9. The polynomial $f(x) = x^3 - 2x + 2$ is irreducible over $\mathbb{Q}$ by Eisenstein's criterion. The polynomial has a real root θ by application of the intermediate value theorem of analysis. The field $\mathbb{Q}(\theta) \cong \mathbb{Q}[x]/(f)$ consists in elements of the form $a + b\theta + c\theta^2$, where a, b, $c \in \mathbb{Q}$. Multiplication is performed by using the distributive law and then reducing using the rule $\theta^3 = 2\theta - 2$ (whence $\theta^4 = 2\theta^2 - 2\theta$). To find the inverse of an element of $\mathbb{Q}(\theta)$, it is convenient to compute in $\mathbb{Q}[x]$. Given $g(\theta) = a + b\theta + c\theta^2$, there exist elements $r(x), s(x) \in \mathbb{Q}[x]$ such that $g(x)r(x) + f(x)s(x) = 1$, since g and f are relatively prime. Furthermore, r and s can be computed by the algorithm implicit in the proof of Theorem 7.2.7. It then follows that $g(\theta)r(\theta) = 1$, so $r(\theta)$ is the desired inverse. Let us compute the inverse of $2+3\theta-\theta^2$ in this way. Put $g(x) = -x^2+3x+2$. Then one can compute that

$$\frac{1}{118}(24+8x-9x^2)g + \frac{1}{118}(35-9x)f = 1,$$

and, therefore,

$$(2+3\theta-\theta^2)^{-1} = \frac{1}{118}(24+8\theta-9\theta^2)$$

From Exercises 8.3.4 and 8.3.6, we have the following proposition.

Proposition 8.3.10. *A field extension $K \subseteq L$ is finite if, and only if, it is generated by finitely many algebraic elements. Furthermore, if $L = K(\alpha_1, \ldots, \alpha_n)$, where the α_i are algebraic over K, then L consists of polynomials in the α_i with coefficients in K.*

Exercises

Exercise 8.3.1. Show that if $K \subseteq L$ are fields, then the identity of K is also the identity of L. Conclude that L is a vector space over K.

Exercise 8.3.2. Fill in the details of the proof of 8.3.2 to show that $\{\lambda_i\mu_j\}$ spans M over K.

Exercise 8.3.3. Let $L \supseteq K$ be a field extension, $f(x) \in K[x]$, and $\alpha \in L$. Check the assertion that the map $\varphi_\alpha : f(x) \mapsto f(\alpha)$ is, in fact, a unital ring homomorphism of $K[x]$ into L.

Exercise 8.3.4.

(a) Show that if $K \subseteq L$ is a finite field extension, then L is algebraic over K. *Hint:* Consider any $a \in L$. Can the powers of a be linearly independent over K?

(b) Show that if $K \subseteq L$ is a finite field extension, then there exist finitely many elements $a_1, \dots, a_n \in L$ such that

$$L = K(a_1, \dots, a_n).$$

Exercise 8.3.5. Show that $\mathbb{R}$ is *not* a finite extension of $\mathbb{Q}$.

Exercise 8.3.6.

(a) Let $L = K(\alpha_1, \dots, \alpha_n)$, where the α_i are algebraic over K. Show that L is a finite extension of K, and, therefore, algebraic. *Hint:* Let $K_0 = K$, and $K_i = K(\alpha_1, \dots, \alpha_i)$, for $1 \leq i \leq n$. Consider the tower of extensions:

$$K \subseteq K_1 \subseteq \cdots \subseteq K_{n-1} \subseteq L.$$

Show that $\dim_{K_i}(K_{i+1})$ is finite for all i, and conclude that $\dim_K(L)$ is finite as well.

(b) Show that $L = K[\alpha_1, \dots, \alpha_n]$, the set of *polynomials* in the α_i with coefficients in K.

Exercise 8.3.7.

(a) Show that the polynomial

$$p(x) = \frac{x^5 - 1}{x - 1} = x^4 + x^3 + x^2 + x + 1$$

is irreducible over $\mathbb{Q}$.

(b) According to the Proposition 8.3.7 , $\mathbb{Q}[\zeta] = \{a\zeta^3 + b\zeta^2 + c\zeta^2 + d : a, b, c, d \in \mathbb{Q}\} \cong \mathbb{Q}[x]/(p(x))$, and $\mathbb{Q}[\zeta]$ is a field. Compute the inverse of $\zeta^2 + 1$ as a polynomial in ζ. *Hint:* Obtain a system of linear equations for the coefficients of the polynomial.

Exercise 8.3.8. Let $\zeta = e^{2\pi i/5}$.

(a) Find the minimal polynomials for $\cos(2\pi/5)$ and $\sin(2\pi/5)$ over $\mathbb{Q}$.

(b) Find the minimal polynomial for ζ over $\mathbb{Q}(\cos(2\pi/5))$.

Exercise 8.3.9. Show that the splitting field E of $x^3 - 2$ has dimension 6 over $\mathbb{Q}$. Refer to Exercise 8.2.8.

Exercise 8.3.10. If α is a fifth root of 2 and β is a seventh root of 3, what is the dimension of $\mathbb{Q}(\alpha, \beta)$ over $\mathbb{Q}$?

Exercise 8.3.11. Find $\dim_{\mathbb{Q}} \mathbb{Q}(\alpha, \beta)$, where

(a) $\alpha^3 = 2$ and $\beta^2 = 2$
(b) $\alpha^3 = 2$ and $\beta^2 = 3$.

Exercise 8.3.12. Show that $f(x) = x^4 + x^3 + x^2 + 1$ is irreducible in $\mathbb{Q}[x]$. For $p(x) \in Q[x]$, let $\overline{p(x)}$ denote the image of $p(x)$ in the field $Q[x]/(f(x))$. Compute the inverse of $\overline{(x^2 + 1)}$.

Exercise 8.3.13. Show that $f(x) = x^3 + 6x^2 - 12x + 3$ is irreducible over $\mathbb{Q}$. Let θ be a real root of $f(x)$, which exists due to the intermediate value theorem. $\mathbb{Q}(\theta)$ consists of elements of the form $a_0 + a_1\theta + a_2\theta^2$. Explain how to compute the product in this field and find the product $(7 + 2\theta + \theta^2)(1 + \theta^2)$. Find the inverse of $(7 + 2\theta + \theta^2)$.

Exercise 8.3.14. Show that $f(x) = x^5 + 4x^2 - 2x + 2$ is irreducible over $\mathbb{Q}$. Let θ be a real root of $f(x)$, which exists due to the intermediate value theorem. $\mathbb{Q}(\theta)$ consists of elements of the form $a_0 + a_1\theta + a_2\theta^2 + a_3\theta^3 + a_4\theta^4$. Explain how to compute the product in this field and find the product $(7 + 2\theta + \theta^3)(1 + \theta^4)$. Find the inverse of $(7 + 2\theta + \theta^3)$.

8.4. Splitting Field of a Cubic Polynomial

In Section 8.2, we considered a field K contained in the complex numbers $\mathbb{C}$ and a cubic polynomial $f(x) = x^3 + px + q \in K[x]$. We obtained explicit expressions involving extraction of square and cube roots for the three roots $\alpha_1, \alpha_2, \alpha_3$ of $f(x)$ in $\mathbb{C}$, and we were beginning to study the splitting field extension $E = K(\alpha_1, a_2, a_3)$.

If $f(x)$ factors in $K[x]$, then either all the roots are in K or exactly one of them (say α_3) is in K and the other two are roots of an irreducible quadratic polynomial in $K[x]$. In this case, $E = K(\alpha_1)$ is a field extension of dimension 2 over K.

Henceforth, we assume that $f(x)$ is irreducible in $K[x]$. Let us first notice that the roots of $f(x)$ are necessarily distinct (exercise).

If α_1 denotes one of the roots, we know that $K(\alpha_1) \cong K[x]/(f(x))$ is a field extension of dimension $3 = \deg(f)$ over K. Since we have $K \subseteq K(\alpha_1) \subseteq E$, it follows from the multiplicativity of dimension (Proposition 8.3.2) that 3 divides the dimension of E over K.

Recall the element $\delta = (\alpha_1 - \alpha_2)(\alpha_1 - \alpha_3)(\alpha_2 - \alpha_3) \in E$; since $\delta^2 = -4p^3 - 27q^2 \in K$, either $\delta \in K$ or $K(\delta)$ is an extension field of dimension

2 over K. In the latter case, since $K \subseteq K(\delta) \subseteq E$, it follows that 2 also divides $\dim_K(E)$.

We will show that there are only two possibilities:

1. $\delta \in K$ and $\dim_K(E) = 3$, or
2. $\delta \notin K$ and $\dim_K(E) = 6$.

We're going to do some algebraic tricks to solve for α_2 in terms of α_1 and δ. The identity $\sum_i \alpha_i = 0$ gives:

$$\alpha_3 = -\alpha_1 - \alpha_2. \tag{8.4.1}$$

Eliminating α_3 in $\sum_{i<j} \alpha_i \alpha_j = p$ gives

$$\alpha_2^2 = -\alpha_1^2 - \alpha_1\alpha_2 - p. \tag{8.4.2}$$

Since $f(\alpha_i) = 0$, we have

$$\alpha_i^3 = -p\alpha_i - q \quad (i = 1, 2). \tag{8.4.3}$$

In the exercises, you are asked to show that

$$\alpha_2 = \frac{\delta + 2\alpha_1 p + 3q}{2\left(3\alpha_1{}^2 + p\right)}. \tag{8.4.4}$$

Proposition 8.4.1. *Let K be a subfield of $\mathbb{C}$, $f(x) = x^3 + px + q \in K[x]$ an irreducible cubic polynomial, and let E denote the splitting field of $f(x)$ in $\mathbb{C}$. Let $\delta = (\alpha_1 - \alpha_2)(\alpha_1 - \alpha_3)(\alpha_2 - \alpha_3)$, where α_i are the roots of $f(x)$. If $\delta \in K$, then* $\dim_K(E) = 3$. *Otherwise,* $\dim_K(E) = 6$.

Proof. Suppose that $\delta \in K$. Then equation (8.4.4) shows that α_2 and, therefore, also α_3 is contained in $K(\alpha_1)$. Thus $E = K(\alpha_1, \alpha_2, \alpha_3) = K(\alpha_1)$, and E has dimension 3 over K.

On the other hand, if $\delta \notin K$, then we have seen that $\dim_K(E)$ is divisible by both 2 and 3, so $\dim_K(E) \geq 6$. Consider the field extension $K(\delta) \subseteq E$. If $f(x)$ were not irreducible in $K(\delta)[x]$, then we would have $\dim_{K(\delta)}(E) \leq 2$, so:

$$\dim_K(E) = \dim_K(K(\delta)) \dim_{K(\delta)}(E) \leq 4,$$

a contradiction. So $f(x)$ must remain irreducible in $K(\delta)[x]$. But then it follows from the previous paragraph (replacing K by $K(\delta)$) that $\dim_{K(\delta)}(E) = 3$. Therefore, $\dim_K(E) = 6$. □

The structure of the splitting field can be better understood if we consider the possible intermediate fields between K and E, and introduce symmetry into the picture as well.

Proposition 8.4.2. *Let F be a field, let $f(x) \in F[x]$ be irreducible, and suppose α and β are two roots of $f(x)$ in some extension field L. Then there is an isomorphism of fields $\sigma : K(\alpha) \to K(\beta)$ such that $\sigma(k) = k$ for all $k \in K$ and $\sigma(\alpha) = \beta$.*

Proof. According to Proposition 8.3.7, there is an isomorphism

$$\sigma_\alpha : K[x]/(f(x)) \to K(\alpha)$$

which takes $[x]$ to α and fixes each element of K. So the desired isomorphism $K(\alpha) \cong K(\beta)$ is $\sigma_\beta \circ \sigma_\alpha^{-1}$. □

Applying this result to the cubic equation, we obtain: *For any two roots α_i and α_j of the irreducible cubic polynomial $f(x)$, there is an isomorphism $K(\alpha_i) \cong K(\alpha_j)$ which fixes each element of K and takes α_i to α_j.*

Now, suppose that $\delta \in K$, so $\dim_K(E) = 3$. Then $E = K(\alpha_i)$ for each i, so for any two roots α_i and α_j of $f(x)$ there is an automorphism of E (i.e., an isomorphism of E onto itself) which fixes each element of K and takes α_i to α_j. Let us consider an automorphism σ of E which fixes K pointwise and maps α_1 to α_2. What is $\sigma(\alpha_2)$? The following general observation shows that $\sigma(\alpha_2)$ is also a root of $f(x)$. Surely, $\sigma(\alpha_2) \neq \alpha_2$, so $\sigma(\alpha_2) \in \{\alpha_1, \alpha_3\}$.

Proposition 8.4.3. *Suppose $F \subseteq L$ is any field extension, $f(x) \in F[x]$, and β is a root of $f(x)$ in L. If σ is an automorphism of L which leaves F fixed pointwise, then $\sigma(\beta)$ is also a root of $f(x)$.*

Proof. If $f(x) = \sum f_i x^i$, then $\sum f_i \sigma(\beta)^i = \sigma(\sum f_i \beta^i) = \sigma(0) = 0$. □

An automorphism of a field L is a field isomorphism from L to itself. The set of all automorphisms, denoted $\mathrm{Aut}(L)$, is a group. If $F \subseteq L$ is a subfield, an automorphism of L which leaves F fixed pointwise is called a *F-automorphism of L.* The set of F-automorphisms of L is denoted $\mathrm{Aut}_F(L)$. $\mathrm{Aut}_F(L)$ is a subgroup of $\mathrm{Aut}(L)$ (exercise).

Return to the irreducible cubic $f(x) \in K[x]$, and suppose that $\delta \in K$ so that the splitting field E has dimension 3 over K. We have seen that the group $\mathrm{Aut}_K(E)$ acts as permutations of the roots of $f(x)$, and the action is transitive; that is, for any two roots α_i and α_j, there is a $\sigma \in \mathrm{Aut}_K(E)$ such that $\sigma(\alpha_i) = \alpha_j$. However, not every permutation of the roots can arise as the restriction of a K-automorphism of E! In fact, any odd permutation of the roots would map $\delta = (\alpha_1 - \alpha_2)(\alpha_1 - \alpha_3)(\alpha_2 - \alpha_3)$ to $-\delta$, so cannot arise from a K-automorphism of E. The group of permutations of the roots induced by $\mathrm{Aut}_K(E)$ is a transitive subgroup of even permutations, so must coincide with A_3. We have proven:

Proposition 8.4.4. *If K is a subfield of $\mathbb{C}$, $f(x) \in K[x]$ is an irreducible cubic polynomial, E is the splitting field of $f(x)$ in $\mathbb{C}$, and* $\dim_K(E) = 3$, *then* $\mathrm{Aut}_K(E) \cong A_3 \cong \mathbb{Z}_3$.

In particular, we have the answer to the question posed above: In the case that $\delta \in K$, if σ is a K-automorphism of the splitting field E such that $\sigma(\alpha_1) = \alpha_2$, then necessarily, $\sigma(\alpha_2) = \alpha_3$ and $\sigma(\alpha_3) = \alpha_1$.

Now, consider the case that $\delta \notin K$. Consider any of the roots of $f(x)$, say α_1. In $K(\alpha_1)[x]$, $f(x)$ factors as $(x - \alpha_1)(x^2 + \alpha_1 x - q/\alpha_1)$, and $p(x) = x^2 + \alpha_1 x - q/\alpha_1$ is irreducible in $K(\alpha_1)[x]$. Now, E is obtained by adjoining a root of $p(x)$ to $K(\alpha_1)$, $E = K(\alpha_1)(\alpha_2)$, so by Proposition 8.4.2, there is an automorphism of E which fixes $K(\alpha_1)$ pointwise and which interchanges α_2 and α_3. Similarly, for any j, $\mathrm{Aut}_K(E)$ contains an automorphism which fixes α_j and interchanges the other two roots. It follows that $\mathrm{Aut}_K(E) \cong S_3$.

Proposition 8.4.5. *If K is a subfield of $\mathbb{C}$, $f(x) \in K[x]$ is an irreducible cubic polynomial, E is the splitting field of $f(x)$ in $\mathbb{C}$, and* $\dim_K(E) = 6$, *then* $\mathrm{Aut}_K(E) \cong S_3$. *Every permutation of the three roots of $f(x)$ is the restriction of a K-automorphism of E.*

The rest of this section will be devoted to working out the *Galois correspondence* between subgroups of $\mathrm{Aut}_K(E)$ and intermediate fields $K \subseteq M \subseteq E$. In the general theory of field extensions, it is crucial to obtain bounds relating the size of the groups $\mathrm{Aut}_M(E)$ and the dimensions $\dim_M(E)$. But for the cubic polynomial, these bounds can be bypassed by *ad hoc* arguments. *I have done this in order to reconstruct something that Galois would have known before he constructed a general theory; Galois would certainly have been intimately familiar with the cubic polynomial as well as with the quartic polynomial before coming to terms with the general case.*

Let us assume that $E \supseteq K$ is the splitting field of an irreducible cubic polynomial in $K[x]$. For each subgroup H of $\mathrm{Aut}_K(E)$, consider $\mathrm{Fix}(H) = \{a \in E : \sigma(a) = a \text{ for all } \sigma \in H\}$. $\mathrm{Fix}(H)$ is always a field intermediate between K and E (exercise). Furthermore, $\mathrm{Fix}(\mathrm{Aut}_K(E)) = K$ (exercise).

In case $\dim_K(E) = 3$, there are no proper intermediate fields between K and E because of multiplicativity of dimensions, and there are no nontrivial subgroups of $\mathrm{Aut}_K(E) \cong \mathbb{Z}_3$. Let us now consider the case $\dim_K(E) = 6$.

Recall that the subgroups of $\mathrm{Aut}_K(E) \cong S_3$ are

- S_3 itself,
- the three copies of S_2, $H_i = \{\sigma \in S_3 : \sigma(i) = i\}$,
- the cyclic group A_3, and

- the trivial subgroup $\{e\}$.

Now, suppose $K \subseteq M \subseteq E$ is some intermediate field. Let

$$H = \mathrm{Aut}_M(E) \subseteq \mathrm{Aut}_K(E) \cong S_3.$$

Then H must be one of the subgroups listed above. Put $\bar{M} = \mathrm{Fix}(H)$; then $M \subseteq \bar{M}$. I want to show that M must be one of the "known" intermediate fields. The first step is the following proposition:

Proposition 8.4.6. *If $K \subseteq M \underset{\neq}{\subset} E$, then* $\mathrm{Aut}_M(E) \neq \{e\}$.

Proof. It is no loss of generality to assume $K \neq M$. Because $\dim_M(E)$ divides $\dim_K(E) = 6$, it follows that $\dim_M(E)$ is either 2 or 3. Let $a \in E \setminus M$. Then $E = M(a)$; there are no more intermediate fields because of the multiplicativity of dimension.

Consider the polynomial

$$g(x) = \prod_{\sigma \in \mathrm{Aut}_K(E)} (x - \sigma(a)).$$

This polynomial has coefficients in E which are invariant under $\mathrm{Aut}_K(E)$; because $\mathrm{Fix}(\mathrm{Aut}_K(E)) = K$, *the coefficients are in* K. Now we can regard $g(x)$ as an element of $M[x]$; since $g(a) = 0$, $g(x)$ has an irreducible factor $h(x) \in M[x]$ such that $h(a) = 0$. Because $g(x)$ splits into linear factors in $E[x]$, so does $h(x)$. Thus E is a splitting field for $h(x)$. Since $\deg(h) = \dim_M(E) \geq 2$, and since the roots of $h(x)$ are distinct by Exercise 8.4.1, $h(x)$ has at least one root in E other than a and, by Proposition 8.4.2, there is an M-automorphism of E which takes a to b. Therefore, $\mathrm{Aut}_M(E) \neq \{e\}$. □

Proposition 8.4.7. *Let $K \subseteq M \subseteq E$ be an intermediate field. Let $H = \mathrm{Aut}_M(E)$ and let $\bar{M} = \mathrm{Fix}(H)$.*

(a) *If H is one of the H_i then $M = \bar{M} = K(\alpha_i)$.*
(b) *If $H = A_3$, then $M = \bar{M} = K(\delta)$.*
(c) *If $H = \mathrm{Aut}_K(E)$, then $M = K$.*
(d) *If $H = \{e\}$, then $M = E$ by the previous Proposition.*

Proof. Exercise 8.4.12. □

We have proved the following:

Theorem 8.4.8. *Let K be a subfield of $\mathbb{C}$, let $f(x) \in K[x]$ be an irreducible cubic polynomial, and let E be the splitting field of $f(x)$ in $\mathbb{C}$. Then there is a bijection between subgroups of* $\mathrm{Aut}_K(E)$ *and intermediate fields* $K \subseteq M \subseteq E$. *Under the bijection, a subgroup H corresponds to the intermediate field* $\mathrm{Fix}(H)$, *and an intermediate field M corresponds to the subgroup* $\mathrm{Aut}_M(E)$.

This is a remarkable result, even for the cubic polynomial. The splitting field E is an infinite set. For any subset $S \subseteq E$, we can form the intermediate field $K(S)$. It is certainly not evident that there are at most six possibilities for $K(S)$ and, without introducing symmetries, this fact would remain obscure.

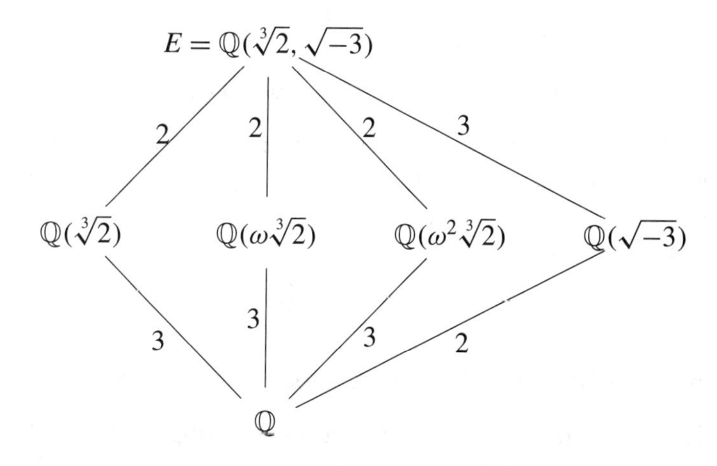

Figure 8.4.1. Intermediate fields for the splitting field of $f(x) = x^3 - 2$ over $\mathbb{Q}$

Example 8.4.9. Consider $f(x) = x^3 - 2$, which is irreducible over $\mathbb{Q}$. The three roots of f in $\mathbb{C}$ are $\sqrt[3]{2}$, $\omega\sqrt[3]{2}$, and $\omega^2\sqrt[3]{2}$, where $\omega = -1/2 + \sqrt{-3}/2$ is a primitive cube root of 1. Let E denote the splitting field of f in $\mathbb{C}$. The discriminant of f is $\delta^2 = -108$, which has square root $\delta = 6\sqrt{-3}$. It follows that the Galois group $G = \mathrm{Aut}_{\mathbb{Q}}(E)$ is of order 6, and $\dim_{\mathbb{Q}}(E) = 6$. Each of the fields $\mathbb{Q}(\alpha)$, where α is one of the roots of f, is a cubic extension of $\mathbb{Q}$, and is the fixed field of the order 2 subgroup of G which exchanges the other two roots. The only other intermediate field between $\mathbb{Q}$ and E is $\mathbb{Q}(\delta) = \mathbb{Q}(\omega) = \mathbb{Q}(\sqrt{-3})$, which is the fixed field of the alternating group $A_3 \cong \mathbb{Z}_3$. A diagram of intermediate fields, with the dimensions of the field extensions indicated is shown as Figure 8.4.1.

Exercises

Exercise 8.4.1. Let $K \subseteq \mathbb{C}$ be a field.

(a) Suppose $f(x) \in K[x]$ is an irreducible quadratic polynomial. Show that the two roots of $f(x)$ in $\mathbb{C}$ are distinct.

(b) Suppose $f(x) \in K[x]$ is an irreducible cubic polynomial. Show that the three roots of $f(x)$ in $\mathbb{C}$ are distinct. *Hint:* One can assume without loss of generality that $f(x) = x^3 + px + q$ has no quadratic term. If f has a double root $\alpha_1 = \alpha_2 = \alpha$, then the third root is $\alpha_3 = -2\alpha$. Now, observe that the relation

$$\sum_{i<j} \alpha_i \alpha_j = p$$

shows that α satisfies a quadratic polynomial over K.

Exercise 8.4.2. Expand the expression $\delta = (\alpha_1 - \alpha_2)(\alpha_1 - \alpha_3)(\alpha_2 - \alpha_3)$, and reduce the result using equations (8.4.1) - (8.4.3) to eliminate α_3 and to reduce higher powers of α_1 and α_2. Show that $\delta = 6\alpha_1^2\alpha_2 - 2\alpha_1 p + 2\alpha_2 p - 3q$. Solve for α_2 to get:

$$\alpha_2 = \frac{\delta + 2\alpha_1 p + 3q}{2\left(3\alpha_1{}^2 + p\right)}.$$

Explain why the denominator in this expression is not zero.

Exercise 8.4.3. (a) Confirm the details of Example 8.4.9.

(b) Consider the irreducible polynomial $x^3 - 3x + 1$ in $\mathbb{Q}[x]$. Show that the dimension over $\mathbb{Q}$ of the splitting field is 3. Conclude that there are no fields intermediate between $\mathbb{Q}$ and the splitting field.

Exercise 8.4.4.

(a) If $\dim_K(E) = 3$, then there are no intermediate fields between K and E.

(b) Suppose that $\dim_K(E) = 6$. Show that $K(\alpha_1)$, $K(\alpha_1)$, $K(\alpha_1)$, and $K(\delta)$ are distinct intermediate fields between K and E.

Exercise 8.4.5. Let $f(x)$ be an irreducible cubic polynomial over a subfield K of $\mathbb{C}$. Let α_1, α_2, α_3 denote the three roots of f in $\mathbb{C}$ and let E denote the splitting field $E = K(\alpha_1, \alpha_2, \alpha_3)$. Let δ denote the square root of the discriminant of f, and suppose that $\delta \notin K$. Show that the lattice of intermediate fields between K and E is as shown in Figure 8.4.2.

Exercise 8.4.6. Show that $\mathrm{Aut}_F(L)$ is a subgroup of $\mathrm{Aut}(L)$.

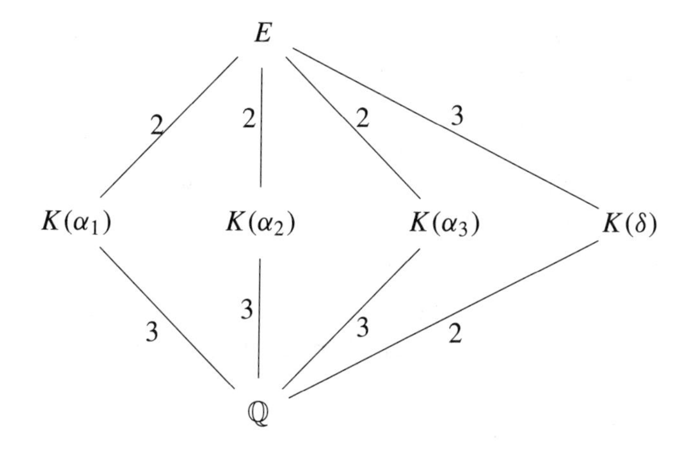

Figure 8.4.2. Intermediate fields for a splitting field with Galois group S_3

Exercise 8.4.7. Suppose $F \subseteq L$ is any field extension, $f(x) \in F[x]$, and $\beta_1, \dots, \beta_r$ are the distinct roots of $f(x)$ in L.

(a) If σ is an automorphism of L which leaves F fixed pointwise, then $\sigma_{|\{\beta_1,\dots,\beta_r\}}$ is a permutation of $\{\beta_1, \dots, \beta_r\}$.

(b) $\sigma \mapsto \sigma_{|\{\beta_1,\dots,\beta_r\}}$ is a homomorphism of $\mathrm{Aut}_F(L)$ into the group of permutations $\mathrm{Sym}(\{\beta_1, \dots, \beta_r\})$.

(c) If L is a splitting field of $f(x)$, $L = K(\beta_1, \dots, \beta_r)$, then the homomorphism $\sigma \mapsto \sigma_{|\{\beta_1,\dots,\beta_r\}}$ is injective.

Exercise 8.4.8. Show that $\mathrm{Fix}(H)$ is always a field intermediate between K and E.

Exercise 8.4.9. Let $f(x) \in K[x]$ be an irreducible cubic polynomial, with splitting field E. Show that $\mathrm{Fix}(\mathrm{Aut}_K(E)) = K$.

Exercise 8.4.10. Let $K \subseteq E$ be the splitting field of an irreducible cubic polynomial in $K[x]$ and that $\dim_K(E) = 6$.

(a) Evidently, $\mathrm{Fix}(\{e\}) = E$. Observe that

$$E \underset{\neq}{\supset} \mathrm{Fix}(H_i) \supseteq K(\alpha_i).$$

Conclude $\mathrm{Fix}(H_i) = K(\alpha_i)$.

(b) Show similarly that $\mathrm{Fix}(A_3) = K(\delta)$.

Exercise 8.4.11.

(a) Show that if M is any of the fields $K(\alpha_i)$, then $\mathrm{Aut}_M(E) = H_i$, and $\mathrm{Fix}(\mathrm{Aut}_M(E)) = M$.

(b) Show that if M is $K(\delta)$, then $\mathrm{Aut}_M(E) = A_3$, and $\mathrm{Fix}(\mathrm{Aut}_M(E)) = M$. *Hint:* Replace K by $K(\delta)$, and use the case $\delta \in K$, which is already finished.

(c) It is trivial that $\mathrm{Fix}(\mathrm{Aut}_E(E)) = E$, and it was shown in the Exercise 8.4.9 that $\mathrm{Fix}(\mathrm{Aut}_K(E)) = K$. Thus $\mathrm{Fix}(\mathrm{Aut}_M(E)) = M$ holds for all the "known" intermediate fields $K \subseteq M \subseteq E$.

Exercise 8.4.12. Prove Proposition 8.4.7.

Exercise 8.4.13. For each of the following polynomials over $\mathbb{Q}$, find the splitting field E and all fields intermediate between $\mathbb{Q}$ and E.

(a) $x^3 + x^2 + x + 1$
(b) $x^3 - 3x^2 + 3$
(c) $x^3 - 3$

8.5. Splitting Fields of Polynomials in $\mathbb{C}[x]$

Our goal in this section will be to state a generalization of Theorem 8.4.8 for arbitrary polynomials in $K[x]$ for K a subfield of $\mathbb{C}$, and to sketch some ideas involved in the proof. This material will be treated systematically and in a more general context in the next chapter.

Let K be any subfield of $\mathbb{C}$, and let $f(x) \in K[x]$. According to Gauss' Fundamental Theorem of Algebra, $f(x)$ factors into linear factors in $\mathbb{C}[x]$. The smallest subfield of $\mathbb{C}$ which contains all the roots of $f(x)$ in $\mathbb{C}$ is called the *splitting field* of $f(x)$. As for the cubic polynomial, in order to understand the structure of the splitting field, it is useful to introduce its symmetries over K. An automorphism of E is said to be a K-*automorphism* if it fixes every point of K. The set of K-automorphisms forms a group (Exercise 8.4.6).

Here is the statement of the main theorem concerning subfields of a splitting field and the symmetries of a splitting field:

Theorem 8.5.1. *Suppose K is a subfield of $\mathbb{C}$, $f(x) \in K[x]$, and E is the splitting field of $f(x)$ in $\mathbb{C}$.*

(a) $\dim_K(E) = \#\mathrm{Aut}_K(E)$.

(b) *The map $M \mapsto \mathrm{Aut}_M(E)$ is a bijection between the set of intermediate fields $K \subseteq M \subseteq E$ and the set of subgroups of $\mathrm{Aut}_K(E)$. The inverse map is $H \mapsto \mathrm{Fix}(H)$. In particular, there are only finitely many intermediate fields between K and E.*

This theorem asserts, in particular, that for any intermediate field $K \subseteq M \subseteq E$, there are sufficiently many M-automorphisms of E so that

$\text{Fix}(\text{Aut}_M(E)) = M$. So the first task in proving the theorem is to see that there is an abundance of M-automorphisms of E. We will maintain the notation: K is a subfield of $\mathbb{C}$, $f(x) \in K[x]$, and E is the splitting field of $f(x)$ in $\mathbb{C}$.

Lemma 8.5.2. *Suppose $\sigma : M \to M'$ is an isomorphism of fields. Then σ extends to an isomorphism of rings from $M[x]$ to $M'[x]$, defined by*

$$\sigma(\sum m_i x^i) = \sum \sigma(m_i) x^i$$

Proof. Exercise 8.5.1. □

Proposition 8.5.3. *Suppose $p(x) \in K[x]$ is an irreducible factor of $f(x)$ and α and α' are two roots of $p(x)$ in E. Then there is a $\tau \in \text{Aut}_K(E)$ such that $\tau(\alpha) = \alpha'$.*

Sketch of proof. By Proposition 8.4.2, there is an isomorphism $\sigma : K(\alpha) \to K(\alpha')$ which fixes K pointwise and sends α to α'. Using an inductive argument based on the fact that E is a splitting field and a variation on the theme of 8.4.2, one shows that the isomorphism σ can be extended to an automorphism of E. This gives an automorphism of E taking α to α' and fixing K pointwise. □

Corollary 8.5.4. $\text{Aut}_K(E)$ *acts faithfully by permutations on the roots of $f(x)$ in E. The action is transitive on the roots of each irreducible factor of $f(x)$.*

Proof. By Exercise 8.4.7, $\text{Aut}_K(E)$ acts faithfully by permutations on the roots of $f(x)$ and, by the previous corollary, this action is transitive on the roots of each irreducible factor. □

Theorem 8.5.5. *Suppose that $K \subseteq \mathbb{C}$ is a field, $f(x) \in K[x]$, and E is the splitting field of $f(x)$ in $\mathbb{C}$. Then* $\text{Fix}(\text{Aut}_K(E)) = K$.

Sketch of proof. One has *a priori* that $K \subseteq \text{Fix}(\text{Aut}_K(E))$. One has to show that if $a \in L \setminus K$, then there is an automorphism of E which leaves K fixed pointwise but does not fix a. □

Corollary 8.5.6. *If $K \subseteq M \subseteq E$ is any intermediate field, then* $\text{Fix}(\text{Aut}_M(E)) = M$.

Proof. E is also the splitting field of f over M, so the previous result applies with K replaced by M. □

Now, we consider a converse:

Proposition 8.5.7. *Suppose $K \subseteq E \subseteq \mathbb{C}$ are fields,* $\dim_K(E)$ *is finite, and* $\text{Fix}(\text{Aut}_K(E)) = K$.

(a) *For any $\beta \in E$, β is algebraic over K, and the minimal polynomial for β over K splits in $E[x]$.*

(b) *For $\beta \in E$, let $\beta = \beta_1, \dots, \beta_r$ be a list of the distinct elements of $\{\sigma(\beta) : \sigma \in \text{Aut}_K(E)\}$. Then $(x - \beta_1)(\dots)(x - \beta_r)$ is the minimal polynomial for β over K.*

(c) *E is the splitting field of a polynomial in $K[x]$.*

Proof. Since $\dim_K(E)$ is finite, E is algebraic over K.

Let $\beta \in E$, and let $p(x)$ denote the minimal polynomial of β over K. Let $\beta = \beta_1, \dots, \beta_r$ be the distinct elements of $\{\sigma(\beta) : \sigma \in \text{Aut}_K(E)\}$. Define $g(x) = (x - \beta_1)(\dots)(x - \beta_r) \in E[x]$. Every $\sigma \in \text{Aut}_K(L)$ leaves $g(x)$ invariant, so the coefficients of $g(x)$ lie in $\text{Fix}(\text{Aut}_K(E)) = K$. Since β is a root of $g(x)$, it follows that $p(x)$ divides $g(x)$. On the other hand, every root of $g(x)$ is of the form $\sigma(\beta)$ for $\sigma \in \text{Aut}_K(L)$ and therefore is also a root of $p(x)$. Since the roots of $g(x)$ are simple (i.e., each root α occurs only once in the factorization $g(x) = \prod(x - \alpha)$,) it follows that $g(x)$ divides $p(x)$. Hence $p(x) = g(x)$. In particular, $p(x)$ splits into linear factors over E. This proves parts (a) and (b).

Since E is finite dimensional over K, it is generated over K by finitely many algebraic elements $\alpha_1, \dots, \alpha_s$. It follows from part (a) that E is the splitting field of $f = f_1 f_2 \cdots f_s$, where f_i is the minimal polynomial of α_i over K. □

Definition 8.5.8. A finite dimensional field extension $K \subseteq E \subseteq \mathbb{C}$ is said to be *Galois* if $\text{Fix}(\text{Aut}_K(E)) = K$.

With this terminology, the previous results say:

Theorem 8.5.9. *For fields $K \subseteq E \subseteq \mathbb{C}$, with* $\dim_K(E)$ *finite, the following are equivalent:*

(a) *The extension E is Galois over K.*

(b) *For all $\alpha \in E$, the minimal polynomial of α over K splits into linear factors over E.*

(c) *E is the splitting field of a polynomial in $K[x]$.*

Corollary 8.5.10. *If $K \subseteq E \subseteq \mathbb{C}$ and E is Galois over K, then E is Galois over every intermediate field $K \subseteq M \subseteq E$.*

Thus far we have sketched "half" of Theorem 8.5.1, namely, the map from intermediate fields M to subgroups of $\text{Aut}_K(E)$, $M \mapsto \text{Aut}_M(E)$, is injective, since $M = \text{Fix}(\text{Aut}_M(E))$. It remains to show that this map is surjective. The key to this result is the equality of the order of subgroup with the dimension of E over its fixed field: If H is a subgroup of $\text{Aut}_K(E)$ and $F = \text{Fix}(H)$, then

$$\dim_F(E) = \#H. \tag{8.5.1}$$

The details of the proof of the equality (8.5.1) will be given in Section 9.5, in a more general setting.

Now, consider a subgroup H of $\text{Aut}_K(E)$, let F be its fixed field, and let $\bar{H}$ be $\text{Aut}_F(E)$. Then we have $H \subseteq \bar{H}$, and

$$\text{Fix}(\bar{H}) = \text{Fix}(\text{Aut}_F(E)) = F = \text{Fix}(H).$$

By the equality (8.5.1) $\#\bar{H} = \dim_F(E) = \#H$, so $H = \bar{H}$. This shows that the map $M \mapsto \text{Aut}_M(E)$ has as its range all subgroups of $\text{Aut}_K(E)$. This completes the sketch of the proof of 8.5.1. Theorem 8.5.1 is known as the Fundamental Theorem of Galois theory.

Example 8.5.11. The field $E = \mathbb{Q}(\sqrt{2}, \sqrt{3})$ is the splitting field of the polynomial $f(x) = (x^2 - 2)(x^2 - 3)$, whose roots are $\pm\sqrt{2}, \pm\sqrt{3}$. The Galois group $G = \text{Aut}_{\mathbb{Q}}(E)$ is isomorphic to $\mathbb{Z}_2 \times \mathbb{Z}_2$; $G = \{e, \alpha, \beta, \alpha\beta\}$, where α sends $\sqrt{2}$ to its opposite and fixes $\sqrt{3}$, and β sends $\sqrt{3}$ to its opposite and fixes $\sqrt{2}$. The subgroups of G are three copies of $\mathbb{Z}_2$ generated by α, β, and $\alpha\beta$. Therefore there are also exactly three intermediate fields between $\mathbb{Q}$ and E, which are the fixed fields of α, β, and $\alpha\beta$. The fixed field of α is $\mathbb{Q}(\sqrt{3})$, the fixed field of β is $\mathbb{Q}(\sqrt{2})$, and the fixed field of $\alpha\beta$ is $Q(\sqrt{6})$. Figure 8.5.3 shows the lattice of intermediate fields, with the dimensions of the extensions indicated.

There is a second part of the Fundamental Theorem which describes the special role of *normal* subgroups of $\text{Aut}_K(E)$.

Theorem 8.5.12. *Suppose K is a subfield of $\mathbb{C}$, $f(x) \in K[x]$, and E is the splitting field of $f(x)$ in $\mathbb{C}$. A subgroup N of* $\text{Aut}_K(E)$ *is normal if, and only if, its fixed field* $\text{Fix}(N)$ *is Galois over K. In this case,* $\text{Aut}_K(\text{Fix}(N)) \cong \text{Aut}_K(E)/N$.

This result is also proved in a more general setting in Section 9.5.

Example 8.5.13. In example 8.5.11, all the field extensions are Galois, since the Galois group is abelian.

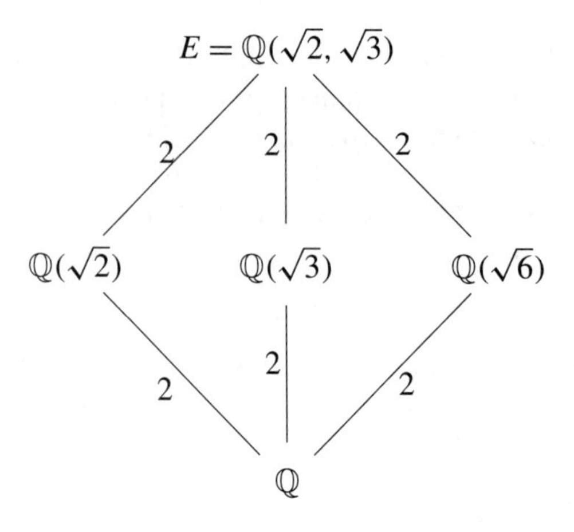

Figure 8.5.3. Lattice of intermediate fields for $\mathbb{Q} \subseteq \mathbb{Q}(\sqrt{2}, \sqrt{3})$

Example 8.5.14. In Example 8.4.9, the subfield $\mathbb{Q}(\delta)$ is Galois; it is the splitting field of a quadratic polynomial, and the fixed field of the normal subgroup A_3 of the Galois group S_3.

Example 8.5.15. Consider $f(x) = x^4 - 2$, which is irreducible over $\mathbb{Q}$ by the Eisenstein criterion. The roots of f in $\mathbb{C}$ are $\pm\sqrt[4]{2}$, $\pm i\sqrt[4]{2}$. Let E denote the splitting field of f in $\mathbb{C}$; evidently $E = \mathbb{Q}(\sqrt[4]{2}, i)$.

The intermediate field $\mathbb{Q}(\sqrt[4]{2})$ is of degree 4 over $\mathbb{Q}$ and $E = \mathbb{Q}(\sqrt[4]{2}, i)$ is of degree 2 over $\mathbb{Q}(\sqrt[4]{2})$, so E is of degree 8 over $\mathbb{Q}$, using Proposition 8.3.2. Therefore, it follows from the equality of dimensions in the Galois correspondence that the Galois group $G = \mathrm{Aut}_{\mathbb{Q}}(E)$ is of order 8.

Since E is generated as a field over $\mathbb{Q}$ by $\sqrt[4]{2}$ and i, a $\mathbb{Q}$-automorphism of E is determined by its action on these two elements. Furthermore, for any automorphism σ of E over $\mathbb{Q}$, $\sigma(\sqrt[4]{2})$ must be one of the roots of f, and $\sigma(i)$ must be one of the roots of $x^2 + 1$, namely $\pm i$. There are exactly eight possibilities for the images of $\sqrt[4]{2}$ and i, namely

$$\sqrt[4]{2} \mapsto i^r\sqrt[4]{2} \quad (0 \le r \le 3), \quad i \mapsto \pm i.$$

As the size of the Galois group is also 8, each of these assignments must determine an element of the Galois group.

In particular, we single out two $\mathbb{Q}$-automorphisms of E:

$$\sigma : \sqrt[4]{2} \mapsto i\sqrt[4]{2}, \quad i \mapsto i,$$

and

$$\tau : \sqrt[4]{2} \mapsto \sqrt[4]{2}, \quad i \mapsto -i.$$

The automorphism τ is complex conjugation restricted to E.

Evidently, σ is of order 4, and τ is of order 2. Furthermore, one can compute that $\tau\sigma\tau = \sigma^{-1}$. It follows that the Galois group is generated by σ and τ, and is isomorphic to the dihedral group D_4. You are asked to check this in the exercises.

We identify the Galois group D_4 as a subgroup of S_4, acting on the roots $\alpha_r = i^{r-1}\sqrt[4]{2}$, $1 \le r \le 4$. With this identification, $\sigma = (1234)$ and $\tau = (24)$.

D_4 has 10 subgroups (including D_4 and $\{e\}$). The lattice of subgroups is show in Figure 8.5.4; all of the inclusions in this diagram are of index 2.

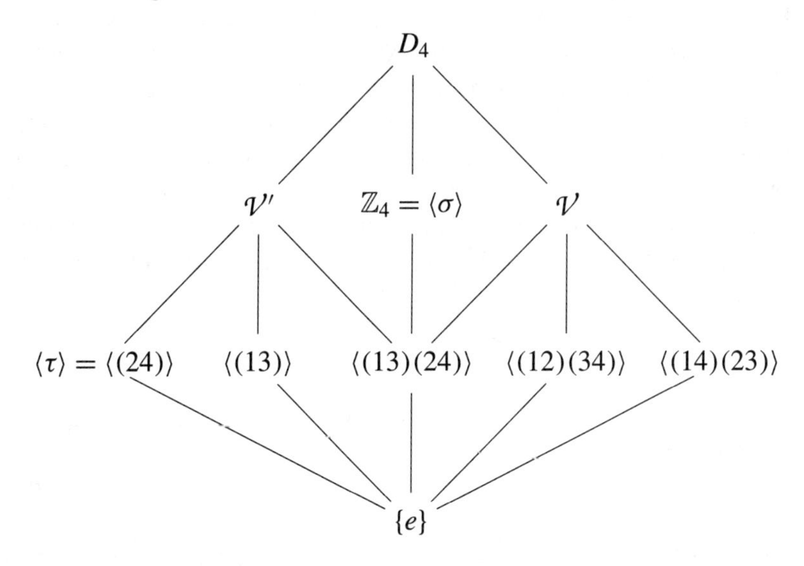

Figure 8.5.4. Lattice of subgroups of D_4

Here $\mathcal{V}$ denotes the group

$$\mathcal{V} = \{e, (12)(34), (13)(24), (14)(23)\},$$

and $\mathcal{V}'$ the group

$$\mathcal{V}' = \{e, (24), (13), (13)(24) = \sigma^2\}.$$

By the Galois correspondence, there are ten intermediate fields between $\mathbb{Q}$ and E, including the the ground field and the splitting field. Each intermediate field is the fixed field of a subgroup of G. The three subgroups of D_4 of index 2 ($\mathcal{V}$, $\mathcal{V}'$, and $\mathbb{Z}_4$) are normal, so the corresponding intermediate fields are Galois over the ground field $\mathbb{Q}$.

It is possible to determine each fixed field rather explicitly. In order to do so, one can find one or more elements which are fixed by the subgroup, and which generate a field of the proper dimension. For example, the fixed field of $\mathcal{V}$ is $\mathbb{Q}(\sqrt{-2})$. You are asked in the exercises to identify all of the intermediate fields as explicitly as possible.

Finally, I want to mention one more useful result, whose proof is also deferred to Section 9.5.

Theorem 8.5.16. *If $K \subseteq F \subseteq \mathbb{C}$ are fields and $\dim_K(F)$ is finite, then there is an $\alpha \in F$ such that $F = K(\alpha)$.*

It is rather easy to show that a finite dimensional field extension is algebraic; that is, every element in the extension field is algebraic over the ground field K. Hence the extension field is certainly obtained by adjoining a finite number of algebraic elements. The tricky part is to show that if a field is obtained by adjoining two algebraic elements, then it can also be obtained by adjoining one algebraic element, $K(\alpha, \beta) = K(\gamma)$ for some γ. Then, it follows by induction that a field obtained by adjoining finitely many algebraic elements can also be obtained by adjoining a single algebraic element.

Exercises

Exercise 8.5.1. Suppose $\sigma : M \to M'$ is an isomorphism of fields. Then σ extends to an isomorphism of rings from $M[x]$ to $M'[x]$, defined by

$$\sigma(\sum m_i x^i) = \sum \sigma(m_i) x^i$$

Exercise 8.5.2. Check the details of Example 8.5.11.

Exercise 8.5.3. In Example 8.5.15, check that the $\mathbb{Q}$-automorphisms σ and τ generate the Galois group, and that the Galois group is isomorphic to D_4.

Exercise 8.5.4. Verify that the diagram of subgroups of D_4 in Example 8.5.15 is correct.

Exercise 8.5.5. Identify the fixed fields of the subgroups of the Galois group in Example 8.5.15. According to the Galois correspondence, these fixed fields are all the intermediate fields between $\mathbb{Q}$ and $\mathbb{Q}(\sqrt[4]{2}, i)$.

Exercise 8.5.6. Let p be a prime. Show that the splitting field of $x^p - 1$ over $\mathbb{Q}$ is $E = \mathbb{Q}(\zeta)$ where ζ is a primitive p-th root of unity in $\mathbb{C}$. Show that $\dim_{\mathbb{Q}}(E) = p - 1$ and that the Galois group is cyclic of order $p - 1$. For $p = 7$, analyze the fields intermediate between $\mathbb{Q}$ and E.

Exercise 8.5.7. Verify that the following field extensions of $\mathbb{Q}$ are Galois. Find a polynomial for which the field is the splitting field. Find the Galois group, the lattice of subgroups, and the lattice of intermediate fields.

(a) $\mathbb{Q}(\sqrt{2}, \sqrt{7})$
(b) $\mathbb{Q}(i, \sqrt{7})$
(c) $\mathbb{Q}(\sqrt{2} + \sqrt{7})$
(d) $\mathbb{Q}(e^{2\pi i/7})$
(e) $\mathbb{Q}(e^{2\pi i/6})$

Exercise 8.5.8. The Galois group G and splitting field E of $f(x) = x^8 - 2$ can be analyzed in a manner quite similar to Example 8.5.15. The Galois group G has order 16 and has a (normal) cyclic subgroup of order 8; however, G is not the dihedral group D_8. Describe the group by generators and relations (2 generators, one relation). Find the lattice of subgroups. G has a normal subgroup isomorphic to D_4; in fact the splitting field E contains the splitting field of $x^4 - 2$. The subgroups of D_4 already account for a large number of the subgroups of G. Describe as explicitly as possible the lattice of intermediate fields between $\mathbb{Q}$ and the splitting field.

Exercise 8.5.9. Suppose E is a Galois extension of a field K (both fields assumed, for now, to be contained in $\mathbb{C}$). Suppose that the Galois group $\text{Aut}_K(E)$ is abelian, and that an irreducible polynomial $f(x) \in K[x]$ has one root $\alpha \in E$. Show that $K(\alpha)$ is then the splitting field of f. Show by examples that this is not true if the Galois group is not abelian or if the polynomial f is not irreducible.

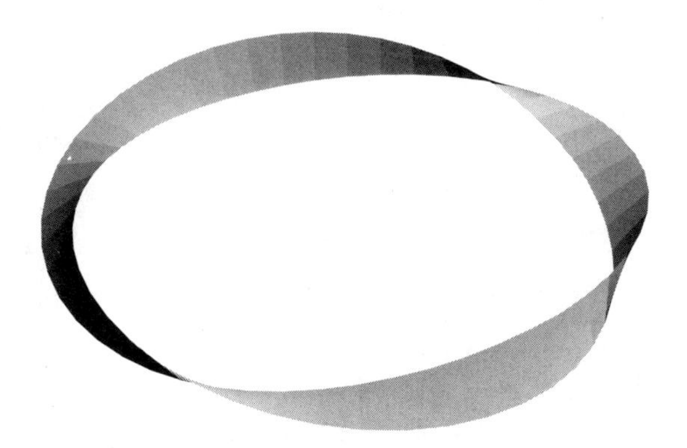

Part II: Topics

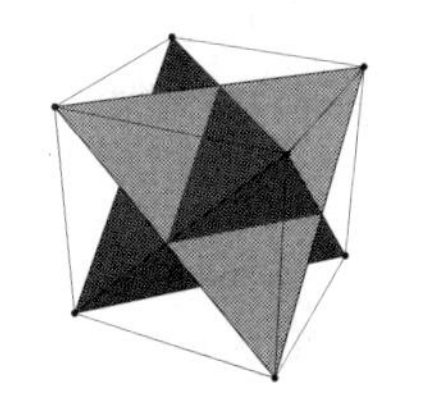

Chapter 9

Field Extensions – Second Look

This chapter contains a systematic introduction to Galois' theory of field extensions and symmetry. The fields in this chapter are general, not necessarily subfields of the complex numbers nor even of characteristic 0. We will restrict ours attention, however, to so called *separable algebraic* extensions.

9.1. Finite and Algebraic Extensions

In this section, we continue, by means of exercises, the exploration of finite and algebraic field extensions, which was begun in Section 8.3. Recall that a field extension $K \subseteq L$ is said to be *finite* if $\dim_K(L)$ is finite and is called *algebraic* in case each element of L satisfies a polynomial equation with coefficients in K.

Exercise 9.1.1.

(a) Let $K \subseteq L$ be a field extension. Suppose that $\alpha_1, \dots, \alpha_s$ are elements of L which are algebraic over K. Suppose β is an element of L which is algebraic over $K(\alpha_1, \dots, \alpha_s)$. Show that β is algebraic over K. *Hint:* Show that $\dim_K(K(\alpha_1, \dots, \alpha_s, \beta))$ is finite.

(b) Suppose that $K \subseteq M \subseteq L$ are field extensions, M is algebraic over K and $\beta \in L$ is algebraic over M. Show that β is algebraic over K. *Hint:* Find $\alpha_1, \dots, \alpha_s$ such that $\alpha_i \in M$ and β is algebraic over $K(\alpha_1, \dots, \alpha_s)$.

(c) Conclude that if M is algebraic over K, and L is algebraic over M, then L is algebraic over K.

Exercise 9.1.2. Let $K \subseteq L$ be a field extension. The set of elements of L which are algebraic over K form a subfield of L. In particular, the set of algebraic numbers (complex numbers which are algebraic over $\mathbb{Q}$) is a countable field. *Hint:* If α and β are algebraic over K, then $K(\alpha, \beta)$ is algebraic over K.

Definition 9.1.1. Suppose that $K \subseteq L$ is a field extension and that E and F are intermediate fields, $K \subseteq E \subseteq L$ and $K \subseteq F \subseteq L$. The *composite* $E \cdot F$ of E and F is the smallest subfield of L containing E and F.

$$\begin{array}{ccccc} F & \subseteq & E \cdot F & \subseteq & L \\ \cup\!| & & \cup\!| & & \\ K & \subseteq & E & & \end{array}$$

In the following exercises, E and F denote fields intermediate between K and L.

Exercise 9.1.3. If E and F are both algebraic over K, then $E \cdot F$ is algebraic over K.

Exercise 9.1.4. If E is algebraic over K and F is arbitrary, then $E \cdot F$ is algebraic over F. *Hint:* Let $a \in E \cdot F$. Then there exist $\alpha_1, \dots, \alpha_n \in E$ such that $a \in F(\alpha_1, \dots, \alpha_n)$.

Exercise 9.1.5. Show that $\dim_F(E \cdot F) \le \dim_K(E)$. *Hint:* In case $\dim_K(E)$ is infinite, there is nothing to be done. So assume the dimension is finite and let $\alpha_1, \dots, \alpha_n$ be a basis of E over K. Conclude successively that $E \cdot F = F(\alpha_1, \dots, \alpha_n)$, then that $E \cdot F = F[\alpha_1, \dots, \alpha_n]$, and finally that $E \cdot F = \mathrm{span}_F\{\alpha_1, \dots, \alpha_n\}$.

Exercise 9.1.6. What is the dimension of $\mathbb{Q}(\sqrt{2} + \sqrt{3})$ over $\mathbb{Q}$?

9.2. Splitting Fields

Now, we are going to turn our point of view around regarding algebraic extensions. Given a polynomial $f(x) \in K[x]$, we *produce* extension fields $K \subseteq L$ in which f has a root, or in which f has a complete set of roots.

Proposition 8.3.7 tells us that if $f(x)$ is irreducible in $K[x]$ and α is a root of $f(x)$ in some extension field, then the field generated by K and the root α is isomorphic to $K[x]/(f(x))$; but $K[x]/(f(x))$ is a field, so we might as well choose our extension field to be $K[x]/(f(x))$! What should α be then? It will have to be the image in $K[x]/(f(x))$ of x, namely $[x] = x + (f(x))$. (We are using $[g(x)]$ to denote the image of $g(x)$ in $K[x]/(f(x))$.) Indeed, $[x]$ is a root of $f(x)$ in $K[x]/(f(x))$, since $f([x]) = [f(x)] = 0$ in L.

Proposition 9.2.1. *Let K be a field and let $f(x)$ be a monic irreducible element of $K[x]$. Then*

(a) *There is an extension field L and an element $\alpha \in L$ such that $f(\alpha) = 0$.*

(b) *For any such extension field and any such α, $f(x)$ is the minimal polynomial for α.*

(c) *If L and L' are extension fields of K containing elements α and α' satisfying $f(\alpha) = 0$ and $f(\alpha') = 0$, then there is an isomorphism $\psi : K(\alpha) \to K(\alpha')$ such that $\psi(k) = k$ for all $k \in K$ and $\psi(\alpha) = \alpha'$.*

Proof. The existence of the field extension containing a root of $f(x)$ was already shown above. The minimal polynomial for α divides f and, therefore, equals f, since f is monic irreducible. For the third part, we have $K(\alpha) \cong K[x]/(f(x)) \cong K(\alpha')$, by isomorphisms which leave K pointwise fixed. □

We will need a technical variation of part (c) of the proposition. Recall from Exercise 8.5.1 that if M and M' are fields and $\sigma : M \to M'$ is a field isomorphism, then σ extends to an isomorphism of rings $M[x] \to M'[x]$ by $\sigma(\sum m_i x^i) = \sum \sigma(m_i) x^i$.

Corollary 9.2.2. *Let K and K' be fields, $\sigma : K \to K'$ be a field isomorphism, and $f(x)$ an irreducible element of $K[x]$. Suppose that α is a root of $f(x)$ in an extension $L \supseteq K$ and that α' is a root of $\sigma(f(x))$ in an extension field $L' \supseteq K'$. Then there is an isomorphism $\psi : K(\alpha) \to K'(\alpha')$ such that $\psi(k) = \sigma(k)$ for all $k \in K$ and $\psi(\alpha) = \alpha'$.*

Proof. The ring isomorphism $\sigma : K[x] \to K'[x]$ induces a field isomorphism $\tilde{\sigma} : K[x]/(f) \to K'[x]/(\sigma(f))$, satisfying $\tilde{\sigma}(k + (f)) = \sigma(k) + (\sigma(f))$ for $k \in K$. Now, use $K(\alpha) \cong K[x]/(f) \cong K'[x]/(\sigma(f)) \cong K'(\alpha')$. □

Now, consider a monic polynomial $f(x) \in K[x]$ of degree $d > 1$, not necessarily irreducible. Factor $f(x)$ into irreducible factors. If any of these factors have degree greater than 1, then choose such a factor and adjoin a root α_1 of this factor to K, as above. Now, regard $f(x)$ as a polynomial over the field $K(\alpha_1)$, and write it as a product of irreducible factors in $K(\alpha_1)[x]$. If any of these factors has degree greater than 1, then choose such a factor and adjoin a root α_2 of this factor to $K(\alpha_1)$. After repeating this procedure at most d times, one obtains a field in which $f(x)$ factors into linear factors. Of course, a proper proof goes by induction (exercise).

Definition 9.2.3. A *splitting field* for a polynomial $f(x) \in K[x]$ is an extension field L such that $f(x)$ factors into linear factors over L, and L is generated by K and the roots of $f(x)$ in L.

If a polynomial $p(x) \in K[x]$ factors into linear factors over a field $M \supseteq K$, and if $\{\alpha_1, \dots \alpha_r\}$ are the distinct roots of $p(x)$ in M, then $K(\alpha_1, \dots, \alpha_r)$, the subfield of M generated by K and $\{\alpha_1, \dots \alpha_r\}$, is a splitting field for $p(x)$. By Exercise 9.2.1, splitting fields always exist.

For polynomials over $\mathbb{Q}$, for example, it is unnecessary to refer to Exercise 9.2.1 for the existence of splitting fields; a rational polynomial splits into linear factors over $\mathbb{C}$, so a splitting field is obtained by adjoining the complex roots of the polynomial to $\mathbb{Q}$.

One consequence of the existence of splitting fields is the existence of many finite fields. See Exercises 9.2.4 and 9.2.5

The following result says that a splitting field is unique up to isomorphism.

Proposition 9.2.4. *Let $K \subseteq L$ and $\tilde{K} \subseteq \tilde{L}$ be field extensions, $\sigma : K \to \tilde{K}$ a field isomorphism, $p(x) \in K[x]$, $\tilde{p}(x) \in \tilde{K}[x]$ polynomials with $\tilde{p}(x) = \sigma(p(x))$. Suppose L is a splitting field for $p(x)$ and $\tilde{L}$ is a splitting field for $\tilde{p}(x)$. Then there is a field isomorphism $\tau : L \to \tilde{L}$ such that $\tau(k) = \sigma(k)$ for all $k \in K$.*

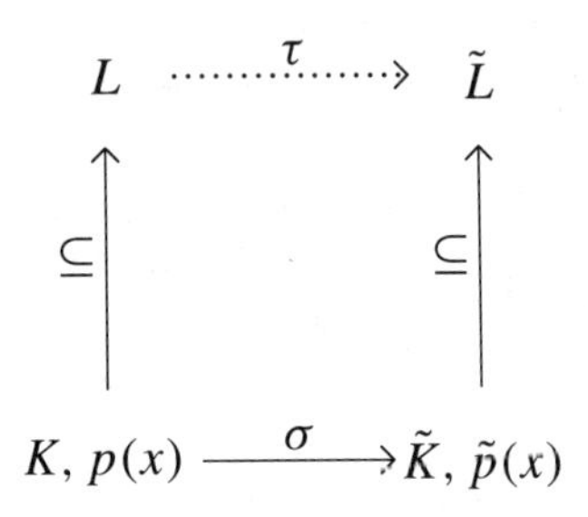

Proof. The idea is to use Proposition 9.2.2 and induction on $\dim_K(L)$. If $\dim_K(L) = 1$, there is nothing to do: The polynomial $p(x)$ factors into linear factors over K, so also $\tilde{p}(x)$ factors into linear factors over $\tilde{K}$, $K = L$, $\tilde{K} = \tilde{L}$, and σ is the required isomorphism.

We make the following induction assumption: Suppose $K \subseteq M \subseteq L$ and $\tilde{K} \subseteq \tilde{M} \subseteq \tilde{L}$ are intermediate field extensions, $\tilde{\sigma} : M \to \tilde{M}$ is a field isomorphism extending σ, and $\dim_M L < n = \dim_K(L)$. Then there is a field isomorphism $\tau : L \to \tilde{L}$ such that $\tau(m) = \tilde{\sigma}(m)$ for all $m \in M$.

Now, since $\dim_K(L) = n > 1$, at least one of the irreducible factors of $p(x)$ in $K[x]$ has degree greater than 1. Choose such an irreducible factor $p_1(x)$, and observe that $\sigma(p_1(x))$ is an irreducible factor of $p(x)$. Let $\alpha \in L$ and $\tilde{\alpha} \in \tilde{L}$ be roots of $p_1(x)$ and $\sigma(p_1(x))$ respectively. By Proposition 9.2.2, there is an isomorphism $\tilde{\sigma} : K(\alpha) \to \tilde{K}(\tilde{\alpha})$ taking α to $\tilde{\alpha}$ and extending σ.

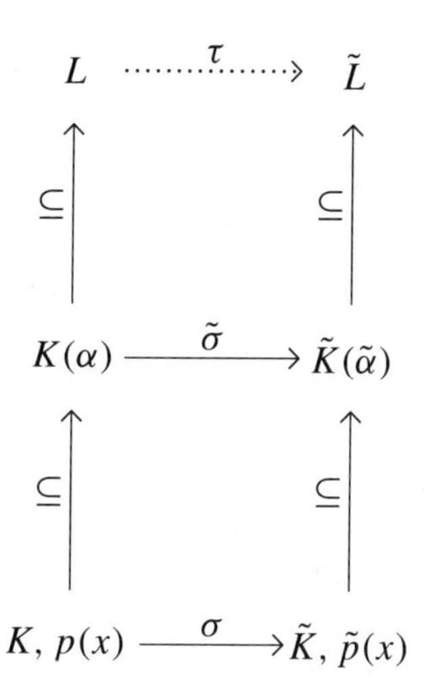

Now, the result follows by applying the induction hypothesis with $M = K(\alpha)$ and $\tilde{M} = \tilde{K}(\tilde{\alpha})$. □

Corollary 9.2.5. *Let $p(x) \in K[x]$, and suppose L and $\tilde{L}$ are two splitting fields for $p(x)$. Then there is an isomorphism $\tau : L \to \tilde{L}$ such that $\tau(k) = k$ for all $k \in K$.*

Proof. Take $K = \tilde{K}$ and $\sigma = \text{id}$ in the proposition. □

Exercises

Exercise 9.2.1. For any polynomial $f(x) \in K[x]$ there is an extension field L of K such that $f(x)$ factors into linear factors in $L[x]$. Give a proof by induction on the degree of f.

Exercise 9.2.2. Verify the following statements: The rational polynomial $f(x) = x^6 - 3$ is irreducible over $\mathbb{Q}$. It factors over $\mathbb{Q}(3^{1/6})$ as

$$(x - 3^{1/6})(x + 3^{1/6})(x^2 - 3^{1/6}x + 3^{1/3})(x^2 + 3^{1/6}x + 3^{1/3}).$$

If $\omega = e^{\pi i/3}$, then the irreducible factorization of $f(x)$ over $\mathbb{Q}(3^{1/6}, \omega)$ is:

$$(x - 3^{1/6})(x + 3^{1/6})(x - \omega 3^{1/6})(x - \omega^2 3^{1/6})(x - \omega^4 3^{1/6})(x - \omega^5 3^{1/6}).$$

Exercise 9.2.3. L is a splitting field for $f(x) \in K[x]$ if, and only if, $f(x)$ factors into linear factors over L, and $f(x)$ does not factor into linear factors over any intermediate field $K \subseteq M \subseteq L$.

Exercise 9.2.4. Show that Euclid's proof of the existence of infinitely many primes in the integers (Theorem C.3) also shows that for any field K, the polynomial ring $K[x]$ has infinitely many irreducible elements.

Exercise 9.2.5.

(a) Show that if K is a finite field, then $K[x]$ has irreducible elements of arbitrarily large degree.

(b) Show that if K is a finite field, then K admits field extensions of arbitrarily large finite degree.

(c) Show that a finite dimensional extension of a finite field is a finite field.

9.3. The Derivative and Multiple Roots

In this section, we examine, by means of exercises, multiple roots of polynomials and their relation to the formal derivative.

One says that a polynomial $f(x) \in K[x]$ has a root *a with multiplicity m* in an extension field L if $f(x) = (x - a)^m g(x)$ in $L[x]$, and $g(a) \neq 0$. A root of multiplicity greater than 1 is called a *multiple root*. A root of multiplicity 1 is called a *simple root*.

The formal derivative in $K[x]$ is defined by the usual rule from calculus: One defines $D(x^n) = nx^{n-1}$ and extends linearly. Thus, $D(\sum k_n x^n) = \sum nk_n x^{n-1}$. The formal derivative satisfies the usual rules for differentiation:

Exercise 9.3.1. Show that $D(f(x) + g(x)) = Df(x) + Dg(x)$ and $D(f(x)g(x)) = D(f(x))g(x) + f(x)D(g(x))$.

Exercise 9.3.2.

(a) Suppose that the field K is of characteristic zero. Show that $Df(x) = 0$ if, and only if, $f(x)$ is a constant polynomial.

(b) Suppose that the field has characteristic p. Show that $Df(x) = 0$ if, and only if, there is a polynomial $g(x)$ such that $f(x) = g(x^p)$.

Exercise 9.3.3. Suppose $f(x) \in K[x]$, L is an extension field of K, and $f(x)$ factors as $f(x) = (x - a)g(x)$ in $L[x]$. Show that the following are equivalent:

(a) a is a multiple root of $f(x)$.

(b) $g(a) = 0$.

(c) $Df(a) = 0$.

Exercise 9.3.4. Let $K \subseteq L$ be a field extension and let $f(x), g(x) \in K[x]$. Then the greatest common divisor of $f(x)$ and $g(x)$ in $L[x]$ is the same as the greatest common divisor in $K[x]$. *Hint:* Review the algorithm for computing the g.c.d., using division with remainder.

Exercise 9.3.5. Suppose $f(x) \in K[x]$, L is an extension field of K, and a is a multiple root of $f(x)$ in L. Show that if $Df(x)$ is not identically zero, then $f(x)$ and $Df(x)$ have a common factor of positive degree in $L[x]$ and, therefore, by the previous exercise, also in $K[x]$.

Exercise 9.3.6. Suppose that K is a field and $f(x) \in K[x]$ is irreducible.

(a) If f has a multiple root in some field extension, then $Df(x) = 0$.

(b) If $\mathrm{Char}(K) = 0$, then $f(x)$ has only simple roots in any field extension.

The exercises above establish the following theorem:

Theorem 9.3.1. *If the characteristic of a field K is zero, then any irreducible polynomial in $K[x]$ has only simple roots in any field extension.*

Exercise 9.3.7. If K is a field of characteristic p and $a \in K$, then $(x + a)^p = x^p + a^p$. *Hint:* The binomial coefficient $\binom{p}{k}$ is divisible by p if $0 < k < p$.

Now, suppose K is a field of characteristic p and that $f(x)$ is an irreducible polynomial in $K[x]$. If $f(x)$ has a multiple root in some extension field, then $Df(x)$ is identically zero, by Exercise 9.3.6. Therefore, there is a $g(x) \in K[x]$ such that $f(x) = g(x^p) = a_0 + a_1x^p + \dots a_rx^{rp}$. Suppose that for each a_i there is a $b_i \in K$ such that $b_i^p = a_i$. Then $f(x) =$

$(b_0 + b_1x + \cdots b_rx^r)^p$, which contradicts the irreducibility of $f(x)$. This proves the following theorem:

Theorem 9.3.2. *Suppose K is a field of characteristic p in which each element has a p-th root. Then any irreducible polynomial in $K[x]$ has only simple roots in any field extension.*

Proposition 9.3.3. *Suppose K is a field of characteristic p. The map $a \mapsto a^p$ is a field isomorphism of K into itself. If K is a finite field, then $a \mapsto a^p$ is an automorphism of K.*

Proof. Clearly $(ab)^p = a^pb^p$ for $a, b \in K$. But also $(a+b)^p = a^p + b^p$ by Exercise 9.3.7. Therefore, the map is a homomorphism. The homomorphism is not identically zero, since $1^p = 1$; since K is simple, the homomorphism must, therefore, be injective. If K is finite, an injective map is bijective. □

Corollary 9.3.4. *Suppose K is a finite field. Then any irreducible polynomial in $K[x]$ has only simple roots in any field extension.*

Proof. K must have some prime characteristic p. By Proposition 9.3.3, any element of K has a p-th root in K and, therefore, the result follows from Theorem 9.3.2. □

9.4. Splitting Fields and Automorphisms

Recall that an *automorphism* of a field L is a field isomorphism of L onto L, and that the set of all automorphisms of L forms a group denoted by $\text{Aut}(L)$. If $K \subseteq L$ is a field extension, we denote the set of automorphisms of L which leave each element of K fixed by $\text{Aut}_K(L)$; we call such automorphisms *K-automorphisms of L*. Recall from Exercise 8.4.6 that $\text{Aut}_K(L)$ is a subgroup of $\text{Aut}(L)$.

Proposition 9.4.1. *Let $f(x) \in K[x]$, let L be a splitting field for $f(x)$, let $p(x)$ be an irreducible factor of $f(x)$, and finally let α and β be two roots of $p(x)$ in L. Then there is an automorphism $\sigma \in \text{Aut}_K(L)$ such that $\sigma(\alpha) = \beta$.*

Proof. Using Proposition 9.2.1, we get an isomorphism from $K(\alpha)$ onto $K(\beta)$ which sends α to β and fixes K pointwise. Now, applying Corollary 9.2.5 to $K(\alpha) \subseteq L$ and $K(\beta) \subseteq L$ gives the result. □

Proposition 9.4.2. *Let L be a splitting field for $p(x) \in K[x]$, let M, M' be intermediate fields, $K \subseteq M \subseteq L$, $K \subseteq M' \subseteq L$, and let σ be an isomorphism of M onto M' which leaves K pointwise fixed. Then σ extends to a K-automorphism of L.*

Proof. This follows from applying Proposition 9.2.4 to:

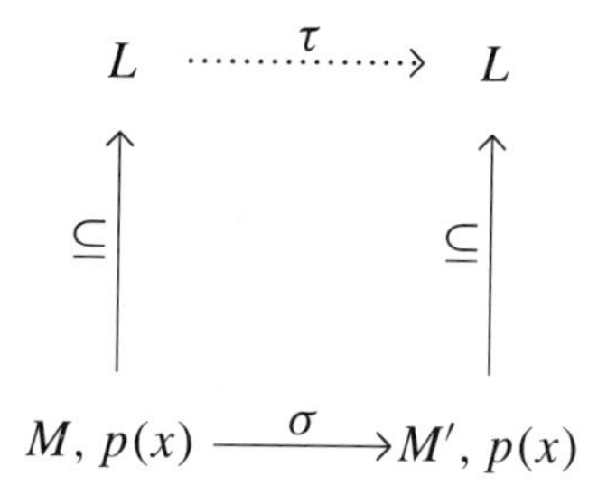

□

Corollary 9.4.3. *Let L be a splitting field for $p(x) \in K[x]$, let M be an intermediate field, $K \subseteq M \subseteq L$. Write $\mathrm{Iso}_K(M, L)$ for the set of field isomorphisms of M into L which leave K fixed pointwise.*

(a) *There is a bijection from the set of left cosets of $\mathrm{Aut}_M(L)$ in $\mathrm{Aut}_K(L)$ onto $\mathrm{Iso}_K(M, L)$.*

(b) $\#\mathrm{Iso}_K(M, L) = [\mathrm{Aut}_K(L) : \mathrm{Aut}_M(L)]$.

Proof. According to Proposition 9.4.2 the map $\tau \mapsto \tau_{|M}$ is a surjection of $\mathrm{Aut}_K(L)$ onto $\mathrm{Iso}_K(M, L)$. Check that $(\tau_1)_{|M} = (\tau_2)_{|M}$ if, and only if, τ_1 and τ_2 are in the same left coset of $\mathrm{Aut}_M(L)$ in $\mathrm{Aut}_K(L)$. This proves part (a), and part (b) follows. □

Proposition 9.4.4. *Let $K \subseteq L$ be a field extension and let $f(x) \in K[x]$.*

(a) *If $\sigma \in \mathrm{Aut}_K(L)$, then σ permutes the roots of $f(x)$ in L.*

(b) *If L is a splitting field of $f(x)$, then $\mathrm{Aut}_K(L)$ acts faithfully on the roots of f in L. Furthermore, the action is transitive on the roots of each irreducible factor of $f(x)$ in $K[x]$.*

Proof. Suppose $\sigma \in \mathrm{Aut}_K(L)$,

$$f(x) = k_0 + k_1 x + \cdots + k_n x^n \in K[x],$$

and α is a root of $f(x)$ in L. Then

$$\begin{aligned} f(\sigma(\alpha)) &= k_0 + k_1\sigma(\alpha) + \cdots + k_n\sigma(\alpha^n)) \\ &= \sigma(k_0 + k_1\alpha + \cdots + k_n\alpha^n) = 0. \end{aligned}$$

Thus, $\sigma(\alpha)$ is also a root of $f(x)$. If A is the set of distinct roots of $f(x)$ in L, then $\sigma \mapsto \sigma_{|A}$ is an action of $\mathrm{Aut}_K(L)$ on A. If L is a splitting field for $f(x)$, then in particular, $L = K(A)$, so the action of $\mathrm{Aut}_K(L)$ on A is faithful. Proposition 9.4.1 says that if L is a splitting field for $f(x)$, then the action is transitive on the roots of each irreducible factor of $f(x)$. □

Definition 9.4.5. If $f \in K[x]$, and L is a splitting field of $f(x)$, then $\mathrm{Aut}_K(L)$ is called the *Galois group of f*, or the *Galois group of the field extension $K \subseteq L$.*

We have seen that the Galois group of an irreducible polynomial f is isomorphic to a transitive subgroup of the group of permutations the roots of f in L. At least for small n, it is possible to classify the transitive subgroups of S_n, and thus to list the possible isomorphism classes for the Galois groups of irreducible polynomials of degree n. For $n = 3, 4$, and 5, we have found all transitive subgroups of S_n, in Exercises 5.1.9 and 5.1.20 and Section 5.4.

Let us quickly recall our investigation of splitting fields of irreducible cubic polynomials in Chapter 8, where we found the properties of the Galois group corresponded to properties of the splitting field. The only possibilities for the Galois group are $A_3 = \mathbb{Z}_3$ and S_3. The Galois group is A_3 if, and only if, the field extension $K \subseteq L$ is of dimension 3, and this occurs if, and only if, the element δ defined in Chapter 8 belongs to the ground field K; in this case there are no intermediate fields between K and L. The Galois group is S_3 if, and only if, the field extension $K \subseteq L$ is of dimension 6. In this case subgroups of the Galois group correspond one-to-one with fields intermediate between K and L.

We are aiming at obtaining similar results in general.

Definition 9.4.6. Let H be a subgroup of $\mathrm{Aut}(L)$. Then the fixed field of H is $\mathrm{Fix}(H) = \{a \in L : \sigma(a) = a \text{ for all } \sigma \in H\}$.

Proposition 9.4.7. *Let L be a field, H a subgroup of* $\mathrm{Aut}(L)$ *and $K \subset L$ a subfield. Then:*

(a) $\mathrm{Fix}(H)$ *is a subfield of L.*
(b) $\mathrm{Aut}_{\mathrm{Fix}(H)}(L) \supseteq H$.
(c) $\mathrm{Fix}(\mathrm{Aut}_K(L)) \supseteq K$.

Proof. Exercise. □

Proposition 9.4.8. *Let L be a field, H a subgroup of* $\mathrm{Aut}(L)$ *and $K \subset L$ a subfield. Introduce the notation $H^\circ = \mathrm{Fix}(H)$ and $K' = \mathrm{Aut}_K(L)$. The previous exercise showed that $H^{\circ\prime} \supseteq H$ and $L'^\circ \supseteq L$.*

(a) *If $H_1 \subseteq H_2 \subseteq \mathrm{Aut}(L)$ are subgroups, then $H_1^\circ \supseteq H_2^\circ$.*
(b) *If $K_1 \subseteq K_2 \subseteq L$ are fields, then $K_1' \supseteq K_2'$.*

Proof. Exercise. □

Proposition 9.4.9. *Let L be a field, H a subgroup of* $\mathrm{Aut}(L)$*, and $K \subset L$ a subfield.*

(a) $(H^\circ)'^\circ = H^\circ$.
(b) $(K')^{\circ\prime} = K'$.

Proof. Exercise. □

Definition 9.4.10. A polynomial in $K[x]$ is said to be *separable* if each of its irreducible factors has only simple roots in some (hence any) splitting field. An algebraic element a in a field extension of K is said to be *separable over K* if its minimal polynomial is separable. An algebraic field extension L of K is said to be *separable over K* if each of its elements is separable over K.

Remark 9.4.11. Separability is automatic if the characteristic of K is zero or if K is finite, by Theorems 9.3.1 and 9.3.4.

Theorem 9.4.12. *Suppose L is a splitting field for a separable polynomial $f(x) \in K[x]$. Then* $\mathrm{Fix}(\mathrm{Aut}_K(L)) = K$.

Proof. Let $\beta_1, \dots, \beta_r$ be the distinct roots of $f(x)$ in L. Consider the tower of fields:

$$\begin{aligned} M_0 = K \subseteq \cdots \subseteq M_j &= K(\beta_1, \dots, \beta_j) \\ &\subseteq \cdots \subseteq M_r = K(\beta_1, \dots, \beta_r) = L. \end{aligned}$$

A priori, $\mathrm{Fix}(\mathrm{Aut}_K(L)) \supseteq K$. We have to show that if $a \in L$ is fixed by all elements of $\mathrm{Aut}_K(L)$, then $a \in K$. I claim that if $a \in M_j$ for some $j \geq 1$, then $a \in M_{j-1}$. It will follow from this claim that $a \in M_0 = K$.

Suppose that $a \in M_j$. If $M_{j-1} = M_j$, there is nothing to show. Otherwise, let $\ell > 1$ denote the degree of the minimal polynomial $p(x)$ for β_j in $M_{j-1}[x]$. Then $\{1, \beta_j, \cdots, \beta_j^{\ell-1}\}$ is a basis for M_j over M_{j-1}. In particular,

$$a = m_0 + m_1\beta_j + \cdots + m_{\ell-1}\beta_j^{\ell-1} \tag{9.4.1}$$

for certain $m_i \in M_{j-1}$.

Since $p(x)$ is a factor of $f(x)$ in $M_{j-1}[x]$, p is separable, and the ℓ distinct roots $\{\alpha_1 = \beta_j, \alpha_2, \dots, \alpha_\ell\}$ of $p(x)$ lie in L. According to Proposition 9.4.1, for each s, there is a $\sigma_s \in \mathrm{Aut}_{M_{j-1}}(L) \subseteq \mathrm{Aut}_K(L)$ such that $\sigma_s(\alpha_1) = \alpha_s$. Applying σ_s to the expression for a and taking into account that a and the m_i are fixed by σ_s, we get:

$$a = m_0 + m_1\alpha_s + \cdots + m_{\ell-1}\alpha_s^{\ell-1} \tag{9.4.2}$$

for $1 \le s \le \ell$. Thus, the polynomial $(m_0 - a) + m_1 x + \cdots + m_{\ell-1}x^{l-1}$ of degree no more than $\ell - 1$ has at least ℓ distinct roots in L, and, therefore, the coefficients are identically zero. In particular, $a = m_0 \in M_{j-1}$. □

The following is the converse to the previous proposition:

Proposition 9.4.13. *Suppose $K \subseteq L$ is a field extension,* $\dim_K(L)$ *is finite, and* $\mathrm{Fix}(\mathrm{Aut}_K(L)) = K$.

(a) *For any $\beta \in L$, β is algebraic and separable over K, and the minimal polynomial for β over K splits in $L[x]$.*

(b) *For $\beta \in L$, let $\beta = \beta_1, \dots, \beta_n$ be a list of the distinct elements of $\{\sigma(\beta) : \sigma \in \mathrm{Aut}_K(L)\}$. Then $(x - \beta_1)(\dots)(x - \beta_n)$ is the minimal polynomial for β over K.*

(c) *L is the splitting field of a separable polynomial in $K[x]$.*

Proof. Since $\dim_K(L)$ is finite, L is algebraic over K.

Let $\beta \in L$, and let $\beta = \beta_1, \dots, \beta_r$ be the distinct elements of $\{\sigma(\beta) : \sigma \in \mathrm{Aut}_K(L)\}$. Define $g(x) = (x - \beta_1)(\dots)(x - \beta_r) \in L[x]$. Every $\sigma \in \mathrm{Aut}_K(L)$ leaves $g(x)$ invariant, so the coefficients of $g(x)$ lie in $\mathrm{Fix}(\mathrm{Aut}_K(L) = K$.

Let $p(x)$ denote the minimal polynomial of β over K. Since β is a root of $g(x)$, it follows that $p(x)$ divides $g(x)$. On the other hand, every root of $g(x)$ is of the form $\sigma(\beta)$ for $\sigma \in \mathrm{Aut}_K(L)$ and, therefore, is also a root of $p(x)$. Since the roots of $g(x)$ are simple, it follows that $g(x)$ divides $p(x)$. Hence $p(x) = g(x)$, as both are monic. In particular, $p(x)$ splits into linear factors over L, and the roots of $p(x)$ are simple. This proves parts (a) and (b).

Since L is finite dimensional over K, it is generated over K by finitely many algebraic elements $\alpha_1, \dots, \alpha_s$. It follows from part (a) that L is the splitting field of $f = f_1 f_2 \cdots f_s$, where f_i is the minimal polynomial of α_i over K. □

Recall that a finite dimensional field extension $K \subseteq L$ is said to be *Galois* if $\mathrm{Fix}(\mathrm{Aut}_K(L)) = K$.

Combining the last results gives:

Theorem 9.4.14. *For a finite dimensional field extension $K \subseteq L$, the following are equivalent:*

(a) *The extension is Galois.*

(b) *The extension is separable, and for all $\alpha \in L$ the minimal polynomial of α over K splits into linear factors over L.*

(c) *L is the splitting field of a separable polynomial in $K[x]$.*

Corollary 9.4.15. *If $K \subseteq L$ is a finite dimensional Galois extension and $K \subseteq M \subseteq L$ is an intermediate field, then $M \subseteq L$ is a Galois extension.*

Proof. L is the splitting field of a separable polynomial over K, and, therefore, also over M. □

Proposition 9.4.16. *If $K \subseteq L$ is a finite dimensional Galois extension, then:*

$$\dim_K L = \#\mathrm{Aut}_K(L). \tag{9.4.3}$$

Proof. The result is evident if $K = L$. Assume inductively that if $K \subseteq M \subseteq L$ is an intermediate field and $\dim_M L < \dim_K L$, then $\dim_M L = \#\mathrm{Aut}_M(L)$. Let $\alpha \in L \setminus K$ and let $p(x) \in K[x]$ be the minimal polynomial of α over K. Since L is Galois over K, p is separable and splits over L, by Theorem 9.4.14. If $\varphi \in \mathrm{Iso}_K(K(\alpha), L)$, then $\varphi(\alpha)$ is a root of p, and φ is determined by $\varphi(\alpha)$. Therefore,

$$\deg(p) = \#\mathrm{Iso}_K(K(\alpha), L) = [\mathrm{Aut}_K(L) : \mathrm{Aut}_{K(\alpha)}(L)], \tag{9.4.4}$$

where the last equality comes from 9.4.3. By the induction hypothesis applied to $K(\alpha)$, $\#\mathrm{Aut}_{K(\alpha)}(L) = \dim_{K(\alpha)} L$ is finite. Therefore, $\mathrm{Aut}_K(L)$ is also finite, and

$$\begin{aligned}\#\mathrm{Aut}_K(L) &= \deg(p)\, \#\mathrm{Aut}_{K(\alpha)}(L) \\ &= \dim_K(K(\alpha))\, \dim_{K(\alpha)}(L) = \dim_K L,\end{aligned}$$

where the first equality comes from equation 9.4.4, the second from the induction hypothesis and the irreducibility of p, and the final equality from the multiplicativity of dimensions, Proposition 8.3.2. □

Corollary 9.4.17. *Let $K \subseteq L$ be a finite dimensional Galois extension and M an intermediate field. Then*

$$\#\mathrm{Iso}_K(M, L) = \dim_K M. \tag{9.4.5}$$

Proof.

$$\begin{aligned}\#\mathrm{Iso}_K(M, L) &= [\mathrm{Aut}_K(L) : \mathrm{Aut}_M(L)] \\ &= \frac{\dim_K(L)}{\dim_M(L)} = \dim_K M,\end{aligned}$$

using Corollary 9.4.3, Proposition 9.4.16, and the multiplicativity of dimension, Proposition 8.3.2. □

Corollary 9.4.18. *Let $K \subseteq M$ be a finite dimensional separable field extension. Then*

$$\#\mathrm{Aut}_K(M) \leq \dim_K M$$

Proof. There is a field extension $K \subseteq M \subseteq L$ such that L is finite dimensional and Galois over K. (In fact, M is obtained from K by adjoining finitely many separable algebraic elements; let L be a splitting field of the product of the minimal polynomials over K of these finitely many elements.) Now, we have $\#\mathrm{Aut}_K(M) \leq \#\mathrm{Iso}_K(M, L) = \dim_K M$. □

Exercises

Exercise 9.4.1. Prove Proposition 9.4.7.

Exercise 9.4.2. Prove Proposition 9.4.8.

Exercise 9.4.3. Prove Proposition 9.4.9.

Exercise 9.4.4. Suppose that $f(x) \in K[x]$ is separable, and $K \subseteq M$ is an extension field. Show that f is also separable when considered as an element in $M[x]$.

9.5. The Galois Correspondence

In this section, we establish the Fundamental Theorem of Galois Theory, a correspondence between intermediate fields $K \subseteq M \subseteq L$ and subgroups of $\mathrm{Aut}_K(L)$, when L is a Galois field extension of K.

Proposition 9.5.1. *Suppose $K \subseteq L$ is finite dimensional separable field extension. Then there is an element $\gamma \in L$ such that $L = K(\alpha)$.*

Proof. If K is finite, then the finite dimensional field extension L is also a finite field. It follows from Exercise 7.2.6 that the multiplicative group of units of L is cyclic. Then $L = K(\alpha)$, where α is a generator of the multiplicative group of units.

Suppose now that K is infinite (which is always the case if the characteristic is zero). L is generated by finitely many separable algebraic elements over K, $L = K(\alpha_1, \dots, \alpha_s)$. It suffices to show that if $L = K(\alpha, \beta)$, where α and β are separable and algebraic, then there is a γ such that $L = K(\gamma)$, for then the general statement follows by an induction on s.

Suppose then that $L = K(\alpha, \beta)$. Let $K \subseteq L \subseteq E$ be a finite dimensional field extension such that E is Galois over K. Write $n = \dim_K L = \#\mathrm{Iso}_K(L, E)$. (Corollary 9.4.17.) Let $\{\varphi_1 = \mathrm{id}, \varphi_2, \dots, \varphi_n\}$ be a listing of $\mathrm{Iso}_K(L, E)$.

I claim that there is an element $k \in K$ such that the elements $\varphi_j(k\alpha + \beta)$ are all distinct. Suppose this for the moment and put $\gamma = k\alpha + \beta$. Then $K(\gamma) \subseteq L$, but $\dim_K(K(\gamma)) = \#\mathrm{Iso}_K(K(\gamma), E) \geq n = \dim_K L$. Therefore, $K(\gamma) = L$.

Now, to prove the claim: Let

$$p(x) = \prod_{1 \leq i < j \leq n} \left(x(\varphi_i(\alpha) - \varphi_j(\alpha) + (\varphi_i(\beta) - \varphi_j(\beta))\right).$$

The polynomial $p(x)$ is not identically zero since the φ_i are distinct on $K(\alpha, \beta)$, so there is an element k of the infinite field K such that $p(k) \neq 0$. But then the elements $k\varphi_i(\alpha) + \varphi_i(\beta) = \varphi_i(k\alpha + \beta)$, $1 \leq i \leq n$ are distinct. □

Corollary 9.5.2. *Suppose $K \subseteq L$ is a finite dimensional field extension, and the characteristic of K is zero. Then $L = K(\alpha)$ for some $\alpha \in L$.*

Proof. Separability is automatic in case the characteristic is zero. □

Proposition 9.5.3. *Let $K \subseteq L$ be a finite dimensional separable field extension and let H be a subgroup of* $\mathrm{Aut}_K(L)$. *Put* $F = \mathrm{Fix}(H)$. *Then*

(a) $\dim_F(L) = \#H$.

(b) $H = \mathrm{Aut}_F(L)$.

Proof. First note that $\#H \leq \#\mathrm{Aut}_K(L) \leq \dim_K(L)$, by Corollary 9.4.18, so H is necessarily finite. Then $F = \mathrm{Fix}(\mathrm{Aut}_F(L))$, by Proposition 9.4.9, so L is Galois over F and, in particular, separable by Proposition 9.4.14.

By Proposition 9.5.1, there is a β such that $L = F(\beta)$. Let $\{\varphi_1 = \mathrm{id}, \varphi_2, \ldots, \varphi_n\}$ be a listing of the elements of H, and put $\beta_i = \varphi_i(\beta)$. Then, by the argument of Proposition 9.4.13, the minimal polynomial for β over F is $g(x) = (x - \beta_1)(\ldots)(x - \beta_n)$. Therefore,

$$\dim_F L = \deg(g) = \#H \leq \#\mathrm{Aut}_F(L) = \dim_F(L),$$

using 9.4.16. □

We are now ready for the Fundamental Theorem of Galois Theory:

Theorem 9.5.4. *Let $K \subseteq L$ be a Galois field extension.*

(a) *There is an order-reversing bijection between subgroups of* $\mathrm{Aut}_K(L)$ *and intermediate fields $K \subseteq M \subseteq L$, given by $H \mapsto$* $\mathrm{Fix}(H)$.

(b) *The following conditions are equivalent for an intermediate field M:*

  (i) *M is Galois over K.*

  (ii) *M is invariant under* $\mathrm{Aut}_K(L)$

  (iii) $\mathrm{Aut}_M(L)$ *is a normal subgroup of* $\mathrm{Aut}_K(L)$.

In this case,

$$\mathrm{Aut}_K(M) \cong \mathrm{Aut}_K(L)/\mathrm{Aut}_M(L).$$

Proof. If $K \subseteq M \subseteq L$ is an intermediate field, then L is Galois over M; i.e., $M = \mathrm{Fix}(\mathrm{Aut}_M(L))$. On the other hand, if H is a subgroup of $\mathrm{Aut}_K(L)$, then, according to Proposition 9.5.3, $H = \mathrm{Aut}_{\mathrm{Fix}(H)}(L)$. Thus, the two maps $H \mapsto \mathrm{Fix}(H)$ and $M \mapsto \mathrm{Aut}_M(L)$ are inverses, which gives part (a).

Let M be an intermediate field. There is an α such that $M = K(\alpha)$ by Proposition 9.5.1. Let $f(x)$ denote the minimal polynomial of α over K. M is Galois over K if, and only if, M is the splitting field for $f(x)$, by Theorem 9.4.14. But the roots of $f(x)$ in L are the images of α under $\mathrm{Aut}_K(L)$ by Proposition 9.4.4. Therefore, M is Galois over K if, and only if, M is invariant under $\mathrm{Aut}_K(L)$.

If $\sigma \in \mathrm{Aut}_K(L)$, then $\sigma(M)$ is an intermediate field with group:

$$\mathrm{Aut}_{\sigma(M)}(L) = \sigma \mathrm{Aut}_M(L)\sigma^{-1}.$$

By part (a), $M = \sigma(M)$ if, and only if,

$$\mathrm{Aut}_M(L) = \mathrm{Aut}_{\sigma(M)}(L) = \sigma \mathrm{Aut}_M(L)\sigma^{-1}.$$

Therefore, M is invariant under $\mathrm{Aut}_K(L)$ if, and only if, $\mathrm{Aut}_M(L)$ is normal.

If M is invariant under $\mathrm{Aut}_K(L)$, then $\pi : \sigma \mapsto \sigma_{|M}$ is a homomorphism of $\mathrm{Aut}_K(L)$ into $\mathrm{Aut}_K(M)$, with kernel $\mathrm{Aut}_M(L)$. I claim that this homomorphism is surjective. In fact, an element of $\sigma \in \mathrm{Aut}_K(M)$ is determined by $\sigma(\alpha)$ which is necessarily a root of $f(x)$. But by Proposition 9.4.1, there is a $\sigma' \in \mathrm{Aut}_K(L)$ such that $\sigma'(\alpha) = \sigma(\alpha)$; therefore, $\sigma = \sigma'_{|M}$. Now, the homomorphism theorem for groups gives:

$$\mathrm{Aut}_K(M) \cong \mathrm{Aut}_K(L)/\mathrm{Aut}_M(L).$$

This completes the proof of part (b). □

Let $K \subseteq L$ be a field extension and let A, B be fields intermediate between K and L. Consider the composite $A \cdot B$, namely the subfield of L generated by $A \cup B$. We have the following diagram of field extensions:

$$\begin{array}{ccccc} & & A & \subseteq & A \cdot B & \subseteq & L \\ & & \cup\!| & & \cup\!| & & \\ K & \subseteq & A \cap B & \subseteq & B & & \end{array}$$

The following is an important technical result which is used in the sequel.

Proposition 9.5.5. *Let $K \subseteq L$ be a finite dimensional field extension and let A, B be fields intermediate between K and L. Suppose that B is Galois over K. Then $A \cdot B$ is Galois over A and* $\mathrm{Aut}_A(A \cdot B) \cong \mathrm{Aut}_{A \cap B}(B)$.

Proof. Exercise 9.5.3. □

The remainder of this section can be omitted without loss of continuity.

It is not hard to obtain the inequality of Corollary 9.4.18 without the separability assumption. It follows that the separability assumption can also be removed in Proposition 9.5.3. Although we consider only separable field extensions in this text, the argument is nevertheless worth knowing.

Proposition 9.5.6. *Let L be a field. Any collection of distinct automorphisms of L is linearly independent.*

Proof. Let $\{\sigma_1, \dots, \sigma_n\}$ be a collection of distinct automorphisms of L. We show by induction on n that the collection is linearly independent (in the vector space of functions from L to L.) If $n = 1$ there is nothing to show, since an automorphism cannot be identically zero. So assume $n > 1$ and assume any smaller collection of distinct automorphisms is linearly independent. Suppose

$$\sum_{i=1}^{n} \lambda_i \, \sigma_i = 0, \tag{9.5.1}$$

where $\lambda_i \in L$. Choose $\lambda \in L$ such that $\sigma_1(\lambda) \neq \sigma_n(\lambda)$. Then for all $\mu \in L$,

$$0 = \sum_{i=1}^{n} \lambda_i \, \sigma_i(\lambda\mu) = \sum_{i=1}^{n} \lambda_i \sigma_i(\lambda) \, \sigma_i(\mu). \tag{9.5.2}$$

In other words,

$$\sum_{i=1}^{n} \lambda_i \sigma_i(\lambda) \, \sigma_i = 0. \tag{9.5.3}$$

We can now eliminate σ_1 between equations 9.5.1 and 9.5.3 to give

$$\sum_{i=2}^{n} \lambda_i (\sigma_1(\lambda) - \sigma_i(\lambda)) \, \sigma_i = 0. \tag{9.5.4}$$

By the inductive assumption, all the coefficients of this equation must be zero. In particular, since $(\sigma_1(\lambda) - \sigma_n(\lambda)) \neq 0$, we have $\lambda_n = 0$. Now, the inductive assumption applied to the original equation 9.5.1 gives that all the coefficients λ_i are zero. □

Proposition 9.5.7. *Let $K \subseteq L$ be a field extension with* $\dim_K(L)$ *finite. Then* $\#\mathrm{Aut}_K(L) \leq \dim_K(L)$.

Proof. Suppose that $\dim_K(L) = n$ and $\{\lambda_1, \dots, \lambda_n\}$ is a basis of L over K. Suppose also that $\{\sigma_1, \dots, \sigma_{n+1}\}$ is a subset of $\mathrm{Aut}_K(L)$. (We *do not* assume that the σ_i are all distinct!) The n-by-$n+1$ matrix

$$\left[\sigma_j(\lambda_i)\right]_{1 \leq i \leq n, \ 1 \leq j \leq n+1}$$

has a non-trivial kernel by basic linear algebra. Thus, there exist $b_1, \dots, b_{n+1}$ in L, not all zero, such that:

$$\sum_j \sigma_j(\lambda_i) b_j = 0$$

for all i. Now, if $k_1, \dots, k_n$ are any elements of K,

$$0 = \sum_i k_i (\sum_j \sigma_j(\lambda_i) \, b_j) = \sum_j b_j \sigma_j (\sum_i k_i \lambda_i).$$

But $\sum_i k_i \lambda_i$ represents an arbitrary element of L, so the last equation gives $\sum_j b_j \sigma_j = 0$. Thus, the collection of σ_j is linearly dependent. By the previous proposition, the σ_j cannot be all distinct. That is, the cardinality of $\mathrm{Aut}_K(L)$ is no more than n. □

Exercises

Exercise 9.5.1. Suppose that $K \subseteq L$ is an algebraic field extension. Show that $L = \cup\{M : K \subseteq M \subseteq L$ and M is finite dimensional$\}$. If there is an $N \in \mathbb{N}$ such that $\dim_K(M) \le N$ whenever $K \subseteq M \subseteq L$ and M is finite dimensional, then also $\dim_K(L) \le N$.

Exercise 9.5.2. Let us consider a statement which is almost identical to that of Proposition 9.5.3. Suppose L is a field, H is a finite subgroup of $\mathrm{Aut}(L)$ and $F = \mathrm{Fix}(H)$. Then $\dim_F(L) \le \#H$. We cannot apply the proposition directly, because we do not know at the outset that $\dim_F(L)$ is finite. So it is necessary to arrange the elements of the proof of the proposition a little more artfully.

(a) Show that L is algebraic and separable over F and, moreover, for each $\beta \in L$, the minimal polynomial for β over F splits into linear factors over L.

(b) Show that if $F \subseteq M \subseteq L$, and M is finite dimensional over F, then $\dim_F(M) \le \#H$.

(c) Using the previous exercise, conclude that $\dim_F(L) \le \#H$.

Exercise 9.5.3. This exercise gives the proof of Proposition 9.5.5. We suppose that $K \subseteq L$ is a finite dimensional field extension, that A, B are fields intermediate between K and L, and that B is Galois over K. Let α be an element of B such that $B = K(\alpha)$ (Proposition 9.5.1). Let $p(x) \in K[x]$ be the minimal polynomial for α. Then B is a splitting field for $p(x)$ over K, and the roots of $p(x)$ are distinct, by 9.4.14.

(a) Show that $A \cdot B$ is Galois over A. *Hint:* $A \cdot B = A(\alpha)$; show that $A \cdot B$ is a splitting field for $p(x) \in A[x]$.

(b) Show that $\tau \mapsto \tau_{|B}$ is an injective homomorphism of $\mathrm{Aut}_A(A \cdot B)$ into $\mathrm{Aut}_{A \cap B}(B)$.

(c) Check surjectivity of $\tau \mapsto \tau_{|B}$ as follows: Let

$$G' = \{\tau_{|B} : \tau \in \mathrm{Aut}_A(A \cdot B)\}.$$

Then

$$\mathrm{Fix}(G') = \mathrm{Fix}(\mathrm{Aut}_A(A \cdot B)) \cap B = A \cap B$$

Therefore, by the Galois correspondence, $G' = \mathrm{Aut}_{A\cap B}(B)$.

9.6. Symmetric Functions

Let K be any field, and let $x_1, \dots, x_n$ be variables. For $\alpha = (\alpha_1, \dots, \alpha_n)$ a vector with non-negative integer entries, let $x^\alpha = x_1^{\alpha_1} \dots x_n^{\alpha_n}$. The *total degree* of the *monic monomial* x^α is $|\alpha| = \sum \alpha_i$. A polynomial is said to be *homogeneous of total degree d* if it is a linear combination of monomials x^α of total degree d.

The symmetric group S_n acts on polynomials and rational functions in n variables over K by $\sigma(f)(x_1, \dots, x_n) = f(x_{\sigma(1)}, \dots, x_{\sigma(n)})$. For $\sigma \in S_n$, $\sigma(x^\alpha) = x_{\sigma(1)}^{\alpha_1} \cdots x_{\sigma(n)}^{\alpha_n}$. A polynomial or rational function is called *symmetric* if it is fixed by the S_n action. The set of symmetric polynomials is denoted $K^S[x_1, \dots, x_n]$, and the set of symmetric rational functions is denoted $K^S(x_1, \dots, x_n)$.

Lemma 9.6.1.

(a) *The action of S_n on $K[x_1, \dots, x_n]$ is an action by ring automorphisms; the action of of S_n on $K(x_1, \dots, x_n)$ is an action by by field automorphisms.*

(b) *$K^S[x_1, \dots, x_n]$ is a subring of $K[x_1, \dots, x_n]$ and $K^S(x_1, \dots, x_n)$ is a subfield of $K(x_1, \dots, x_n)$.*

(c) *The field of symmetric rational functions is the field of fractions of the ring of symmetric polynomials in n-variables.*

Proof. Exercise. □

Proposition 9.6.2. *The field $K(x_1, \dots, x_n)$ of rational functions is Galois over the field $K^S(x_1, \dots, x_n)$ of symmetric rational functions, and the Galois group $\mathrm{Aut}_{K^S(x_1,\dots,x_n)}(K(x_1, \dots, x_n))$ is S_n.*

Proof. By Exercise 9.6.1, S_n acts on $K(x_1, \dots, x_n)$ by field automorphisms and $K^S(x_1, \dots, x_n)$ is the fixed field. Therefore, by Exercise 9.4.9, the extension is Galois, and by 9.5.3, the Galois group is S_n. □

We define a distinguished family of symmetric polynomials, the *elementary symmetric functions* as follows:

$$\begin{aligned}
\epsilon_0(x_1, \dots, x_n) &= 1 \\
\epsilon_1(x_1, \dots, x_n) &= x_1 + x_2 + \cdots + x_n \\
\epsilon_2(x_1, \dots, x_n) &= \sum_{1 \le i < j \le n} x_i x_j \\
&\dots \\
\epsilon_k(x_1, \dots, x_n) &= \sum_{1 \le i_1 < i_2 < \cdots < i_k \le n} x_{i_1} \cdots x_{i_k} \\
&\dots \\
\epsilon_n(x_1, \dots, x_n) &= x_1 x_2 \cdots x_n
\end{aligned}$$

Lemma 9.6.3.

$$\begin{aligned}
&(x - x_1)(x - x_2)(\cdots)(x - x_n) \\
&= x^n - \epsilon_1 x^{n-1} + \epsilon_2 x^{n-2} - \cdots + (-1)^n \epsilon_n \\
&= \sum_{k=0}^{n} (-1)^k \epsilon_k x^{n-k},
\end{aligned}$$

where ϵ_k is short for $\epsilon_k(x_1, \dots, x_n)$.

Proof. Exercise. □

Corollary 9.6.4.

(a) *Let $f(x) = x^n + a_{n-1}x^{n-1} + \cdots + a_0$ be a monic polynomial in $K[x]$ and let $\alpha_1, \dots, \alpha_n$ be the roots of f in a splitting field. Then $a_i = (-1)^{n-i}\epsilon_{n-i}(\alpha_1, \dots, \alpha_n)$.*

(b) *$f(x) = a_n x^n + a_{n-1}x^{n-1} + \cdots + a_0 \in K[x]$ be of degree n, and let $\alpha_1, \dots, \alpha_n$ be the roots of f in a splitting field. Then $a_i/a_n = (-1)^{n-i}\epsilon_{n-i}(\alpha_1, \dots, \alpha_n)$.*

Proof. For part (a),

$$\begin{aligned}
f(x) &= (x - \alpha_1)(x - \alpha_2) \cdots (x - \alpha_n) \\
&= x^n - \epsilon_1(\alpha_1, \dots, \alpha_n)x^{n-1} + \epsilon_2(\alpha_1, \dots, \alpha_n)x^{n-2} - \\
&\qquad \cdots + (-1)^n \epsilon_n(\alpha_1, \dots, \alpha_n).
\end{aligned}$$

For part (b), apply part (a) to $(x-\alpha_1)(x-\alpha_2)\cdots(x-\alpha_n) = \sum_i (a_i/a_n)x^i$. □

Definition 9.6.5. Let K be a field and $\{u_1, \dots, u_n\}$ a set of elements in an extension field. One says that $\{u_1, \dots, u_n\}$ is *algebraically independent* over K if there is no polynomial $f \in K[x_1, \dots, x_n]$ such that $f(u_1, \dots, u_n) = 0$.

The following is called the Fundamental Theorem of Symmetric Functions:

Theorem 9.6.6. *The set of elementary symmetric functions $\{\epsilon_1, \dots, \epsilon_n\}$ in $K[x_1, \dots, x_n]$ is algebraically independent over K, and generates $K^S[x_1, \dots, x_n]$ as a ring. Consequently, $K(\epsilon_1, \dots, \epsilon_n) = K^S(x_1, \dots, x_n)$.*

The *algebraic independence* of the ϵ_i is the same as *linear independence* of the monic monomials in the ϵ_i. First, we establish an indexing system for the monic monomials: A *partition* is a finite decreasing sequence of non-negative integers, $\lambda = (\lambda_1, \dots, \lambda_k)$. One can picture a partition by means of an M-by-N matrix Λ of zeroes and ones, where $M \geq k$ and $N \geq \lambda_1$; $\Lambda_{rs} = 1$ if $r \leq k$ and $s \leq \lambda_r$, and $\Lambda_{rs} = 0$ otherwise. Here is a matrix representing the partition $\lambda = (5, 4, 4, 2)$:

$$\Lambda = \begin{bmatrix} 1 & 1 & 1 & 1 & 1 & 0 \\ 1 & 1 & 1 & 1 & 0 & 0 \\ 1 & 1 & 1 & 1 & 0 & 0 \\ 1 & 1 & 0 & 0 & 0 & 0 \\ 0 & 0 & 0 & 0 & 0 & 0 \\ 0 & 0 & 0 & 0 & 0 & 0 \end{bmatrix}.$$

The size of a partition λ is $|\lambda| = \sum \lambda_i$. The non-zero entries in λ are referred to as the *parts* of λ. The number of parts is called the *length* of λ. The *conjugate partition* λ^* is that represented by the transposed matrix Λ^*, namely, $\lambda_r^* = \#\{i : \lambda_i \geq r\}$. For $\lambda = (5, 4, 4, 2)$, one has $\lambda^* = (4, 4, 3, 3, 1)$, corresponding to the matrix

$$\Lambda^* = \begin{bmatrix} 1 & 1 & 1 & 1 & 0 & 0 \\ 1 & 1 & 1 & 1 & 0 & 0 \\ 1 & 1 & 1 & 0 & 0 & 0 \\ 1 & 1 & 1 & 0 & 0 & 0 \\ 1 & 0 & 0 & 0 & 0 & 0 \\ 0 & 0 & 0 & 0 & 0 & 0 \end{bmatrix}.$$

For $\lambda = (\lambda_1, \lambda_2, \dots, \lambda_n)$, define

$$\epsilon_\lambda(x_1, \dots, x_n) = \prod_i \epsilon_{\lambda_i}(x_1, \dots, x_n).$$

For example, $\epsilon_{(5,4,4,2)} = (\epsilon_5)(\epsilon_4)^2(\epsilon_2)$.

We will show that the set of ϵ_λ with $|\lambda| = d$ is a basis of $K_d^S[x_1, \ldots, x_n]$. In order to do this, we first produce a more obvious basis.

For a partition $\lambda = (\lambda_1, \ldots \lambda_n)$ with no more than n non-zero parts, define the *monomial symmetric function*

$$m_\lambda(x_1, \ldots, x_n) = (1/f) \sum_{\sigma \in S_n} \sigma(x^\lambda),$$

where f is the size of the stabilizer of x^λ under the action of the symmetric group. (Thus, m_λ is a sum of monic monomials, each occurring exactly once.) For example,

$$m_{(5,4,4,2)}(x_1, \ldots, x_4) = x_1^5 x_2^4 x_3^4 x_4^2 + x_1^5 x_2^4 x_3^2 x_4^4 + x_1^5 x_2^2 x_3^4 x_4^4 + \ldots,$$

a sum of 12 monic monomials.

In the exercises, you are asked to check that the monomial symmetric functions m_λ with $\lambda = (\lambda_1, \ldots, \lambda_n)$ and $|\lambda| = d$ form a linear basis of $K_d^S[x_1, \ldots, x_n]$.

Define a total order on n-tuples of non-negative integers and, in particular, on partitions by $\alpha > \beta$ if the first non-zero difference $\alpha_i - \beta_i$ is positive. Note that $\lambda \geq \sigma(\lambda)$ for any permutation σ. This total order on n-tuples α induces a total order on monomials x^α. This order is called *lexicographic order*.

Lemma 9.6.7.

$$\epsilon_{\lambda^*}(x_1, \ldots, x_n) = \sum_{|\mu| = |\lambda|} T_{\lambda\mu} m_\mu(x_1, \ldots, x_n),$$

where $T_{\lambda\mu}$ is a non-negative integer, $T_{\lambda\lambda} = 1$, and $T_{\lambda\mu} = 0$ if $\mu > \lambda$.

Proof. This is probably best explained by example: Take $n = 4$ and $\lambda = (5, 4, 4, 2)$. Then $\lambda^* = (4, 4, 3, 3, 2)$, and:

$$\begin{aligned}
\epsilon_{\lambda^*} &= (\epsilon_4)^2(\epsilon_3)^2\epsilon_1 \\
&= (x_1x_2x_3x_4)^2(x_1x_2x_3 + x_1x_3x_4 + x_2x_3x_4)^2(x_1 + \ldots x_4) \\
&= x_1^5x_2^4x_3^4x_4^2 + \ldots,
\end{aligned}$$

where the remaining monomials are less than $x_1^5x_2^4x_3^4x_4^2$ in lexicographic order. It follows that $\epsilon_{\lambda^*} = m_\lambda +$ a sum of m_μ, where $|\mu| = |\lambda|$ and $\mu < \lambda$.

In general, ϵ_{λ^*} is a symmetric polynomial with leading term x^λ; therefore, $\epsilon_{\lambda^*} = m_\lambda +$ a sum of m_μ for $\mu < \lambda$. □

Proof of Theorem 9.6.6. It is immediate from the previous lemma that the ϵ_λ of fixed degree are linearly independent (and contained in the integer linear span of the m_μ of the same degree.)

Moreover, a triangular integer matrix with 1's on the diagonal has an inverse of the same sort. Therefore, the monomial symmetric functions m_μ are integer linear combinations of the elementary symmetry functions ϵ_λ of the same degree.

It follows that the ϵ_λ of a fixed degree d are a basis for the linear space $K_d^S[x_1, \dots, x_n]$, and the symmetric functions ϵ_λ of arbitrary degree are a basis for $K^S[x_1, \dots, x_n]$.

Moreover, for any ring A, the ring of symmetric polynomials in $A[x_1, \dots, x_n]$ equals $A[\epsilon_1, \dots, \epsilon_n]$. □

Algorithm for expansion of symmetric polynomials in the elementary symmetric polynomials. The matrix $T_{\lambda\mu}$ was convenient for showing that the ϵ_λ are linearly independent. It is neither convenient nor necessary, however, to compute the matrix and to invert it in order to expand symmetric polynomials as linear combinations of the ϵ_λ. This can be done by the following algorithm instead:

Let $p = \sum a_\beta x^\beta$ be a homogeneous symmetric polynomial of degree d in variables $x_1, \dots, x_n$. Let x^α be the greatest monomial (in lexicographic order) appearing in p; since p is symmetric, α is necessarily a partition. Then

$$p = a_\alpha m_\alpha + \text{ lexicographically lower terms},$$

and

$$\epsilon_{\alpha^*} = m_\alpha + \text{ lexicographically lower terms}.$$

Therefore $p_1 = p - a_\alpha e_{\alpha^*}$ is a homogeneous symmetric polynomial of the same degree which contains only monomials lexicographically lower than x^α, or $p_1 = 0$. Now, iterate this procedure. The algorithm must terminate after finitely many steps because there are only finitely many monic monomials of degree d.

One can formalize the structure of this proof by introducing the notion of induction on a totally ordered set. Let $\mathcal{P}(d, n)$ denote the totally ordered set of partitions of size d with no more than n parts. Suppose one has a predicate $P(\lambda)$ depending on a partition. In order to prove that $P(\lambda)$ holds for all $\lambda \in \mathcal{P}(d, n)$, it suffices to show

1. $P(\kappa)$, where κ is the least partition in $\mathcal{P}(d, n)$, and
2. For all $\lambda \in \mathcal{P}(d, n)$, if $P(\mu)$ holds for all $\mu \in \mathcal{P}(d, n)$ such that $\mu < \lambda$, then $P(\lambda)$ holds.

To apply this idea to our situation, take $P(\lambda)$ to be the predicate: If p is a symmetric homogeneous polynomial of degree d in n variables, in which the lexicographically highest monomial is x^λ, then p is a linear combination of ϵ^*_μ with $\mu \in \mathcal{P}(d,n)$ and $\mu \le \lambda$.

Let κ be the least partition in $\mathcal{P}(d,n)$. Then,

$$m_\kappa(x_1, \dots, x_n) = \epsilon_\kappa(x_1, \dots, x_n).$$

Moreover, if p is a symmetric homogeneous polynomial of degree d in n variables whose *highest* monomial is x^κ, then p is a multiple of $m_\kappa(x_1, \dots, x_n)$. This shows that $P(\kappa)$ holds.

Now, fix a partition $\lambda \in \mathcal{P}(d,n)$, and suppose that $P(\mu)$ holds for all partitions $\mu \in \mathcal{P}(d,n)$ such that $\mu < \lambda$. (This is the *inductive hypothesis.*) Let p be a homogeneous symmetric polynomial of degree d whose leading term is $a_\lambda x^\lambda$. As observed previously, $p_1 = p - a_\lambda \epsilon_{\lambda^*}$ is either zero or a homogeneous symmetric polynomial of degree d with leading monomial x^β for some $\beta < \lambda$. According to the inductive hypothesis, p_1 is a linear combination of ϵ^*_μ with $\mu \le \beta < \lambda$. Therefore $p = p_1 + a_\lambda \epsilon_{\lambda^*}$ is a linear combination of ϵ^*_μ with $\mu \le \lambda$.

Example 9.6.8. We illustrate the algorithm by an example. Consider polynomials in three variables. Take $p = x^3 + y^3 + z^3$. Then,

$$\begin{aligned} p_1 &= p - \epsilon_{1^3} \\ &= -3x^2y - 3xy^2 - 3x^2z - 6xyz - 3y^2z - 3xz^2 - 3yz^2, \\ p_2 &= p_1 + 3\epsilon_{(2,1)} \\ &= 3xyz \\ &= 3\epsilon_3. \end{aligned}$$

Thus, $p = \epsilon_{1^3} - 3\epsilon_{(2,1)} + 3\epsilon_3$.

Of course, such computations can be automated. A program in *Mathematica* for expanding symmetric polynomials in elementary symmetric functions is available on my world wide web site.

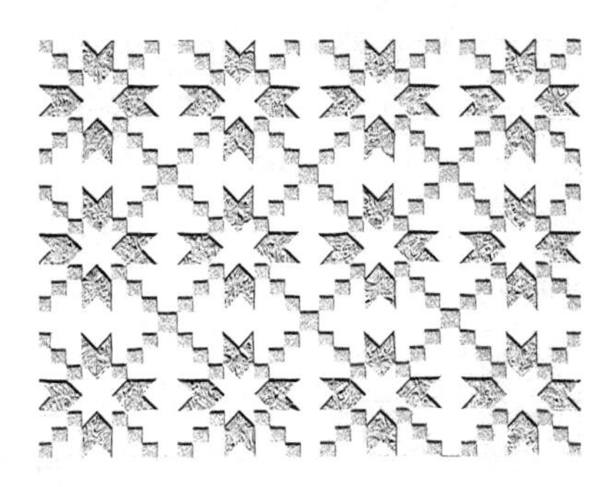

Exercises

Exercise 9.6.1.

(a) Show that the set $K_d[x_1, \dots, x_n]$ of polynomials which are homogeneous of total degree d or identically zero is a finite dimensional vector subspace of the K-vector space $K[x_1, \dots, x_n]$.

(b) Find the dimension of $K_d[x_1, \dots, x_n]$.

(c) Show that $K[x_1, \dots, x_n]$ is the direct sum of $K_d[x_1, \dots, x_n]$, where d ranges over the non-negative integers.

Exercise 9.6.2. Prove Lemma 9.6.1.

Exercise 9.6.3.

(a) For each d, $K_d[x_1, \dots, x_n]$ is invariant under the action of S_n.

(b) $K_d^S[x_1, \dots, x_n] = K_d[x_1, \dots, x_n] \cap K^S[x_1, \dots, x_n]$ is a vector subspace of $K^S[x_1, \dots, x_n]$.

(c) $K^S[x_1, \dots, x_n]$ is the direct sum of the subspaces $K_d^S[x_1, \dots, x_n]$ for $d \geq 0$.

Exercise 9.6.4. Prove Lemma 9.6.3.

Exercise 9.6.5. Show that every monic monomial $\epsilon_n^{m_n} \epsilon_{n-1}^{m_{n-1}} \dots \epsilon_1^{m_1}$ in the ϵ_i is an ϵ_λ, and relate λ to the multiplicities m_i.

Exercise 9.6.6. ϵ_λ is homogeneous of total degree $|\lambda|$.

Exercise 9.6.7. The monomial symmetric functions m_λ with $\lambda = (\lambda_1, \dots, \lambda_n)$ and $|\lambda| = d$ form a linear basis of $K_d^S[x_1, \dots, x_n]$.

Exercise 9.6.8. An symmetric function ϵ_λ of degree d is an integer linear combination of monomials x^α of degree d, and therefore an integer linear combination of monomial symmetric functions m_μ with $|\mu| = d$. (Substitute $\mathbb{Z}_q$-linear combinations in case the characteristic is q.)

Exercise 9.6.9. An upper triangular matrix T with 1's on the diagonal and integer entries has an inverse of the same type.

Exercise 9.6.10. Write out the monomial symmetric functions $m_{3,3,1}(x_1, x_2, x_3)$ and $m_{3,2,1}(x_1, x_2, x_3)$, and note that they have different numbers of summands.

Exercise 9.6.11. Consult the *Mathematica* notebook **Symmetric- Functions.nb** which is available on my world wide web site. Use the *Mathematica* function **monomialSymmetric[ ]** to compute the monomial symmetric functions m_λ in n variables for:

(a) $\lambda = [2, 2, 1, 1]$, $n = 5$
(b) $\lambda = [3, 3, 2]$, $n = 5$
(c) $\lambda = [3, 1]$, $n = 5$.

Exercise 9.6.12. Use the algorithm described in this section to expand the following symmetric polynomials as polynomials in the elementary symmetric functions.

(a) ${x_1}^2{x_2}^2x_3 + {x_1}^2x_2{x_3}^2 + x_1{x_2}^2{x_3}^2$
(b) $x_1^3 + x_2^3 + x_3^3$

Exercise 9.6.13. Consult the *Mathematica* notebook **Symmetric- Functions.nb** which is available on my world wide web site. Use the *Mathematica* function **elementaryExpand[ ]** to compute the expansion of the following symmetric functions as polynomials in the elementary symmetric functions.

(a) $[(x_1 - x_2)(x_1 - x_3)(x_1 - x_4)(x_2 - x_3)(x_2 - x_4)(x_3 - x_4)]^2$
(b) m_λ, for $\lambda = [4, 3, 3, 1]$, in 4 variables.

Exercise 9.6.14. Suppose that f is an *antisymmetric* polynomial in n variables; that is for each $\sigma \in S_n$, $\sigma(f) = \epsilon(\sigma)f$, where ϵ denotes the parity homomorphism. Show that f has a factorization $f = \delta(x_1, \dots, x_n)g$, where $\delta(x_1, \dots, x_n) = \prod_{i<j}(x_i - x_j)$, and g is symmetric.

9.7. The General Equation of Degree n.

Consider the quadratic formula, or Cardano's formulas for solutions of a cubic equation, which calculate the roots of a polynomials in terms of the coefficients; in these formulas, the coefficients may be regarded as variables and the roots as functions of these variables. This observation suggests the notion of the *general polynomial of degree n*, which is defined as follows:

Let $t_1, \dots, t_n$ be variables. The *general (monic) polynomial of degree n over K* is:

$$P_n(x) = x^n - t_1x^{n-1} + \cdots + (-1)^{n-1}t_n \in K(t_1, \dots, t_n)(x)$$

Let $u_1, \dots, u_n$ denote the roots of this polynomial in a splitting field E. Then $(x - u_1)\cdots(x - u_n) = P_n(x)$, and $t_j = \epsilon_j(u_1, \dots, u_n)$ for $1 \le j \le n$, by Corollary 9.6.4. We shall now show that the Galois group of the general polynomial of degree n is the symmetric group S_n.

Theorem 9.7.1. *Let E be a splitting field of the general polynomial $P_n(x) \in K(t_1, \dots, t_n)(x)$. The Galois group $\mathrm{Aut}_{K(t_1,\dots,t_n)}(E)$ is the symmetric group S_n.*

Proof. Introduce a new set of variables $v_1, \dots, v_n$ and let

$$f_j = \epsilon_j(v_1, \dots, v_n) \quad \text{for } 1 \le j \le n,$$

where the ϵ_j are the elementary symmetric functions. Consider the polynomial

$$\tilde{P}_n(x) = (x - v_1) \cdots (x - v_n) = x^n + \sum_j (-1)^j f_j x^{n-j}.$$

The coefficients lie in $K(f_1, \dots, f_n)$, which is equal to $K^S(v_1, \dots, v_n)$ by Theorem 9.6.6. According to Proposition 9.6.2, $K(v_1, \dots, v_n)$ is Galois over $K(f_1, \dots, f_n)$ with Galois group S_n. Furthermore, $K(v_1, \dots, v_n)$ is the splitting field over $K(f_1, \dots, f_n)$ of $\tilde{P}_n(x)$.

Let $u_1, \dots, u_n$ be the roots of $P_n(x)$ in E. Then $t_j = \epsilon_j(u_1, \dots, u_n)$ for $1 \le j \le n$, so $E = K(t_1, \dots, t_n)(u_1, \dots, u_n) = K(u_1, \dots, u_n)$.

Since $\{t_i\}$ are variables, and the $\{f_i\}$ are algebraically independent over K, according to Theorem 9.6.6, there is a ring isomorphism $K[t_1, \dots, t_n] \to K[f_1, \dots, f_n]$ fixing K and taking t_i to f_i. This ring isomorphism extends to an isomorphism of fields of fractions $K(t_1, \dots, t_n) \cong K(f_1, \dots, f_n)$, and to the polynomial rings $K(t_1, \dots, t_n)[x] \cong K(f_1, \dots, f_n)[x]$; the isomorphism of polynomial rings carries $P_n(x)$ to $\tilde{P}_n(x)$. Therefore, by Proposition 9.2.4, there is an isomorphism of splitting fields $E \cong K(v_1, \dots, v_n)$ extending the isomorphism $K(t_1, \dots, t_n) \cong K(f_1, \dots, f_n)$.

It follows that the Galois groups are isomorphic:

$$\mathrm{Aut}_{K(t_1,\dots,t_n)}(E) \cong \mathrm{Aut}_{K(f_1,\dots,f_n)}(K(v_1, \dots, v_n)) \cong S_n.$$

□

We shall see in Section 10.6 that this result implies that *there can be no analogue of the quadratic and cubic formulas for equations of degree 5 or more.* (We shall work out formulas for quartic equations in Section 9.8.)

The discriminant. We now consider some symmetric polynomials which arise in the study of polynomials and Galois groups.

Write

$$\delta = \delta(x_1, \dots, x_n) = \prod_{1 \le i < j \le n} (x_i - x_j).$$

We know that δ distinguishes even and odd permutations. Every permutation σ satisfies $\sigma(\delta) = \pm\delta$ and σ is even if, and only if, $\sigma(\delta) = \delta$. The symmetric polynomial δ^2 is called the *discriminant polynomial.*

Now let $f = \sum_i a_i x^i \in K[x]$ be polynomial of degree n. Let $\alpha_1, \dots, \alpha_n$ be the roots of $f(x)$ in a splitting field E. The element

$$\delta^2(f) = a_n^{2n-2} \delta^2(\alpha_1, \dots, \alpha_n)$$

is called the *discriminant of f*.

Now suppose in addition that f is irreducible and separable. Since $\delta^2(f)$ is invariant under the Galois group of f, it follows that $\delta^2(f)$ is an element of the ground field K. Since the roots of f are distinct, the element $\delta(f) = a_n^{n-1} \delta(\alpha_1, \dots, \alpha_n)$ is non-zero, and an element $\sigma \in \mathrm{Aut}_K(E)$ induces an even permutation of the roots of f if, and only if, $\sigma(\delta(f)) = \delta(f)$. Thus we have the following result:

Proposition 9.7.2. *Let $f \in K[x]$ be an irreducible separable polynomial. Let E be a splitting field for f, and let $\alpha_1, \dots, \alpha_n$ be the roots of $f(x)$ in E. The Galois group* $\mathrm{Aut}_K(E)$*, regarded as a group of permutations of the set of roots, is contained in the alternating group A_n if, and only if, the discriminant $\delta^2(f)$ of f has a square root in K.*

Proof. The discriminant has a square root in K if, and only if, $\delta(f) \in K$. But $\delta(f) \in K$ if, and only if, it is fixed by the Galois group, if, and only if, the Galois group consists of even permutations. □

One does not need to know the roots of f in order to compute the discriminant. Because the discriminant of $f = \sum_i a_i x^i$ is the discriminant of the monic polynomial $(1/a_n) f$ multiplied by a_n^{2n-1}, it suffices to give a method for computing the discriminant of a monic polynomial. Suppose, then, that f is monic.

The discriminant is a symmetric polynomial in the roots, and therefore, a polynomial in the coefficients of f, by Theorem 9.6.6 and Corollary 9.6.4. For fixed n, one can expand the discriminant polynomial $\delta^2(x_1, \dots, x_n)$ as a polynomial in the elementary symmetric functions, say $\delta^2(x_1, \dots, x_n) = d_n(\epsilon_1, \dots, \epsilon_n)$. Then

$$\delta^2(f) = d_n(-a_{n-1}, a_{n-2}, \dots, (-1)^n a_0).$$

This is not the most efficient method of calculating the discriminant of a polynomial, but it works well for polynomials of low degree. (A program for computing the discriminant by this method is available on my web page.)

Example 9.7.3. (Galois group of a cubic.) The Galois group of a monic cubic irreducible polynomial f is either $A_3 = \mathbb{Z}_3$ or S_3, according to whether

$\delta^2(f) \in K$ or not. (These are the only transitive subgroups of S_3.) One can compute that

$$\delta^2(x_1, x_2, x_3) = \epsilon_1{}^2\,\epsilon_2{}^2 - 4\,\epsilon_2{}^3 - 4\,\epsilon_1{}^3\,\epsilon_3 + 18\,\epsilon_1\,\epsilon_2\,\epsilon_3 - 27\,\epsilon_3{}^2.$$

Therefore, a cubic polynomial $f(x) = x^3 + ax^2 + bx + c$ has

$$\delta^2(f) = a^2\,b^2 - 4\,b^3 - 4\,a^3\,c + 18\,a\,b\,c - 27\,c^2.$$

In particular, a polynomial of the special form $f(x) = x^3 + px + q$ has

$$\delta^2(f) = -4\,p^3 - 27\,q^2,$$

as already computed in Chapter 8.

Consider for example $f(x) = x^3 - 4x^2 + 2x + 13 \in \mathbb{Z}[x]$. To test a cubic for irreducibility, it suffices to show it has no root in $\mathbb{Q}$, which follows from the rational root test. The discriminant of f is $\delta^2(f) = -3075$. Since this is not a square in $\mathbb{Q}$, it follows that the Galois group of f is S_3.

Example 9.7.4. The Galois group of an irreducible quartic polynomial f must be one of the following, as these are the only transitive subgroups of S_4, according to Exercise 5.1.20:

- A_4, $\mathcal{V} \subseteq A_4$, or
- S_4, D_4, $\mathbb{Z}_4 \not\subseteq A_4$.

One can compute that the discriminant polynomial has the expansion:

$$\begin{aligned}\delta^2(x_1, \dots, x_4) =& \epsilon_1{}^2\,\epsilon_2{}^2\,\epsilon_3{}^2 - 4\,\epsilon_2{}^3\,\epsilon_3{}^2 - 4\,\epsilon_1{}^3\,\epsilon_3{}^3 + 18\,\epsilon_1\,\epsilon_2\,\epsilon_3{}^3 - \\ & 27\,\epsilon_3{}^4 - 4\,\epsilon_1{}^2\,\epsilon_2{}^3\,\epsilon_4 + 16\,\epsilon_2{}^4\,\epsilon_4 + 18\,\epsilon_1{}^3\,\epsilon_2\,\epsilon_3\,\epsilon_4 - \\ & 80\,\epsilon_1\,\epsilon_2{}^2\,\epsilon_3\,\epsilon_4 - 6\,\epsilon_1{}^2\,\epsilon_3{}^2\,\epsilon_4 + 144\,\epsilon_2\,\epsilon_3{}^2\,\epsilon_4 - 27\,\epsilon_1{}^4\,\epsilon_4{}^2 + \\ & 144\,\epsilon_1{}^2\,\epsilon_2\,\epsilon_4{}^2 - 128\,\epsilon_2{}^2\,\epsilon_4{}^2 - 192\,\epsilon_1\,\epsilon_3\,\epsilon_4{}^2 + 256\,\epsilon_4{}^3\end{aligned}$$

Therefore, for $f(x) = x^4 + ax^3 + bx^2 + cx + d$,

$$\begin{aligned}\delta^2(f) =& a^2\,b^2\,c^2 - 4\,b^3\,c^2 - 4\,a^3\,c^3 + 18\,a\,b\,c^3 - 27\,c^4 - 4\,a^2\,b^3\,d + \\ & 16\,b^4\,d + 18\,a^3\,b\,c\,d - 80\,a\,b^2\,c\,d - 6\,a^2\,c^2\,d + 144\,b\,c^2\,d - \\ & 27\,a^4\,d^2 + 144\,a^2\,b\,d^2 - 128\,b^2\,d^2 - 192\,a\,c\,d^2 + 256\,d^3.\end{aligned}$$

For example, take $f(x) = x^4 + 3x^3 + 4x^2 + 7x - 5$. The reduction of f mod 3 is $x^4 + x^2 + x + 1$, which can be shown to be irreducible over $\mathbb{Z}_3$ by an ad hoc argument. Therefore, $f(x)$ is also irreducible over $\mathbb{Q}$. One computes that $\delta^2(f) = -212836$, which is not a square in $\mathbb{Q}$, so the Galois group must be S_4, D_4, or $\mathbb{Z}_4$.

Resultants. In the remainder of this section, we will discuss the *resultant* of two polynomials. *This material can be omitte of continuity.*

The resultant of polynomials $f = \sum_{i=0}^{n} a_i x^i$, $g = \sum_{i=0}^{m} b_i x^i \in K[x]$, of degrees n and m, is defined to be the product

$$R(f, g) = a_n^m b_m^n \prod_i \prod_j (\xi_i - \eta_j), \tag{9.7.1}$$

where the ξ_i and η_j are the roots of f and g, respectively, in a common splitting field.

The product

$$R_0(f, g) = \prod_i \prod_j (\xi_i - \eta_j), \tag{9.7.2}$$

is evidently symmetric in the roots of f and in the roots of g, and therefore, is a polynomial in the quantities (a_i/a_n) and (b_j/b_n). One can show that the total degree as a polynomial in the (a_i/a_n) is m and the total degree as a polynomial in the (b_j/b_m) is m (exercise). Therefore, the resultant (9.7.1) is a polynomial in the a_i and b_j, homogeneous of degree m in the a_i and of degree n in the b_j.

Furthermore, $R(f, g) = 0$ precisely when f and g have a common root, that is, if, and only if, f and g have a a non-constant common factor in $K[x]$. We shall find an expression for $R(f, g)$ by exploiting this observation.

If the polynomials f and g have a common divisor $q(x)$ in $K[x]$, then there exist polynomials $\varphi(x)$ and $\psi(x)$ of degrees no more than $n - 1$ and $m - 1$, respectively, such that:

$$\begin{aligned} f(x) &= q(x)\varphi(x), \text{ and} \\ g(x) &= q(x)\psi(x), \text{ so} \\ f(x)\psi(x) &= g(x)\varphi(x) = q(x)\varphi(x)\psi(x). \end{aligned}$$

Conversely, the existence of polynomials $\varphi(x)$ of degree no more than $n - 1$ and $\psi(x)$ of degree no more than $m - 1$ such that $f(x)\psi(x) = g(x)\varphi(x)$ implies that f and g have a non-constant common divisor (exercise).

Now, write $\psi(x) = \sum_{i=0}^{m-1} \alpha_i x^i$ and $\varphi(x) = \sum_{i=0}^{n-1} \beta_i x^i$, and match coefficients in the equation $f(x)\psi(x) = g(x)\varphi(x)$ to get a system of linear equations in the variables

$$(\alpha_0, \ldots, \alpha_{m-1}, -\beta_0, \ldots, -\beta_{n-1}).$$

Assuming without loss of generality that $n \geq m$, the matrix $\mathcal{R}(f, g)$ of the system of linear equations has the form displayed in Figure 9.7.1.

$$
\begin{bmatrix}
a_0 & & & & & b_0 & & & & & & \\
a_1 & a_0 & & & & b_1 & b_0 & & & & & \\
a_2 & a_1 & a_0 & & & b_2 & b_1 & b_0 & & & & \\
\vdots & \vdots & \vdots & \ddots & & \vdots & \vdots & \vdots & \ddots & & & \\
\vdots & \vdots & \vdots & \cdots & a_0 & \vdots & \vdots & \vdots & \cdots & & & \\
\vdots & \vdots & \vdots & \cdots & a_1 & b_m & b_{m-1} & b_{m-2} & \cdots & & & \\
\vdots & \vdots & \vdots & \cdots & a_2 & & b_m & b_{m-1} & \cdots & \cdots & \ddots & \\
\vdots & \vdots & \vdots & \cdots & \vdots & & & b_m & \cdots & \cdots & \cdots & b_0 \\
a_n & a_{n-1} & & \cdots & a_{n-m+1} & & & & \ddots & \cdots & \cdots & b_1 \\
 & a_n & \vdots & \cdots & \vdots & & & & & & \cdots & \vdots \\
 & & a_n & \cdots & \vdots & & & & & & \cdots & \vdots \\
 & & & \ddots & \vdots & & & & & & \ddots & \vdots \\
 & & & & a_n & & & & & & & b_m
\end{bmatrix}
$$

Figure 9.7.1. The matrix $\mathcal{R}(f, g)$

The matrix has m columns with a_i's and n columns with b_i's. It follows from the discussion above that f and g have a non-constant common divisor if, and only if, the matrix $\mathcal{R}(f, g)$ has determinant zero.

We want to show that $\det(\mathcal{R}(f, g)) = (-1)^{n+m} R(f, g)$. Since $\det(\mathcal{R}(f, g)) = a_n^m b_m^n \det(\mathcal{R}(a_n^{-1} f, b_m^{-1} g))$, and similarly, $R(f, g) = a_n^m b_m^n R(a_n^{-1} f, b_m^{-1} g)$, it suffices to consider the case where both f and g are monic.

At this point we need to shift our point of view slightly. Regard the roots ξ_i and η_j as variables, the coefficients a_i as symmetric polynomials in the ξ_i, and the coefficients and b_j as symmetric polynomials in the η_j. Then $\det(\mathcal{R}(f, g))$ is a polynomial in the ξ_i and η_j, symmetric with respect to each set of variables, which is zero when any ξ_i and η_j agree. It follows that $\det(\mathcal{R}(f, g))$ is divisible by $\prod_i \prod_j (\xi_i - \eta_j) = R(f, g)$ (exercise). But as both $\det(\mathcal{R})$ and $R(f, g)$ are polynomials in the a_i and b_j of total degree $n + m$, they are equal up to a scalar factor, and it remains to show that the scalar factor is $(-1)^{n+m}$. In fact, $\det(\mathcal{R})$ has a summand a_0^m. On the other hand $R(f, g) = (-1)^{n+m} \prod_j f(\beta_j)$, according to Exercise 9.7.10, and so has a summand $(-1)^{n+m} a_0^m$.

Proposition 9.7.5. $R(f, g) = (-1)^{n+m} \det(\mathcal{R}(f, g))$.

We now observe that the discriminant of f can be compu... resultant of f and its formal derivative f'. In fact, from Exercise 9.7.10,

$$R(f, f') = a_n^{n-1} \prod_i f'(\xi_i).$$

Using

$$f(x) = a_n \prod_i (x - \xi_i),$$

one can verify that

$$f'(\xi_i) = a_n \prod_{j \neq i} (\xi_i - \xi_j).$$

Therefore,

$$\begin{aligned} R(f, f') &= a_n^{n-1} \prod_i f'(\xi_i) = a_n^{2n-1} \prod_i \prod_{j \neq i} (\xi_i - \xi_j) \\ &= a_n^{2n-1} (-1)^{n(n-1)/2} \prod_{i<j} (\xi_i - \xi_j)^2 \\ &= a_n (-1)^{n(n-1)/2} \delta^2(f). \end{aligned}$$

Proposition 9.7.6. $\delta^2(f) = a_n^{-1} (-1)^{n(n-1)/2} R(f, f')$.

Although in general, determinants tend to be inefficient for computations, the matrix $\mathcal{R}$ above is sparse, and it appears to be more efficient to calculate the discriminant using the determinant of $\mathcal{R}(f, f')$ than to use the method described earlier in this section.

Exercises

Exercise 9.7.1. Determine the Galois groups of the following cubic polynomials:

(a) $x^3 + 2x + 1$, over $\mathbb{Q}$.
(b) $x^3 + 2x + 1$, over $\mathbb{Z}_3$.
(c) $x^3 - 7x^2 - 7$, over $\mathbb{Q}$.

Exercise 9.7.2. Verify that the Galois group of $f(x) = x^3 + 7x + 7$ over $\mathbb{Q}$ is S_3. Determine, as explicitly as possible, all intermediate fields between the rationals and the splitting field of $f(x)$.

ercise 9.7.3. Show that the discriminant polynomial $\delta^2(x_1,\dots,x_n)$ is a polynomial of degree $2n-2$ in the elementary symmetric polynomials. *Hint:* Show that when $\delta^2(x_1,\dots,x_n)$ is expanded as a polynomial in the elementary symmetric functions,

$$\delta^2(x_1,\dots,x_n) = d_n(\epsilon_1,\dots,\epsilon_n),$$

the monomial of highest total degree in d_n comes from the lexicographically highest monomial in $\delta^2(x_1,\dots,x_n)$. Identify the lexicographically highest monomial in $\delta^2(x_1,\dots,x_n)$, say x^α, and find the degree of the corresponding monomial ϵ_{α^*}.

Exercise 9.7.4. Let $f(x) = \sum_i a_i x^i$ have degree n and roots $\alpha_1,\dots,\alpha_n$. Using the previous exercise, show that $\delta^2(f) = a_n^{2n-2}\delta^2(\alpha_1,\dots,\alpha_n)$ is a homogeneous polynomial of degree $2n-2$ in the coefficients of f.

Exercise 9.7.5. Determine $\delta^2(f)$ as a polynomial in the coefficients of f when the degree of f is $n=2$ or $n=3$.

Exercise 9.7.6. Check that x^4+x^2+x+1 is irreducible over $\mathbb{Z}_3$.

Exercise 9.7.7. Use a computer algebra package (e.g., *Maple* or *Mathematica*) to find the discriminants of the following polynomials. You may refer to the *Mathematica* notebook **Discriminants-and-Resultants.nb**, available on my world wide web site. The *Mathematica* function **Discriminant[ ]**, available in that notebook, computes the discriminant.

(a) x^4-3x^2+2x+5
(b) x^4+10x^3-3x+4
(c) $x^3-14x+10$
(d) x^3-12

Exercise 9.7.8. Each of the polynomials in the previous exercise is irreducible over the rationals. The Eisenstein criterion applies to one of the polynomials, and the others can be checked by computing factorizations of the reductions modulo small primes. For each polynomial, determine which Galois groups are consistent with the computation of the discriminant.

Exercise 9.7.9. Suppose that f and g are polynomials in $K[x]$ of degrees n and m respectively, and that there exist $\varphi(x)$ of degree no more than $n-1$ and $\psi(x)$ of degree no more than $m-1$ such that $f(x)\psi(x) = g(x)\varphi(x)$. Show that f and g have a non-constant common divisor in $K[x]$.

Exercise 9.7.10. Let f and g be polynomials of degree n and m, respectively, with roots ξ_i and η_j.

(a) Show that $R(f, g) = (-1)^{n+m} R(g, f)$.
(b) Show that

$$R(f, g) = a_n^m \prod_i g(\xi_i).$$

(c) Show that

$$R(f, g) = (-1)^{n+m} b_m^n \prod_j f(\eta_j).$$

Exercise 9.7.11. Show that

$$\prod_{i=1}^{n} \prod_{j=1}^{m} (x_i - y_j)$$

is a polynomial of total degree m in the elementary symmetric functions $\epsilon_i(x_1, \dots, x_n)$ and of total degree n in the elementary symmetric functions $\epsilon_j(y_1, \dots, y_m)$.

Exercise 9.7.12. Suppose f and g are polynomials of degree n and m, respectively. Suppose that there exist polynomials $\varphi(x)$ of degree no more than $n - 1$ and $\psi(x)$ of degree no more than $m - 1$ such that $f(x)\psi(x) = g(x)\varphi(x)$. Show that f and g have a non-constant common divisor.

Exercise 9.7.13. Verify the assertion in the text that $\det(\mathcal{R}(f, g))$ is divisible by $\prod_i \prod_j (\xi_i - \eta_j) = R(f, g)$.

Exercise 9.7.14. Use a computer algebra package (e.g., *Maple* or *Mathematica* to find the resultants:

(a) $R(x^3 + 2x + 5, x^2 - 4x + 5)$
(b) $R(3x^4 + 7x^3 + 2x - 5, 12x^3 + 21x^2 + 2)$

9.8. Quartic Polynomials

In this section, we determine the Galois groups of quartic polynomials. Consider a quartic polynomial

$$f(x) = x^4 + ax^3 + bx^2 + cx + d \tag{9.8.1}$$

with coefficients in some field K. It is convenient first to eliminate the cubic term by the linear change of variables $x = y - a/4$. This yields

$$f(x) = g(y) = y^4 + px^2 + qx + r, \tag{9.8.2}$$

with

$$p = -\frac{3a^2}{8} + b,$$
$$(9.8.3) \qquad q = \frac{a^3}{8} - \frac{ab}{2} + c,$$
$$r = -\frac{3a^4}{256} + \frac{a^2 b}{16} - \frac{ac}{4} + d,$$

We can suppose, without loss of generality, that g is irreducible, since otherwise we could analyze g by analyzing its factors. We also suppose that g is separable. Let G denote the Galois group of g over K.

Using one of the techniques for computing the discriminant from the previous section, one computes that the discriminant of g (or f) is:

$$(9.8.4) \qquad -4\,p^3\,q^2 - 27\,q^4 + 16\,p^4\,r + 144\,p\,q^2\,r - 128\,p^2\,r^2 + 256\,r^3,$$

or, in terms of a, b, c, and d,

$$a^2\,b^2\,c^2 - 4\,b^3\,c^2 - 4\,a^3\,c^3 + 18\,a\,b\,c^3 - 27\,c^4 - 4\,a^2\,b^3\,d +$$
$$(9.8.5) \qquad 16\,b^4\,d + 18\,a^3\,b\,c\,d - 80\,a\,b^2\,c\,d - 6\,a^2\,c^2\,d + 144\,b\,c^2\,d -$$
$$27\,a^4\,d^2 + 144\,a^2\,b\,d^2 - 128\,b^2\,d^2 - 192\,a\,c\,d^2 + 256\,d^3 \quad .$$

We can distinguish two cases, according to whether $\delta(g) \in K$ or not:

Case 1. $\delta(g) \in K$. Then the Galois group is isomorphic to A_4 or $\mathcal{V} \cong \mathbb{Z}_2 \times \mathbb{Z}_2$, since these are the transitive subgroups of A_4, according to Exercise 5.1.20.

Case 2. $\delta(g) \notin K$. Then the Galois group is isomorphic to S_4, D_4, or $\mathbb{Z}_4$, since these are the transitive subgroups of S_4 which are not contained in A_4, according to Exercise 5.1.20.

Denote the roots of g by $\alpha_1, \alpha_2, \alpha_3, \alpha_4$, and let E denote the splitting field $K(\alpha_1, \dots, \alpha_4)$. The next idea in analyzing the quartic equation is to introduce the elements:

$$\theta_1 = (\alpha_1 + \alpha_2)(\alpha_3 + \alpha_4)$$
$$(9.8.6) \qquad \theta_2 = (\alpha_1 + \alpha_3)(\alpha_2 + \alpha_4)$$
$$\theta_3 = (\alpha_1 + \alpha_4)(\alpha_2 + \alpha_3),$$

and the cubic polynomial, called the *resolvant cubic*:

$$(9.8.7) \qquad h(y) = (y - \theta_1)(y - \theta_2)(y - \theta_3).$$

Expanding $h(y)$ and identifying symmetric functions in the α_i with coefficients of g gives:

$$(9.8.8) \qquad h(y) = y^3 - 2py^2 + (p^2 - 4r)y + q^2.$$

The discriminant $\delta^2(h)$ turns out to be identical with the discriminant $\delta^2(g)$ (exercise). We distinguish cases according to whether the resolvant cubic is irreducible or not.

Case 1A. $\delta(g) \in K$ and h is irreducible over K. In this case,

$$[K(\theta_1, \theta_2, \theta_3) : K] = 6,$$

and 6 divides the order of G. The only possibility is $G = A_4$.

Case 2A. $\delta(g) \notin K$ and h is irreducible over K. Again, 6 divides the order of G. The only possibility is $G = S_4$.

Case 1B. $\delta(g) \in K$ and h is not irreducible over K. Then G must be $\mathcal{V}$. (In particular, 3 does not divide the order of G, so the Galois group of h must be trivial, and h factors into linear factors over K. Conversely, if h splits over K, then the θ_i are all fixed by G, which implies that $G \subseteq \mathcal{V}$. The only possibility is then that $G = \mathcal{V}$.)

Case 2B. $\delta(g) \notin K$ and h is not irreducible over K. The Galois group must be one of D_4 and $\mathbb{Z}_4$. By the remark under Case 1B, h does not split over K, so it must factor into a linear factor and an irreducible quadratic; i.e., exactly one of the θ_i lies in K.

We can assume without loss of generality that $\theta_1 \in K$. Then, G is contained in the stabilizer of θ_1, which is the copy of D_4 generated by $\{(12)(34), (13)(24)\}$,

$$D_4 = \{e, (1324), (12)(34), (1423), (34), (12), (13)(24), (14)(23)\}.$$

G is either equal to D_4 or to

$$\mathbb{Z}_4 = \{e, (1324), (12)(34), (1423)\}.$$

It remains to distinguish between these cases.

Since $K(\delta) \subseteq K(\theta_1, \theta_2, \theta_3)$ and both are quadratic over K, it follows that $K(\delta) = K(\theta_1, \theta_2, \theta_3)$. $K(\delta)$ is the fixed field of $G \cap A_4$, which is either

$$D_4 \cap A_4 = \{e, (12)(34), (13)(24), (14)(23)\} \cong \mathcal{V}, \quad \text{or}$$

$$\mathbb{Z}_4 \cap A_4 = \{e, (12)(34)\} \cong \mathbb{Z}_2.$$

It follows that the degree of the splitting field E over the intermediate field $K(\delta) = K(\theta_1, \theta_2, \theta_3)$ is either 4 or 2, according to whether G is D_4 or $\mathbb{Z}_4$. So one possibility for finishing the analysis of the Galois group in Case 2B is to compute the dimension of E over $K(\delta)$. The following lemma can be used for this. This result is taken from L. Kappe and B. Warren, *An elementary test for the Galois group of a quartic polynomial*, The American Mathematical Monthly **96** (1989) 133-137 .

Lemma 9.8.1. *Let $g(x) = y^4 + py^2 + qy + r$ be an irreducible quartic polynomial over a field K. Let $h(x)$ denote the resolvant cubic of g. Suppose that $\delta^2 = \delta^2(g)$ is not a square in K and that h is not irreducible over K. Let θ denote the one root of the resolvant cubic which lies in K, and define:*

$$H(x) = (x^2 + \theta)(x^2 + (\theta - p)x + r). \tag{9.8.9}$$

Then the Galois group of g is $\mathbb{Z}_4$ if, and only if, $H(x)$ splits over $K(\delta)$.

Proof. We maintain the notation from the discussion above: the roots of g are denoted by α_i, the splitting field of g by E, and the roots of h by θ_i. Let $L = K(\delta) = K(\theta_1, \theta_1, \theta_3)$. Assume without loss of generality that the one root of h in K is θ_1. Now consider the polynomial

$$\begin{aligned} &(x-\alpha_1\alpha_2)(x - \alpha_3\alpha_4)(x - (\alpha_1 + \alpha_2))(x - (\alpha_3 + \alpha_4) \\ &\quad = (x^2 - (\alpha_1\alpha_2 + \alpha_3\alpha_4)x + r)(x^2 + \theta_1). \end{aligned} \tag{9.8.10}$$

Compute that $p - \theta = \alpha_1\alpha_2 + \alpha_3\alpha_4$. It follows that the polynomial displayed above is none other than $H(x)$.

Suppose that $H(x)$ splits over L, so $\alpha_1\alpha_2$, $\alpha_3\alpha_4$, $\alpha_1 + \alpha_2$, $\alpha_3 + \alpha_4 \in L$. It follows that α_1 satisfies a quadratic polynomial over L,

$$(x - \alpha_1)(x - \alpha_2) = x^2 - (\alpha_1 + \alpha_2)x + \alpha_1\alpha_2 \in L[x],$$

and $[L(\alpha_1) : L] = 2$. But one can check that $L(\alpha_1) = E$. Consequently, $[E : L] = 2$ and $G = \mathbb{Z}_4$, by the discussion preceding the lemma.

Conversely, suppose that the Galois group G is $\mathbb{Z}_4$. Because $\theta_1 = (\alpha_1 + \alpha_2)(\alpha_3 + \alpha_4)$ is in K and is, therefore, fixed by the generator σ of G, one has $\sigma^{\pm 1} = (1324)$. The fixed field of $\sigma^2 = (12)(34)$ is the unique intermediate field between K and E, so L equals this fixed field. Each of the roots of $H(x)$ is fixed by σ^2 and, hence, an element of L. That is, H splits over L. □

Examples of the use of this lemma are given below. We shall now explain how to solve explicitly for the roots α_i of the quartic equation in terms of the roots θ_i of the resolvant cubic.

Note that

$$(\alpha_1 + \alpha_2)(\alpha_3 + \alpha_4) = \theta_1 \quad \text{and} \quad (\alpha_1 + \alpha_2) + (\alpha_3 + \alpha_4) = 0, \tag{9.8.11}$$

which means that $(\alpha_1 + \alpha_2)$ and $(\alpha_3 + \alpha_4)$ are the two square roots of $-\theta_1$. Similarly, $(\alpha_1 + \alpha_3)$ and $(\alpha_2 + \alpha_4)$ are the two square roots of $-\theta_2$, and $(\alpha_1 + \alpha_4)$ and $(\alpha_2 + \alpha_3)$ are the two square roots of $-\theta_3$. It is possible to choose the signs of the square roots consistently, noting that:

$$\sqrt{-\theta_1}\sqrt{-\theta_2}\sqrt{-\theta_2} = \sqrt{-\theta_1\theta_2\theta_3} = \sqrt{q^2} = q. \tag{9.8.12}$$

That is, it is possible to choose the square roots so that their product is q. One can check that

$$(\alpha_1 + \alpha_2)(\alpha_1 + \alpha_3)(\alpha_1 + \alpha_4) = q, \tag{9.8.13}$$

and so put

(9.8.14)

$$(\alpha_1 + \alpha_2) = \sqrt{-\theta_1} \quad (\alpha_1 + \alpha_3) = \sqrt{-\theta_2} \quad (\alpha_1 + \alpha_4) = \sqrt{-\theta_3}.$$

Using this together with $\alpha_1 + \alpha_2 + \alpha_3 + \alpha_4 = 0$, one gets:

$$2\alpha_i = \pm\sqrt{-\theta_1} \pm \sqrt{-\theta_2} \pm \sqrt{-\theta_3}, \tag{9.8.15}$$

with the four choices of signs giving the four roots α_i (as long as the characteristic of the ground field is not 2).

Example 9.8.2. Take $K = \mathbb{Q}$ and $f(x) = x^4 + 3x^3 - 3x - 2$. Applying the linear change of variables $x = y - 3/4$ gives $f(x) = g(y) = -\frac{179}{256} + \frac{3}{8}\,y - \frac{27}{8}\,y^2 + y^4$. The reduction of f modulo 5 is irreducible over $\mathbb{Z}_5$ and, therefore, f is irreducible over $\mathbb{Q}$. The discriminant of f (or g or h) is -2183, which is not a square in $\mathbb{Q}$. Therefore, the Galois group of f is not contained in the alternating group A_4. The resolvant cubic of g is $h(y) = \frac{9}{64} + \frac{227\,y}{16} + \frac{27\,y^2}{4} + y^3$. The reduction of $64h$ modulo 7 is irreducible over $\mathbb{Z}_7$ and, hence, h is irreducible over $\mathbb{Q}$. It follows that the Galois group is S_4.

Example 9.8.3. Take $K = \mathbb{Q}$ and $f(x) = 21 + 12\,x + 6\,x^2 + 4\,x^3 + x^4$. Applying the linear change of variables $x = y - 1$ gives $f(x) = g(y) = 12 + 8\,y + y^4$. The reduction of g modulo 5 factors as:

$$\bar{g}(y) = y^4 + 3y + 2 = (1 + y)\left(2 + y + 4\,y^2 + y^3\right).$$

By the rational root test, g has no rational root; therefore, if it is not irreducible, it must factor into two irreducible quadratics. But this would be inconsistent with the factorization of the reduction modulo 5. Hence, g is irreducible.

The discriminant of g is $331776 = (576)^2$, and therefore, the Galois group of g is contained in the alternating group A_4. The resolvant cubic of g is $h(y) = 64 - 48\,y + y^3$. The reduction of h modulo 5 is irreducible over $\mathbb{Z}_5$, so h is irreducible over $\mathbb{Q}$. It follows that the Galois group of g is A_4.

Example 9.8.4. Take $K = \mathbb{Q}$ and $f(x) = x^4 + 16x^2 - 1$. The reduction of f modulo 3 is irreducible over $\mathbb{Z}_3$, so f is irreducible over $\mathbb{Q}$. The discriminant of f is -1081600, so the Galois group G of f is not contained in the alternating group. The resolvant cubic of f is

$$h(x) = 260\,x - 32\,x^2 + x^3 = x\left(260 - 32\,x + x^2\right)$$

Because h is reducible, it follows that G is either Z_4 or D_4. To determine which, we can use the criterion of Lemma 9.8.1.

Note first that $\delta = \sqrt{-1081600} = 1040\sqrt{-1}$, so $\mathbb{Q}(\delta) = \mathbb{Q}(\sqrt{-1})$. The root of the resolvant cubic in $\mathbb{Q}$ is zero, so the polynomial $H(x)$ of Lemma 9.8.1 is $x^2(x^2 - 16x - 1)$. The non-zero roots of this are $8 \pm \sqrt{65}$, so H does not split over $\mathbb{Q}(\delta)$. Therefore, the Galois group is D_4.

The following lemma also implies that the Galois group is D_4 rather than $\mathbb{Z}_4$:

Lemma 9.8.5. *Let K be a subfield of $\mathbb{R}$ and let f be an irreducible quartic polynomial over K whose discriminant is negative. Then The Galois group of f over K is not $\mathbb{Z}_4$.*

Proof. Complex conjugation, which we denote here by γ, is an automorphism of $\mathbb{C}$ which leaves invariant the coefficients of f. The splitting field E of f can be taken to be a subfield of $\mathbb{C}$, and γ induces a K-automorphism of E, since E is Galois over K.

The square root δ of the discriminant δ^2 is always contained in the splitting field. As δ^2 is assumed to be negative, δ is pure imaginary. In particular, E is not contained in the reals, and the restriction of γ to E has order 2.

Suppose that the Galois group G of f is cyclic of order 4. Then G contains a unique subgroup of index 2, which must be the subgroup generated by γ. By by the Galois Correspondence, there is a unique intermediate field $K \subseteq L \subseteq E$ of dimension 2 over K, which is the fixed field of γ. But $K(\delta)$ is a quadratic extension of K which is *not* fixed pointwise by γ. This is a contradiction. □

Example 9.8.6. Take the ground field to be $\mathbb{Q}$ and $f(x) = x^4 + 5x + 5$. The irreducibility of f follows from the Eisenstein Criterion. The discriminant of f is 15125, which is not a square in Q; therefore, the Galois group G of f is not contained in the alternating group. The resolvant cubic of f is $25 - 20\,x + x^3 = (5 + x)\left(5 - 5\,x + x^2\right)$. Because h is reducible, it follows that the G is either Z_4 or D_4, but this time the discriminant is positive, so the criterion of Lemma 9.8.5 fails.

Note that the splitting field of the resolvant cubic is $\mathbb{Q}(\delta) = \mathbb{Q}(\sqrt{5})$, since $\sqrt{(\delta)} = 55\sqrt{5}$. The one root of the resolvant cubic in $\mathbb{Q}$ is -5. Therefore, the polynomial $H(x)$ of Lemma 9.8.1 is $\left(-5 + x^2\right)\left(5 - 5x + x^2\right)$. Because $H(x)$ splits over $\mathbb{Q}(\sqrt{5})$, the Galois group is $\mathbb{Z}_4$.

Example 9.8.7. Set $f(x) = x^4 + 6x^2 + 4$. This is a so-called biquadratic polynomial, whose roots are $\pm\sqrt{\alpha}, \pm\sqrt{\beta}$, where α and β are the roots of the quadratic polynomial $x^2 + 6x + 4$. Because the quadratic polynomial is irreducible over the rationals, so is the quartic f. The discriminant of f is $25600 = 160^2$, so the Galois group is contained in the alternating group. But the resolvant cubic $20x - 12x^2 + x^3 = (-10 + x)(-2 + x)x$ is reducible over $\mathbb{Q}$, so the Galois group is $\mathcal{V}$.

A *Mathematica* notebook **Quartic.nb** for investigation of Galois groups of quartic polynomials can be found on my web site. This notebook can be used to verify the computations in the examples as well as to help with the exercises.

Exercises

Exercise 9.8.1. Verify Equation 9.8.8.

Exercise 9.8.2. Show that a linear change of variables $y = x + c$ does not alter the discriminant of a polynomial.

Exercise 9.8.3. Show how to modify the treatment in this section to deal with a quartic polynomial which is not monic.

Exercise 9.8.4. Show that the resolvant cubic h of an irreducible quartic polynomial f has the same discriminant as f, $\delta^2(h) = \delta^2(f)$.

Exercise 9.8.5. Let $f(x)$ be an irreducible quartic polynomial over a field K, and let δ^2 denote the discriminant of f. Show that the splitting field of the resolvant cubic of f equals $K(\delta)$ if, and only if, the resolvant cubic is not irreducible over K.

Exercise 9.8.6. Show that $x^4 + 3x + 3$ is irreducible over $\mathbb{Q}$, and determine the Galois group.

Exercise 9.8.7. For p a prime other than 3 and 5, show that $x^4 + px + p$ is irreducible with Galois group S_4. The case $p = 3$ is treated in the previous exercise, and the case $p = 5$ is treated in Example 9.8.6.

Exercise 9.8.8. Find the Galois group of $f(x) = x^4 + 5x^2 + 3$. Compare Example 9.8.7.

Exercise 9.8.9. Show that the biquadratic $f(x) = x^4 + px^2 + r$ has Galois group $\mathcal{V}$ precisely when r is a square in the ground field K. Show that in general $K(\delta) = K(\sqrt{r})$.

Exercise 9.8.10. Find examples of the biquadratic $f(x) = x^4 + px^2 + r$ for which the Galois group is $\mathbb{Z}_4$ and examples for which the Galois group is D_4. Find conditions for the Galois group to be cyclic and for the Galois group to be the dihedral group.

Exercise 9.8.11. Determine the Galois group over $\mathbb{Q}$ for each of the following polynomials. You may need to do computer aided calculations, for example, using the *Mathematica* notebook **Quartic.nb** on my web site.

(a) $21 + 6x - 17x^2 + 3x^3 + 21x^4$
(b) $1 - 12x + 36x^2 - 19x^4$
(c) $-33 + 16x - 39x^2 + 26x^3 - 9x^4$
(d) $-17 - 50x - 43x^2 - 2x^3 - 33x^4$
(e) $-8 - 18x^2 - 49x^4$
(f) $40 + 48x + 44x^2 + 12x^3 + x^4$
(g) $82 + 59x + 24x^2 + 3x^3 + x^4$

9.9. Galois Groups of Higher Degree Polynomials

In this section, we shall discuss the computation of Galois groups of polynomials in $\mathbb{Q}[x]$ of degree 5 or more.

If a polynomial $f(x) \in K[x]$ of degree n has irreducible factors of degrees $m_1 \geq m_2 \geq \cdots \geq m_r$, where $\sum_i m_i = n$, we say that $(m_1, m_2, \ldots, m_r)$ is the *degree partition* of f. Let $f(x) \in \mathbb{Z}[x]$ and let $\tilde{f}(x) \in \mathbb{Z}_p[x]$ be the reduction of f modulo a prime p; if $\tilde{f}$ has degree partition α, we say that *f has degree partition α modulo p.*

The following theorem is fundamental for the computation of Galois groups over $\mathbb{Q}$.

Theorem 9.9.1. *Let f be an irreducible polynomial of degree n with coefficients in $\mathbb{Z}$. Let p be a prime which does not divide the leading coefficient of f nor the discriminant of f. Suppose that f has degree partition α modulo p. Then the Galois group of f contains a permutation of cycle type α.*

If f is an irreducible polynomial of degree n whose Galois group does not contain an n-cycle, then the reduction of f modulo a prime p is *never* irreducible of degree n. If the prime p does not divide the leading coefficient of f or the discriminant, then the reduction of f modulo p factors, according to the theorem. If the prime divides the leading coefficient, then the reduction has degree less than n. Finally, if the prime divides the discriminant, then the discriminant of the reduction is zero; but an irreducible polynomial over $\mathbb{Z}_p$ can never have multiple roots and therefore cannot have zero discriminant.

We shall not prove Theorem 9.9.1 here. One can find a proof in B. L. van der Waerden, *Algebra*, Volume I, Frederick Ungar Publishing Co., 1970, Section 8.10 (Translation of the 7th German edition, Springer Verlag, 1966).

This theorem, together with the computation of the discriminant, often suffices to determine the Galois group.

Example 9.9.2. Consider $f(x) = x^5 + 5x^4 + 3x + 2$ over the ground field $\mathbb{Q}$. The discriminant of f is 4557333, which is not a square in $\mathbb{Q}$ and which has prime factors 3, 11, and 138101. The reduction of f modulo 7 is irreducible, and the reduction modulo 41 has degree partition $(2, 1, 1, 1)$. It follows that f is irreducible over $\mathbb{Q}$, and that its Galois group over $\mathbb{Q}$ contains a 5-cycle and a 2-cycle. But a 5-cycle and a 2-cycle generate S_5, so the Galois group of f is S_5.

Example 9.9.3. Consider $f(x) = 4 - 4x + 9x^3 - 5x^4 + x^5$ over the ground field $\mathbb{Q}$. The discriminant of f is $15649936 = 3956^2$, so the Galois group G of f is contained in the alternating group A_5. The prime factors of the discriminant are 2, 23, and 43. The reduction of f modulo 3 is irreducible and the reduction modulo 5 has degree partition $(3, 1, 1)$. Therefore, f is irreducible over $\mathbb{Q}$, and its Galois group over $\mathbb{Q}$ contains a 5-cycle and a 3-cycle. But a 5-cycle and a 3-cycle generate the alternating group A_5, so the Galois group is A_5.

The situation is quite a bit more difficult if the Galois group is not S_n or A_n. In this case, one has to show that certain cycle types *do not* appear. The rest of this section is devoted to a (by no means definitive) discussion of this question.

Example 9.9.4. Consider $f(x) = -3 + 7x + 9x^2 + 8x^3 + 3x^4 + x^5$ over the ground field $\mathbb{Q}$. The discriminant of f is $1306449 = 1143^2$, so the Galois group G of f is contained in the alternating group A_5. The prime factors of the discriminant are 3 and 127. The transitive subgroups of the alternating group A_5 are A_5, D_5, and $\mathbb{Z}_5$, according to Exercise 5.1.20. The reduction of f modulo 2 is irreducible, and the reduction modulo 5 has degree partition $(2, 2, 1)$. Therefore G contains a 5-cycle and an element of cycle type $(2, 2, 1)$. This eliminates $\mathbb{Z}_5$ as a possibility for the Galois group.

If one computes the factorization of the reduction of f modulo a number of primes, one finds no instances of factorizations with degree partition $(3, 1, 1)$ In fact, for the first 1000 primes, the frequencies of degree partitions modulo p are:

1^5	$2, 1^4$	$2^2, 1$	$3, 1^2$	$3, 2$	$4, 1$	5
.093	0	.5	0	0	0	.4

These data certainly suggest strongly that the Galois group has no 3-cycles and therefore must be D_5 rather than A_5, but the empirical evidence does not yet constitute a proof! I would now like to present some facts which nearly, but not quite, constitute a method for determining that the Galois group is D_5 rather than A_5.

Let us compare our frequency data with the frequencies of various cycle types in the transitive subgroups of S_5. The first table below shows the number of elements of each cycle type in the various transitive subgroups of S_5, and the second table displays the frequencies of the cycle types, that is, the number of elements of a given cycle type divided by the order of the group.

	1^5	$2, 1^4$	$2^2, 1$	$3, 1^2$	$3, 2$	$4, 1$	5
$\mathbb{Z}_5$	1	0	0	0	0	0	4
D_5	1	0	5	0	0	0	4
A_5	1	0	15	20	0	0	24
$\mathbb{Z}_4 \ltimes \mathbb{Z}_5$	1	0	5	0	0	10	4
S_5	1	10	15	20	0	30	24

Numbers of elements of various cycle types

	1^5	$2, 1^4$	$2^2, 1$	$3, 1^2$	$3, 2$	$4, 1$	5
$\mathbb{Z}_5$	.2	0	0	0	0	0	.8
D_5	.1	0	.5	0	0	0	.4
A_5	.017	0	.25	.333	0	0	.4
$\mathbb{Z}_4 \ltimes \mathbb{Z}_5$	.05	0	.25	0	0	.5	.2
S_5	.0083	.083	.125	.167	0	.25	.2

Frequencies of various cycle types

Notice that our empirical data for frequencies of degree partitions modulo p for $f(x) = -3 + 7x + 9x^2 + 8x^3 + 3x^4 + x^5$ are remarkably close to the frequencies of cycle types for the group D_5. Let's go back to our previous example, the polynomial $f(x) = 4 - 4x + 9x^3 - 5x^4 + x^5$, whose Galois group is known to be A_5. The frequencies of degree partitions modulo the first 1000 primes are:

1^5	$2, 1^4$	$2^2, 1^3$	$3, 1^2$	$3, 2$	$4, 1$	5
.019	0	.25	.32	0	0	.41

These data are, again, remarkably close to the data for the distribution of cycle types in the group A_5. The following theorem (conjectured by Frobenius, and proved by Chebotarev in 1926) asserts that the frequencies of degree partitions modulo primes inevitably approximate the frequencies of cycle types in the Galois group:

Theorem 9.9.5 (Chebotarev). *Let $f \in \mathbb{Z}[x]$ be an irreducible polynomial of degree n, and let G denote the Galois group of f over $\mathbb{Q}$. For each partition α of n let d_α be the fraction of elements of G of cycle type α. For each $N \in \mathbb{N}$, let $d_{\alpha,N}$ be the fraction of primes p in the interval $[1, N]$ such that f has degree partition α modulo p. Then*

$$\lim_{N\to\infty} d_{\alpha,N} = d_\alpha.$$

Because the distribution of degree partitions modulo primes of the polynomial $f(x) = -3 + 7x + 9x^2 + 8x^3 + 3x^4 + x^5$ for the first 1000 primes is quite close to the distribution of cycle types for the group D_5 and quite far from the distribution of cycle types for A_5, our belief that the Galois group of f is D_5 is encouraged, but still not rigorously confirmed by this theorem! The difficulty is that we do not know for certain that, if we examine yet more primes, the distribution of degree partitions will not shift towards the distribution of cycle types for A_5.

What we would need in order to turn these observations into a method is a practical error estimate in Chebotarev's theorem. In fact, a number of error estimates have been published in the 1970's and 1980's, but I have not been able to see how to turn these estimates into a practical method, nor have I been able to find any description in the literature of a practical method based on these estimates. (The reader who wishes to look at the literature on this subject could start with a survey by J.-P. Serre, *Quelque applications du théorèm de densité de Chebotarev*, Publications Mathématiques, Institut de Hautes Études Scientifiques, **54** (1981). Be warned that you will find difficult and unfamiliar mathematics there, but you if you don't expect to understand everything, you can still learn something.)

The situation is quite annoying but also intriguing! Empirically, the distribution of degree partitions converge quite rapidly to the distribution of cycle types for the Galois group, so that one can usually identify the Galois group by this method, without having a proof that it is in fact the Galois group! (By the way, there exist pairs of non-isomorphic transitive subgroups of S_n for $n \geq 12$ which have the same distribution of cycle types, so that frequency of degree partitions cannot always determine the Galois group of a polynomial.)

A practical method of computing Galois groups of polynomials of small degree is described in L. Soicher and J. McKay, *Computing Galois groups over the Rationals*, Journal of Number Theory, vol. 20 (1985) pp. 273-281. The method is based on the computation and factorization of certain resolvant polynomials. This method has two great advantages: first, it works, and second, it is based on mathematics which you now know! I am not going to describe the method in full here, however. This method, or a similar one, has been implemented in the computer algebra package *Maple*; the command **galois( )** in *Maple* will compute Galois groups for polynomials of degree no more than 7.

Both the Chebotarev non-method and the Soicher-McKay method are based on having at hand a catalogue of potential Galois groups (i.e., transitive subgroups of S_n) together with certain identifying data for these groups. Transitive subgroups of S_n have been catalogued at least for $n \leq 11$; see G. Butler and J. McKay, *The transitive subgroups of degree up to 11*, Communications in Algebra, vol. 11 (1983), pp. 863-911.

By the way, if you write down a polynomial with integer coefficients "at random," the polynomial will probably be irreducible, will probably have Galois group S_n, and you will probably be able to show that the Galois group is S_n by examining the degree partition modulo p for only a few primes. In fact, just writing down polynomials "at random," you will have a hard time finding one whose Galois group is *not* S_n. Apparently, not all that much is known about the probability distribution of various Galois groups, aside from the predominant occurrence of the symmetric group.

A major unsolved problem is the so-called *inverse Galois problem*: Which groups can occur as Galois groups of polynomials over the rational numbers? A great deal is known about this problem; for example it is known that all solvable groups occur as Galois groups over $\mathbb{Q}$. (See Chapter 10 for a discussion of solvability.) However, the definitive solution to the problem is still out of reach.

Exercises

Find the *probable* Galois groups for each of the following quintic polynomials, by examining reductions of the polynomials modulo primes. In *Mathematica*, the command **Factor[f, Modulus → p]** will compute the factorization of the reduction of a polynomial f modulo a prime p. Using this, you can do a certain amount of computation "by hand." By writing a simple loop in *Mathematica*, you can examine the degree partition modulo p for the first N primes, say for $N = 100$. This will already give you a good idea of the Galois group in most cases. You can find a program for the computation of frequencies of degree partitions on my web site; consult the notebook **Galois-Groups.nb**.

Exercise 9.9.1. $x^5 + 2$.

Exercise 9.9.2. $x^5 + 20x + 16$.

Exercise 9.9.3. $x^5 - 5x + 12$.

Exercise 9.9.4. $x^5 + x^4 - 4x^3 - 3x^2 + 3x + 1$.

Exercise 9.9.5. $x^5 - x + 1$.

Exercise 9.9.6. Write down a few random polynomials of various degrees, and show that the Galois group is S_n.

Exercise 9.9.7. Project: Read the paper of Soicher and McKay. Write a program for determining the Galois group of polynomials over $\mathbb{Q}$ of degree ≤ 5 based on the method of Soicher and McKay.

Exercise 9.9.8. Project: Investigate the probability distribution of Galois groups for polynomials of degree ≤ 4 over the integers. If you have done the previous exercise, extend your investigation to polynomials of degree ≤ 5.

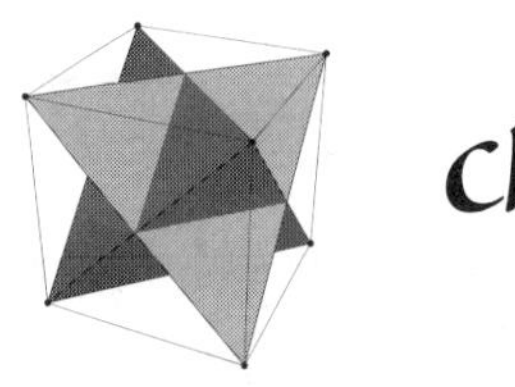

Chapter 10

Solvability

10.1. Composition Series and Solvable Groups

This section treats a decomposition of any finite group into simple pieces.

Definition 10.1.1. A group G which has no proper normal subgroup N ($\{e\} \subsetneq N \subsetneq G$) is called *simple*.

We know that a cyclic group of prime order is simple, as it has no proper subgroups at all. In fact, the cyclic groups of prime order comprise all abelian simple groups, see Exercise 10.1.2. In the next section, we will show that the alternating groups A_n for $n \geq 5$ are simple.

One of the heroic achievements of mathematics in this century is the complete classification of finite simple groups, which was finished in 1981. There are a number of infinite families of finite simple groups, namely, the cyclic groups of prime order, alternating groups, and certain groups of matrices over finite fields (simple groups of Lie type). Aside from the infinite families, there are 26 so-called *sporadic simple groups*. In the 1950's not all of the simple groups of Lie type were known, and only five of the sporadic groups were known. In the period from 1960 to 1980, knowledge of the simple groups of Lie type was completed and systematized, and the remaining sporadic groups were discovered. At the same time, classification theorems of increasing strength eventually showed that the known list of finite simple groups covered all possibilities. This achievement was a collaborative effort of many mathematicians.

Now, consider the chain of groups $\{e\} \subsetneq N \subsetneq G$. If N is not simple, then it is possible to interpose a subgroup N', $\{e\} \subsetneq N' \subsetneq N$, with N' normal in N. If G/N is not simple, then there is a proper normal subgroup $\{e\} \subsetneq \bar{N}'' \subsetneq G/N$; then the inverse image $N'' = \{g \in G : gN \in \bar{N}''\}$ is a normal subgroup of G satisfying $N \subsetneq N'' \subsetneq G$. If one continues in this way to interpose intermediate normal subgroups until no further additions are possible, the result is a *composition series*.

Definition 10.1.2. A *composition series* for a group G is a chain of subgroups $\{e\} = G_0 \subsetneq G_1 \subsetneq \cdots \subsetneq G_n = G$ such that for all i, G_i is normal in G_{i+1} and G_{i+1}/G_i is simple.

An induction based on the observations above shows that a finite group always has a composition series.

Proposition 10.1.3. *Any finite group has a composition series.*

Proof. A group of order 1 is simple, so has a composition series. Let G be a group of order $n > 1$, and assume inductively that whenever G' is a finite group of order less than n, then G' has a composition series. If G is simple, then $\{e\} \subsetneq G$ is already a composition series. Otherwise, there is a proper normal subgroup $\{e\} \subsetneq N \subsetneq G$. By the induction hypothesis, N has a composition series $\{e\} \subsetneq G_1 \subsetneq \cdots \subsetneq G_r = N$. Likewise, G/N has a composition series $\{e\} \subsetneq N_1 \subsetneq \cdots \subsetneq N_s = G/N$. Let $\pi : G \to G/N$ be the quotient map, and let $G_{r+i} = \pi^{-1}(N_i)$ for $0 \leq i \leq s$, and finally put $n = r + s$. Then $\{e\} \subsetneq G_1 \subsetneq \cdots \subsetneq G_n = G$ is a composition series for G. □

For abelian groups one can make a sharper statement: It follows from the structure theory of finite abelian groups that the only simple abelian groups are cyclic of prime order, and that any finite abelian group has a composition series in which the successive quotients are cyclic of prime order (exercise).

Note that there are many choices to be made in the construction of a composition series. For example an abelian group of order 45 has compositions series in which successive quotient groups have order 3, 3, 5, another in which the successive quotient groups have order 3, 5, 3, and another in which successive quotient groups have order 5, 3, 3 (exercise).

There is a theorem of C. Jordan and O. Hölder (Hölder, 1882, based on earlier work of Jordan) which says that the simple groups appearing in any two composition series of a finite group are the same up to order (and isomorphism). I am not going to give a proof of the Jordan-Hölder theorem here, but you can find a proof in any more advanced book on group theory or in many reference works on algebra.

Definition 10.1.4. A finite group is said to be *solvable* if it has a composition series in which the successive quotients are cyclic of prime order.

Proposition 10.1.5. *A finite group G is solvable if, and only if, it has a chain of subgroups $\{e\} \underset{\neq}{\subset} G_1 \underset{\neq}{\subset} \cdots \underset{\neq}{\subset} G_n = G$ in which each G_i is normal in G_{i+1} and the quotients G_{i+1}/G_i are abelian.*

Proof. Exercise 10.1.3. □

Exercises

Exercise 10.1.1. Show that an abelian group G of order 45 has a composition series $\{e\} \subseteq G_1 \subseteq G_2 \subseteq G_3 = G$ in which the successive quotient groups have order $3, 3, 5$, and another in which the successive quotient groups have orders $3, 5, 3$, and yet another in which the successive quotient groups have orders $5, 3, 3$.

Exercise 10.1.2.

(a) Show that any group of order p^n, where p is a prime, has a composition series in which successive quotients are cyclic of order p.

(b) Show that the structure theory for finite abelian groups implies that any finite abelian group has a composition series in which successive quotients are cyclic of prime order. In particular, the only simple abelian groups are the cyclic groups of prime order.

Exercise 10.1.3. Prove Proposition 10.1.5.

Exercise 10.1.4. Show that the symmetric groups S_n for $n \leq 4$ are solvable.

10.2. Commutators and Solvability

In this section, we develop by means of exercises a description of solvability of groups in terms of *commutators*. This section can be skipped without loss of continuity; however, the final three exercises in the section will be referred to in later proofs.

Definition 10.2.1. The *commutator* of elements x, y in a group is $[x, y] = x^{-1}y^{-1}xy$. The commutator subgroup or derived subgroup of a group G is the subgroup generated by all commutators of elements of G. The commutator subgroup is denoted $[G, G]$ or G'.

Exercise 10.2.1. Calculate the commutator subgroup of the dihedral group D_n.

Exercise 10.2.2. Calculate the commutator subgroup of the symmetric groups S_n for $n = 3, 4$.

Exercise 10.2.3. For H and K subgroups of a group G, let $[H, K]$ be the subgroup of G generated by commutators $[h, k]$ with $h \in H$ and $k \in K$.

(a) Show that if H and K are both normal, then $[H, K]$ is normal in G, and that $[H, K] \subseteq H \cap K$.

(b) Conclude that the commutator subgroup $[G, G]$ is normal in G.

(c) Define $G^{(0)} = G$, $G^{(1)} = [G, G]$, and, in general, $G^{(i+1)} = [G^{(i)}, G^{(i)}]$. Show by induction that $G^{(i)}$ is normal in G for all i and $G^{(i)} \supseteq G^{(i+1)}$.

Exercise 10.2.4. Show that G/G' is abelian and that G' is the unique smallest subgroup of G such that G/G' is abelian.

Theorem 10.2.2. *For a group G, the following are equivalent:*

(a) *G is solvable.*

(b) *There is a natural number n such that the n-th commutator subgroup $G^{(n)}$ is the trivial subgroup $\{e\}$.*

Proof. We use the characterization of solvability in Exercise 10.1.5.

If the condition (b) holds, then:

$$\{e\} = G^{(n)} \subseteq G^{(n-1)} \subseteq \cdots \subseteq G^{(1)} \subseteq G$$

is a chain of subgroups, each normal in the next, with abelian quotients. Therefore, G is solvable.

Suppose, conversely, that G is solvable and that

$$\{e\} = H_r \subseteq H_{r-1} \subseteq \cdots \subseteq H_1 \subseteq G$$

is a chain of subgroups, each normal in the next, with abelian quotients. We show by induction that $G^{(k)} \subseteq H_k$ for all k; in particular, $G^{(r)} = \{e\}$. Since G/H_1 is abelian, it follows from Exercise 10.2.4 that $G^{(1)} \subseteq H_1$.

Assume inductively that $G^{(i)} \subseteq H_i$ for some i $(1 \le i < r)$. Since H_i/H_{i+1} is assumed to be abelian, it follows from Exercise 10.2.4 that $[H_i, H_i] \subseteq H_{i+1}$; but then $G^{(i+1)} = [G^{(i)}, G^{(i)}] \subseteq [H_i, H_i] \subseteq H_{i+1}$. This completes the inductive argument. □

In the following exercises, you can use whatever criterion for solvability appears most convenient.

Exercise 10.2.5. Show that any subgroup of a solvable group is solvable.

Exercise 10.2.6. Show that any quotient group of a solvable group is solvable.

Exercise 10.2.7. Show that if $N \underset{\neq}{\subset} G$ is a normal subgroup and both N and G/N are solvable, then also G is solvable.

10.3. Simplicity of the Alternating Groups

In this section, we will prove that the symmetric groups S_n and the alternating groups A_n for $n \ge 5$ are not solvable. In fact, we will see that the alternating group A_n is the unique proper normal subgroup of S_n for $n \ge 5$, and moreover that A_n is a simple group for $n \ge 5$.

Recall that conjugacy classes in the symmetric group S_n are determined by cycle structure. Two elements are conjugate precisely when they have the same cycle structure. This fact is used frequently in the following.

Lemma 10.3.1. *For $n \ge 3$, the alternating group A_n is generated by 3-cycles.*

Proof. A product of two 2-cycles in S_n is conjugate to $(12)(12) = e = (123)(132)$, if the two 2-cycles are equal; or to $(12)(23) = (123)$, if the two 2-cycles have one digit in common; or to $(12)(34) = (132)(134)$, if the two 2-cycles have no digits in common. Thus, any product of two 2-cycles can be written as a product of one or two 3-cycles. Any even permutation is a product of an even number of 2-cycles, and therefore, can be written as a product of 3-cycles. □

Since the center of a group is always a normal subgroup, if one wants to show that a non-abelian group is simple, it makes sense to check first that the center is trivial. You are asked to show in Exercise 10.3.1 that for $n \geq 3$ the center of S_n is trivial and in Exercise 10.3.4 that for $n \geq 4$ the center of A_n is trivial.

Theorem 10.3.2. *If $n \geq 5$ and N is a normal subgroup of S_n such that $N \neq \{e\}$, then $N \supseteq A_n$.*

Proof. By Exercise 10.3.1, the center of S_n consists of e alone. Let $\sigma \neq e$ be an element of N; since σ is not central and since 2-cycles generate S_n there is some 2-cycle τ such that $\sigma\tau \neq \tau\sigma$. Consider the element $\tau\sigma\tau\sigma^{-1}$; writing this as $(\tau\sigma\tau)\sigma^{-1}$ and using the normality of N, we see that this element is in N; on the other hand, writing the element as $\tau(\sigma\tau\sigma^{-1})$, we see that the element is a product of two unequal 2-cycles.

If these two 2-cycles have a digit in common, then the product is a 3-cycle.

If the two 2-cycles have no digit in common, then by normality of N, N contains all elements which are a product of two disjoint 2-cycles. In particular, N contains the elements $(12)(34)$ and $(12)(35)$, and, therefore, the product $(12)(34)(12)(35) = (34)(35) = (435)$. So also in this case, N contains a 3-cycle.

By normality of N, N contains all 3-cycles, and, therefore, by Lemma 10.3.1, $N \supseteq A_n$. □

Lemma 10.3.3. *If $n \geq 5$, then all 3-cycles are conjugate in A_n*

Proof. It suffices to show that any 3-cycle is conjugate in A_n to (123). Let σ be a 3-cycle. Since σ and (123) have the same cycle structure, there is an element of $\tau \in S_n$ such that $\tau(123)\tau^{-1} = \sigma$. If τ is even, there is nothing more to do. Otherwise, $\tau' = \tau(45)$ is even, and $\tau'(123)\tau'^{-1} = \sigma$. □

Theorem 10.3.4. *If $n \geq 5$, then A_n is simple.*

Proof. Suppose $N \neq \{e\}$ is a normal subgroup of A_n and that $\sigma \neq e$ is an element of N. Since the center of A_n is trivial and A_n is generated by 3-cycles, there is a 3-cycle τ which does not commute with σ. Then (as in the proof of the previous theorem) the element $\sigma\tau\sigma^{-1}\tau^{-1}$ is a non-identity element of N which is a product of two 3-cycles.

The rest of the proof consists in showing that N must contain a 3-cycle. Then by Lemma 10.3.3, N must contain all 3-cycles, and, therefore, $N = A_n$ by Lemma 10.3.1.

The product π of two 3-cycles must be of one of the following types:

1. $(a_1a_2a_3)(a_4a_5a_6)$
2. $(a_1a_2a_3)(a_1a_4a_5) = (a_1a_4a_5a_2a_3)$
3. $(a_1a_2a_3)(a_1a_2a_4) = (a_1a_3)(a_2a_4)$
4. $(a_1a_2a_3)(a_2a_1a_4) = (a_1a_4a_3)$
5. $(a_1a_2a_3)(a_1a_2a_3) = (a_1a_3a_2)$

In either of the last two cases, N contains a 3-cycle.

In case (3), since $n \geq 5$, A_n contains an element $(a_2a_4a_5)$. Compute that $(a_2a_4a_5)\pi(a_2a_4a_5)^{-1} = (a_1a_3)(a_4a_5)$, and the product

$$\pi^{-1}(a_2a_4a_5)\pi(a_2a_4a_5)^{-1} = (a_2a_5a_4)$$

is a 3-cycle in N.

In case (2), compute that $(a_1a_4a_3)\pi(a_1a_4a_3)^{-1} = (a_4a_3a_5a_2a_1)$, and $\pi^{-1}(a_1a_4a_3)\pi(a_1a_4a_3)^{-1} = (a_4a_2a_3)$ is a 3-cycle in N.

Finally, in case (1), compute that

$$(a_1a_2a_4)\pi(a_1a_2a_4)^{-1} = (a_2a_4a_3)(a_1a_5a_6),$$

and the product $\pi^{-1}(a_1a_2a_4)\pi(a_1a_2a_4)^{-1}$ is $(a_1a_4a_6a_2a_3)$, namely, a 5-cycle. But then by case (b), N contains a 3-cycle. □

Let me remark, in case these computations look mysterious and unmotivated, that the idea is to take a 3-cycle x and to form the "commutator" $\pi^{-1}x\pi x^{-1}$. This is a way to get a lot of new elements of N. If one experiments just a little, one can find a 3-cycle x such that the commutator is either a 3-cycle or, at the worst, has one of the cycle structures already dealt with.

Here is an alternative way to finish the proof which avoids the computations. I learned this method from I. Herstein, *Abstract Algebra*, 3rd edition, Prentice Hall, 1996.

I claim that the simplicity of A_n for $n > 6$ follows from the simplicity of A_6. Suppose $n > 6$ and that $N \neq \{e\}$ is a normal subgroup of A_n. By the first part of the proof, N contains a non-identity element which is a product of two 3-cycles. These two 3-cycles involve at most 6 of the digits $1 \leq k \leq n$; so there is an isomorphic copy of S_6 in S_n such that $N \cap A_6 \neq \{e\}$. Since $N \cap A_6$ is a normal subgroup of A_6, and A_6 is supposed to be simple, it follows that $N \cap A_6 = A_6$ and, in particular, N contains a 3-cycle. But if N contains a 3-cycle, then $N = A_n$, by an argument used in the original proof.

So it suffices to prove that A_5 and A_6 are simple. Take $n = 5$ or 6, and suppose that $N \neq \{e\}$ is a normal subgroup of A_n which is minimal among all such subgroups; that is, if $\{e\} \subseteq K \underset{\neq}{\subset} N$ is a normal subgroup of A_n, then $K = \{e\}$.

Let X be the set of conjugates of N in S_n, and let S_n act on X by conjugation. What is the stabilizer (centralizer) of N? Since N is normal in A_n, $\mathrm{Cent}_{S_n}(N) \supseteq A_n$. There are no subgroups properly between A_n and S_n, so the centralizer is either A_n or S_n. If the centralizer is all of S_n, then N is normal in S_n, so by Theorem 10.3.2, $N = A_n$.

If the centralizer of N is A_n, then X has $2 = [S_n : A_n]$ elements. Let $M \neq N$ be the other conjugate of N in S_n. Then $M = \sigma N \sigma^{-1}$ for any odd permutation σ. Note that $M \cong N$ and, in particular, M and N have the same number of elements.

I leave it as an exercise to show that M is also a normal subgroup of A_n.

Since $N \neq M$, $N \cap M$ is a normal subgroup of A_n properly contained in N. By the assumption of minimality of N, it follow that $N \cap M = \{e\}$. Therefore, MN is a subgroup of A_n, isomorphic to $M \times N$. MN is normal in A_n, but if σ is an odd permutation, then $\sigma MN \sigma^{-1} = \sigma M \sigma^{-1} \sigma N \sigma^{-1} = NM = MN$, so MN is normal in S_n. Therefore, by Theorem 10.3.2 again, $MN = A_n$. *In particular, the cardinality of A_n is $\#(M \times N) = (\#N)^2$. But neither $5!/2 = 60$ nor $6!/2 = 360$ is a perfect square, so this is impossible.*

This completes the alternative proof of Theorem 10.3.4.

It follows from the simplicity of A_n for $n \geq 5$ that neither S_n nor A_n is solvable for $n \geq 5$ (exercise).

Exercises

Exercise 10.3.1. Show that for $n \geq 3$, the center of S_n is $\{e\}$. *Hint:* Suppose that $\sigma \in Z(A_n)$. On the one hand, any other element of S_n with the same cycle structure is conjugate to σ. But on the other hand, since σ is central, it is conjugate only to itself. Thus, σ must be the unique element of its cycle structure. Show that this implies that $\sigma = e$

Exercise 10.3.2. The subgroup $\mathcal{V} = \{e, (12)(34), (13)(24), (14)(23)\}$ of A_4 is normal in S_4.

Exercise 10.3.3. If A is a normal subgroup of a group G, then the center $Z(A)$ of A is normal in G.

Exercise 10.3.4. This exercise shows that the center of A_n is trivial for all $n \geq 4$.

(a) Compute that the center of A_4 is $\{e\}$.

(b) For $n \geq 4$, A_n is not abelian.

(c) Use Exercise 10.3.3 and Theorem 10.3.2 to show that if $n \geq 5$, then the center of A_n is either $\{e\}$ or A_n. Since A_n is not abelian, the center is $\{e\}$.

Exercise 10.3.5. Show that the subgroup M appearing in the alternative proof of 10.3.4 is normal in A_n.

Exercise 10.3.6. Observe that a simple non-abelian group is not solvable, and conclude that neither A_n nor S_n are solvable for $n \geq 5$.

10.4. Cyclotomic Polynomials and Cyclotomic Field Extensions

In this section, we will study factors of the polynomial $x^n - 1$. Recall that a primitive n-th root of unity in a field K is an element $\zeta \in K$ such that $\zeta^n = 1$ but $\zeta^d \neq 1$ for $d < n$. The primitive n-th roots of 1 in $\mathbb{C}$ are the numbers $e^{2\pi i r/n}$, where r is relatively prime to n. Thus the number of primitive n-th roots is $\varphi(n)$, where φ denotes the the Euler function (see Section 6.1.)

Definition 10.4.1. The n-th cyclotomic polynomial $\Psi_n(x)$ is defined by

$$\Psi_n(x) = \prod \{(x - \zeta) : \zeta \text{ a primitive } n\text{-th root of 1 in } \mathbb{C}\}.$$

A priori $\Psi_n(x)$ is a polynomial in $\mathbb{C}[x]$, but, in fact, it turns out to have integer coefficients. It is evident that $\Psi_n(x)$ is a monic polynomial of degree $\varphi(n)$. Note the factorization:

$$\begin{aligned} x^n - 1 &= \prod \{(x - \zeta) : \zeta \text{ an } n\text{-th root of 1 in } \mathbb{C}\} \\ &= \prod_{d \text{ divides } n} \prod \{(x - \zeta) : \zeta \text{ a primitive } d\text{-th root of 1 in } \mathbb{C}\} \\ &= \prod_{d \text{ divides } n} \Psi_d(x). \end{aligned}$$

Using this, one can compute the polynomials $\Psi_n(x)$ recursively, beginning with $\Psi_1(x) = x - 1$. See Exercise 10.4.2.

Proposition 10.4.2. *For all n, the cyclotomic polynomial $\Psi_n(x)$ is a monic polynomial of degree $\varphi(n)$ with integer coefficients.*

Proof. The assertion is valid for $n = 1$ by inspection. We proceed by induction on n. Fix $n > 1$, and put

$$f(x) = \prod_{\substack{d<n \\ d \text{ divides } n}} \Psi_d(x),$$

so that $x^n - 1 = f(x)\Psi_n(x)$. By the induction hypothesis, $f(x)$ is a monic polynomial in $\mathbb{Z}[x]$. Therefore,

$$\Psi_n(x) \in \mathbb{Q}(x) \cap \mathbb{C}[x] = \mathbb{Q}[x],$$

by Exercise 10.4.3. Since all the polynomials involved in the factorization $x^n - 1 = f(x)\Psi_n(x)$ are monic, it follows from Gauss' lemma that $\Psi_n(x)$ has integer coefficients. □

Note that the splitting field of $x^n - 1$ coincides with the splitting field of $\Psi_n(x)$, and is equal to $\mathbb{Q}(\zeta)$, where ζ is any primitive n-th root of unity (exercise). Next, we wish to show that $\Psi_n(x)$ is irreducible over $\mathbb{Q}$, so $\mathbb{Q}(\zeta)$ is a field extension of degree $\varphi(n)$ over $\mathbb{Q}$. First we note the following:

Lemma 10.4.3. *The following statements are equivalent:*

(a) $\Psi_n(x)$ *is irreducible.*

(b) *If ζ is a primitive n-th root of unity in $\mathbb{C}$, f is the minimal polynomial of ζ over $\mathbb{Q}$, and p is a prime not dividing n, then ζ^p is a root of f.*

(c) *If ζ is a primitive n-th root of unity in $\mathbb{C}$, f is the minimal polynomial of ζ over $\mathbb{Q}$, and r is relatively prime to n, then ζ^r is a root of f.*

(d) *If ζ is a primitive n-th root of unity in $\mathbb{C}$, and r is relatively prime to n, then $\zeta \mapsto \zeta^r$ determines an automorphism of $\mathbb{Q}(\zeta)$ over $\mathbb{Q}$.*

Proof. Exercise. □

Theorem 10.4.4. $\Psi_n(x)$ *is irreducible over* $\mathbb{Q}$.

Proof. Suppose that $\Psi_n(x) = f(x)g(x)$, where $f, g \in \mathbb{Z}[x]$, and f is irreducible. Let ζ be a root of f in $\mathbb{C}$; thus ζ is a primitive n-th root of unity and f is the minimal polynomial of ζ over $\mathbb{Q}$. Suppose that p is a prime not dividing n. According to the previous lemma, it suffice to show that ζ^p is a root of f.

Suppose that ζ^p is not a root of f. Then, necessarily, ζ^p is a root of g, or, equivalently, ζ is a root of $g(x^p)$. It follows that $f(x)$ divides $g(x^p)$, because f is the minimal polynomial of ζ. Reducing all polynomials modulo p, one has that $\tilde{f}(x)$ divides $\tilde{g}(x^p) = (\tilde{g}(x))^p$. In particular, $\tilde{f}(x)$ and $\tilde{g}(x)$

are not relatively prime in $\mathbb{Z}_p[x]$. Now, it follows from $\tilde{\Psi}_n(x) = \tilde{f}(x)\tilde{g}(x)$ that $\tilde{\Psi}_n(x)$ has a multiple root, and therefore $x^n - 1$ has a multiple root over $\mathbb{Z}_p$. But this cannot be so, because the derivative of $x^n - 1$ in $\mathbb{Z}_p[x]$, namely $[n]x^{n-1}$, is not identically zero, since p does not divide n, and has no roots in common with $x^n - 1$. This contradiction shows that, in fact, ζ^p is a root of f. ☐

Corollary 10.4.5. *The splitting field of $x^n - 1$ over $\mathbb{Q}$ is $\mathbb{Q}(\zeta)$, where ζ is a primitive n-th root of unity. The dimension of $\mathbb{Q}(\zeta)$ over $\mathbb{Q}$ is $\varphi(n)$, and the Galois group of $\mathbb{Q}(\zeta)$ over $\mathbb{Q}$ is isomorphic to $\Phi(n)$, the multiplicative group of units in $\mathbb{Z}_n$.*

Proof. The irreducibility of $\Psi_n(x)$ over $\mathbb{Q}$ implies that $\dim_{\mathbb{Q}}(\mathbb{Q}(\zeta)) = \varphi(n)$. One can define an injective homomorphism of $G = \mathrm{Aut}_{\mathbb{Q}}(\mathbb{Q}(\zeta)$ into $\Phi(n)$ by $\sigma \mapsto [r]$ if $\sigma(\zeta) = \zeta^r$. This map is surjective since both groups have the same cardinality. ☐

Proposition 10.4.6. *Let K be any field whose characteristic does not divide n. The the Galois group of $x^n - 1$ over K is isomorphic to a subgroup of $\Phi(n)$. In particular if n is prime, then the Galois group is cyclic of order dividing $n - 1$.*

Proof. Exercise. ☐

Exercises

Exercise 10.4.1. If ζ is any primitive n-th root of unity in $\mathbb{C}$, show that the set of all primitive n-th roots of unity in $\mathbb{C}$ is $\{\zeta^r : r \text{ is relatively prime to } n\}$.

Exercise 10.4.2. Show how to use Equation 10.4 to compute the cyclotomic polynomials recursively. Compute the first several cyclotomic polynomials by this method.

Exercise 10.4.3. Show that $\mathbb{Q}(x) \cap \mathbb{C}[x] = \mathbb{Q}[x]$.

Exercise 10.4.4. Prove Lemma 10.4.3.

Exercise 10.4.5. Prove Proposition 10.4.6. *Hint:* Let E be a splitting field of $x^n - 1$ over K. Show that the hypothesis on characteristics implies that E contains a primitive n-th root of unity ζ. Define an injective homomorphism from $G = \mathrm{Aut}_K(E)$ into $\Phi(n)$ by $\sigma \mapsto [r]$, where $\sigma(\zeta) = \zeta^r$. Show that this is a well defined, injective homomorphism of groups. (In particular, check the proof of Corollary 10.4.5.)

10.5. The Equation $x^n - b = 0$

Consider the polynomial $x^n - b \in K[x]$, where $b \neq 0$, and n is relatively prime to the characteristic of K. The polynomial $x^n - b$ has n distinct roots in a splitting field E, and the ratio of any two roots is an n-th root of unity, so again E contains n distinct roots of unity, and, therefore, a primitive n-th root of unity u.

If a is one root of $x^n - b$ in E, then all the roots are of the form $u^j a$, and $E = K(u, a)$. We consider the extension in two stages $K \subseteq K(u) \subseteq E = K(u, a)$.

A $K(u)$-automorphism τ of E is determined by $\tau(a)$, and $\tau(a)$ must be of the form $\tau(a) = u^i a$.

Lemma 10.5.1. *The map $\tau \mapsto \tau(a)a^{-1}$ is an injective group homomorphism from $\mathrm{Aut}_{K(u)}(E)$ into the cyclic group generated by u. In particular, $\mathrm{Aut}_{K(u)}(E)$ is a cyclic group whose order divides n.*

Proof. Exercise. □

We have proved the following proposition:

Proposition 10.5.2. *If K contains a primitive n-th root of unity (where n is necessarily relatively prime to the characteristic) and E is a splitting field of $x^n - b \in K[x]$, then $\mathrm{Aut}_K(E)$ is cyclic. Furthermore, $E = K(a)$ where a is any root in E of $x^n - b$.*

Corollary 10.5.3. *If K is a field, n is relatively prime to the characteristic of K, and E is a splitting field over K of $x^n - b \in K[x]$, then $\mathrm{Aut}_K(E)$ has a cyclic normal subgroup N such that $\mathrm{Aut}_K(E)/N$ is abelian.*

Proof. As shown above, E contains a primitive n-th root of unity u, and if a is one root of $x^n - b$ in E, then $K \subseteq K(u) \subseteq K(u, a) = E$, and the intermediate field $K(u)$ is a Galois over K. By Proposition 10.5.2, $N = \mathrm{Aut}_{K(u)}(E)$ is cyclic, and by the Fundamental Theorem, N is normal. The quotient $\mathrm{Aut}_K(E)/N \cong \mathrm{Aut}_K(K(u))$ is a subgroup of $\Phi(n)$, and in particular abelian, by Proposition 10.4.6. □

Exercises

Exercise 10.5.1. Prove Lemma 10.5.1.

Exercise 10.5.2. Find the Galois group of $x^{13} - 1$ over $\mathbb{Q}$. Find all intermediate fields in as explicit a form as possible.

Exercise 10.5.3. Find the Galois group of $x^{13} - 2$ over $\mathbb{Q}$. Find all intermediate fields in as explicit a form as possible.

Exercise 10.5.4. Let $n = p_1^{m_1} p_2^{m_2} \cdots p_s^{m_s}$. Let ζ be a primitive n-th root of unity in $\mathbb{C}$ and let ζ_i be a primitive $p_i^{m_i}$ root of unity. Show that the Galois group of $x^n - 1$ over $\mathbb{Q}$ is isomorphic to the direct product of the Galois groups of $x^{p_i^{m_i}} - 1$, $(1 \le i \le s)$. Show that each $\mathbb{Q}(\zeta_i)$ is a subfield of $\mathbb{Q}(\zeta)$, the intersection of the $\mathbb{Q}(\zeta_i)$ is $\mathbb{Q}$, and the composite of all the $\mathbb{Q}(\zeta_i)$ is $\mathbb{Q}(\zeta)$.

10.6. Solvability by Radicals

Definition 10.6.1. A tower of fields $K = K_0 \subseteq K_1 \subseteq \cdots \subseteq K_r$ is called a *radical tower* if there is a sequence of elements $a_i \in K_i$ such that $K_1 = K_0(a_1)$, $K_2 = K_1(a_2)$, $\ldots$, $K_r = K_{r-1}(a_r)$, and there is a sequence of natural numbers n_i such that $a_i^{n_i} \in K_{i-1}$ for $1 \le i \le r$.

We call r the *length* of the radical tower. An extension $K \subseteq L$ is called a *radical extension* if there is a radical tower $K = K_0 \subseteq K_1 \subseteq \cdots \subseteq K_r$ with $K_r = L$.

The idea behind this definition is that the elements of a radical extension L can be computed, starting with elements of K, by rational operations – that is, field operations – and by "extracting roots," that is, by solving equations of the form $x^n = b$.

I am aiming for the following result:

Theorem 10.6.2. (Galois) *If $K \subseteq E \subseteq L$ are field extensions with E Galois over K and L radical over K, then the Galois group* $\mathrm{Aut}_K(E)$ *is a solvable group.*

Lemma 10.6.3. *If $K \subseteq L$ is a radical extension of K, then there is a radical extension $L \subseteq \bar{L}$ such that $\bar{L}$ is Galois over K.*

Proof. Suppose there is a radical tower $K = K_0 \subseteq K_1 \subseteq \cdots \subseteq K_r = L$ of length r. We prove the statement by induction on r. If $r = 0$, there is nothing to do. So suppose $r \geq 1$ and that there is a radical extension $F \supseteq K_{r-1}$ such that F is Galois over K. There is an $a \in L$ and a natural number n such that $L = K_{r-1}(a)$, and $b = a^n \in K_{r-1}$. Let $p(x) \in K[x]$ be the minimal polynomial of b, and let $f(x) = p(x^n)$. Then a is a root of $f(x)$.

Let $\bar{L}$ be a splitting field of $f(x)$ over F which contains $F(a)$. We have the inclusions:

$$\begin{array}{ccccccccc} & & F & \subseteq & & & F(a) & \subseteq & \bar{L} \\ & & \cup\!| & & & & \cup\!| & & \\ K & \subseteq & K_{r-1} & \subseteq & L & = & K_{r-1}(a) & & \end{array}$$

If $g(x) \in K[x]$ is a polynomial such that F is a splitting field of $g(x)$ over K, then $\bar{L}$ is a splitting field of $g(x)f(x)$ over K and, hence, $\bar{L}$ is Galois over K.

Finally, for any root α of $f(x)$, α^n is a root of $p(x)$ and, therefore, $\alpha^n \in F$. It follows that $\bar{L}$ is a radical extension of F and, therefore, of K. □

Lemma 10.6.4. *Suppose $K \subseteq L$ is a field extension such that L is radical and Galois over K. Let $K = K_0 \subseteq K_1 \subseteq \cdots \subseteq K_{r-1} \subseteq K_r = L$ be a radical tower, let $a_i \in K_i$ and $n_i \in \mathbb{N}$ satisfy $K_i = K_{i-1}(a_i)$ and $a_i^{n_i} \in K_{i-1}$ for $1 \leq i \leq r$. Let $n = n_1 n_2 \cdots n_r$, and suppose that K contains a primitive n-th root of unity. Then the Galois group* $\mathrm{Aut}_K(L)$ *is solvable.*

Proof. We prove this by induction on the length r of the radical tower. If $r = 0$, then $K = L$ and the Galois group is $\{e\}$. So suppose $r \geq 1$, and suppose that the result holds for radical Galois extensions with radical towers of length less than r. It follows from this inductive hypothesis that the Galois group $G_1 = \mathrm{Aut}_{K_1}(L)$ is solvable.

We have $K_1 = K(a_1)$ and $a^{n_1} \in K$. Since n_1 divides n, K contains a primitive n_1-th root of unity. Then it follows from Proposition 10.5.2 that K_1 is Galois over K with cyclic Galois group $\mathrm{Aut}_K(K_1)$.

By the Fundamental Theorem, G_1 is normal in the Galois group $G = \mathrm{Aut}_K(L)$, and $G/G_1 \cong \mathrm{Aut}_K(K_1)$. Since G_1 is solvable and normal in G and the quotient G/G_1 is cyclic, it follows that G is solvable. □

Proof of Theorem 10.6.2. Let $K \subseteq E \subseteq L$ be field extensions, where E is Galois over K and L is radical over K. By Lemma 10.6.3, we can assume that L is Galois over K. Since $\mathrm{Aut}_K(E)$ is a quotient group of $\mathrm{Aut}_K(L)$, it suffices to prove that $\mathrm{Aut}_K(L)$ is solvable.

This has been done in Lemma 10.6.4 under the additional assumption that K contains certain roots of unity. The strategy will be to reduce to this case by introducing roots of unity.

Let $K = K_0 \subseteq K_1 \subseteq \cdots \subseteq K_{r-1} \subseteq K_r = L$ be a radical tower, let $a_i \in K_i$ and $n_i \in \mathbb{N}$ satisfy $K_i = K_{i-1}(a_i)$ and $a_i^{n_i} \in K_{i-1}$ for $1 \leq i \leq r$. Let $n = n_1 n_2 \cdots n_r$, and let ζ be a primitive n-th root of unity in an extension field of L.

Consider the inclusions:

$$\begin{array}{ccccccc} & & K(\zeta) & \subseteq & L(\zeta) & = & K(\zeta) \cdot L \\ & & \cup| & & \cup| & & \\ K & \subseteq & K(\zeta) \cap L & \subseteq & L & & \end{array}$$

By Proposition 9.5.5, $L(\zeta) = K(\zeta) \cdot L$ is Galois over $K(\zeta)$ and, furthermore, $\mathrm{Aut}_{K(\zeta)}(L(\zeta)) \cong \mathrm{Aut}_{K(\zeta)\cap L}(L)$. But $L(\zeta)$ is obtained from $K(\zeta)$ by adjoining the elements a_i, so $L(\zeta)$ is radical over $K(\zeta)$, and, furthermore, Lemma 10.6.4 is applicable to the extension $K(\zeta) \subseteq L(\zeta)$. Hence, $\mathrm{Aut}_{K(\zeta)}(L(\zeta)) \cong \mathrm{Aut}_{K(\zeta)\cap L}(L)$ is solvable.

The extension $K \subseteq K(\zeta)$ is Galois with abelian Galois group, by Proposition 10.4.6. Therefore, every intermediate field is Galois over K with abelian Galois group, by the Fundamental Theorem. In particular, $K(\zeta) \cap L$ is Galois over K and $\mathrm{Aut}_K(K(\zeta) \cap L)$ is abelian, so solvable.

Write $G = \mathrm{Aut}_K(L)$ and $N = \mathrm{Aut}_{K(\zeta)\cap L}(L)$. Since $K(\zeta) \cap L$ is Galois over K, by the Fundamental Theorem, N is normal in G and $G/N \cong \mathrm{Aut}_K(K(\zeta) \cap L)$.

Since both N and G/N are solvable, so is G. □

Definition 10.6.5. Let K be a field and $p(x) \in K[x]$. One says that *p(x) is solvable by radicals* over K if there is a radical extension of K which contains a splitting field of $p(x)$ over K.

The idea of this definition is that the roots of $p(x)$ can be obtained, beginning with elements of K, by rational operations and by extraction of roots.

Corollary 10.6.6. *Let $p(x) \in K[x]$, and let E be a splitting field of $p(x)$ over K. If $p(x)$ is solvable by radicals, then the Galois group* $\mathrm{Aut}_K(E)$ *is solvable.*

Corollary 10.6.7. (Abel, Galois) *The general equation of degree n over a field K is not solvable by radicals if $n \geq 5$.*

Proof. According to Theorem 9.7.1, the Galois group of the general equation of degree n is S_n, an it follows from Theorem 10.3.4 that S_n is not solvable when $n \geq 5$. □

This result implies that there can be no analogue of the quadratic (and cubic and quartic) formula for the general polynomial equation of degree 5 or higher. A general method for solving a degree n equation over K is a method of solving the general equation of degree n, of finding the roots u_i in terms of the variables t_i. Such a method can be specialized to any particular equation with coefficients in K. The procedures for solving equations of degrees 2, 3, and 4 are general methods of obtaining the roots in terms of rational operations and extractions of radicals. If such a method existed for equations of degree $n \geq 5$, then the general equation of degree n would have to be solvable by radicals over the field $K(t_1, \dots, t_n)$, which is not so.

10.7. Radical Extensions

This section can be omitted without loss of continuity.

In this section, we will obtain a partial converse to the results of the previous section: If $K \subseteq E$ is a Galois extension with a solvable Galois group, then E is contained in a radical extension. I will prove this only in the case that the ground field has characteristic 0. This restriction is not essential but is made to avoid technicalities. *All the fields in this section are assumed to have characteristic 0.*

We begin with a converse to Proposition 10.5.2.

Proposition 10.7.1. *Suppose the field K contains a primitive n-th root of unity. Let E be a Galois extension of K such that* $\mathrm{Aut}_K(E)$ *is cyclic of order n. Then E is the splitting field of an irreducible polynomial $x^n - b \in K[x]$, and $E = K(a)$, where a is any root of $x^n - b$ in E.*

The proof of this is subtle and requires some preliminary apparatus. Let $K \subseteq E$ be a Galois extension. Recall that E^* denotes the set of non-zero elements of E.

Definition 10.7.2. A function $f : \mathrm{Aut}_K(E) \to E^*$ is called a *multiplicative 1-cocycle* if it satisfies $f(\sigma\tau) = f(\sigma)\sigma(f(\tau))$.

The basic example of a multiplicative 1-cocycle is the following: If a is any non-zero element of E, then the function $g(\sigma) = \sigma(a)a^{-1}$ is a multiplicative 1-cocycle (exercise).

Proposition 10.7.3. *If $K \subseteq E$ is a Galois extension, and $f : \mathrm{Aut}_K(E) \to E^*$ is a multiplicative 1-cocycle, then there is an element $a \in E^*$ such that $f(\sigma) = \sigma(a)a^{-1}$.*

Proof. The elements of the Galois group are linearly independent, by Proposition 9.5.6, so there is an element $b \in E^*$ such that

$$\sum_{\tau \in \mathrm{Aut}_K(E)} f(\tau)\tau(b) \neq 0.$$

Call this non-zero element a^{-1}. In the following computation, the 1-cocycle relation is used in the form $\sigma(f(\tau)) = f(\sigma)^{-1} f(\sigma\tau)$. Now, for any $\sigma \in \mathrm{Aut}_K(E)$,

$$\begin{aligned}
\sigma(a^{-1}) &= \sigma\left(\sum_{\tau \in \mathrm{Aut}_K(E)} f(\tau)\tau(b)\right) \\
&= \sum_{\tau \in \mathrm{Aut}_K(E)} \sigma(f(\tau))\sigma\tau(b) \\
&= \sum_{\tau \in \mathrm{Aut}_K(E)} f(\sigma)^{-1} f(\sigma\tau)\sigma\tau(b) \\
&= f(\sigma)^{-1} \sum_{\tau \in \mathrm{Aut}_K(E)} f(\tau)\tau(b) \\
&= f(\sigma)^{-1} a^{-1}.
\end{aligned}$$

This gives $f(\sigma) = \sigma(a)a^{-1}$. □

Definition 10.7.4. Let $K \subseteq E$ be a Galois extension. For $a \in E$, the *norm* of a is defined by:

$$N(a) = N_K^E(a) = \prod_{\sigma \in \mathrm{Aut}_K(E)} \sigma(a).$$

Note that $N(a)$ is fixed by all $\sigma \in \mathrm{Aut}_K(E)$, so $N(a) \in K$ (because E is a Galois extension). For $a \in K$, $N(a) = a^{\dim_K(E)}$.

Proposition 10.7.5 (Hilbert). *Let $K \subseteq E$ be a Galois extension with cyclic Galois group. Let σ be a generator of $\mathrm{Aut}_K(E)$. If $b \in E^*$ satisfies $N(b) = 1$, then there is an $a \in E^*$ such that $b = \sigma(a)a^{-1}$.*

Proof. Let n be the order of the cyclic group $\mathrm{Aut}_K(E)$. Define $f : \mathrm{Aut}_K(E) \to E^*$ by $f(\mathrm{id}) = 1$, $f(\sigma) = b$, and $f(\sigma^i) = \sigma^{i-1}(b) \cdots \sigma(b)b$, for $i \geq 1$.

One can check that $N(b) = 1$ implies that $f(\sigma^{i+n}) = f(\sigma^i)$, so f is well defined on $\mathrm{Aut}_K(E)$, and that f is a multiplicative 1-cocycle (exercise).

Because f is a multiplicative 1-cocycle, it follows from the previous proposition that there is an $a \in E^*$ such that $b = f(\sigma) = \sigma(a)a^{-1}$. □

Proof of Proposition 10.7.1. Suppose that $\mathrm{Aut}_K(E)$ is cyclic of order n, that σ is a generator of the Galois group, and that $\zeta \in K$ is a primitive n-th root of unity. Because $\zeta \in K$, its norm satisfies $N(\zeta) = \zeta^n = 1$. Hence, by Proposition 10.7.5, there is an element $a \in E^*$ such that $\zeta = \sigma(a)a^{-1}$, or $\sigma(a) = \zeta a$. Let $b = a^n$; one has $\sigma(b) = \sigma(a)^n = (\zeta a)^n = a^n = b$, so b is fixed by $\mathrm{Aut}_K(E)$, and, therefore, $b \in K$, since E is Galois over K. The elements $\sigma^i(a) = \zeta^i a$, $0 \le i \le n-1$ are distinct roots of $x^n - b$ in E, and $\mathrm{Aut}_K(E)$ acts transitively on these roots, so $x^n - b = \prod_{i=0}^{n-1}(x - \zeta^i a)$ is irreducible in $K[x]$. (In fact, if f is an irreducible factor of $x^n - b$ in $K[x]$, and $\zeta^i a$ is one root of f, then for all j, one has that $\sigma^{j-i}(\zeta^i a) = \zeta^j a$ is also a root of f; hence, $\deg f \ge n$ and $f(x) = x^n - b$.) Since $\dim_K(K(a)) = n = \dim_K(E)$, one has $K(a) = E$. □

Theorem 10.7.6. *Suppose $K \subseteq E$ is a Galois field extension. If the Galois group $\mathrm{Aut}_K(E)$ is solvable, then there is a radical extension L of K such that $K \subseteq E \subseteq L$.*

Proof. Let $n = \dim_K(E)$, and let ζ be a primitive n-th root of unity in a field extension of E. Consider the extensions:

$$\begin{array}{ccccc} & & K(\zeta) & \subseteq & E(\zeta) = K(\zeta)\cdot E \\ & & \cup\vert & & \cup\vert \\ K & \subseteq & K(\zeta)\cap E & \subseteq & E \end{array}$$

By Proposition 9.5.5, $E(\zeta)$ is Galois over $K(\zeta)$ with Galois group

$$\mathrm{Aut}_{K(\zeta)}(E(\zeta)) \cong \mathrm{Aut}_{K(\zeta)\cap E}(E) \subseteq \mathrm{Aut}_K(E).$$

Therefore, $\mathrm{Aut}_{K(\zeta)}(E(\zeta))$ is solvable.

Let $G = G_0 = \mathrm{Aut}_{K(\zeta)}(E(\zeta))$, and let $G_0 \supseteq G_1 \supseteq \cdots \supseteq G_r = \{e\}$ be a composition series of G with cyclic quotients. Define $K_i = \mathrm{Fix}(G_i)$ for $0 \le i \le r$; thus:

$$K(\zeta) = K_0 \subseteq K_1 \subseteq \cdots \subseteq K_r = E(\zeta).$$

By the Fundamental Theorem, each extension $K_{i-1} \subseteq K_i$ is Galois with Galois group $\mathrm{Aut}_{K_{i-1}}(K_i) \cong G_{i-1}/G_i$, which is cyclic. Since K_{i-1} contains a primitive d-th root of unity, where $d = \dim_{K_{i-1}}(K_i)$, it follows from

Proposition 10.5.2 that $K_i = K_{i-1}(a_i)$, where a_i satisfies an irreducible polynomial $x^d - b_i \in K_{i-1}[x]$.

Therefore, $E(\zeta)$ is a radical extension of $K(\zeta)$. Since also $K(\zeta)$ is a radical extension of K, $E(\zeta)$ is a radical extension of K containing E, as required. □

Corollary 10.7.7. *If E is the splitting field of a polynomial $p(x) \in K[x]$, and the Galois group* $\mathrm{Aut}_K(E)$ *is solvable, then $p(x)$ is solvable by radicals*

Exercises

Exercise 10.7.1. If $a \in E^*$, then the function $g(\sigma) = \sigma(a)a^{-1}$ is a multiplicative 1-cocycle.

Exercise 10.7.2. With notation as in the proof of 10.7.5, check that if $N(b) = 1$, then f is well-defined on $\mathrm{Aut}_K(E)$, and that f is a multiplicative 1-cocycle.

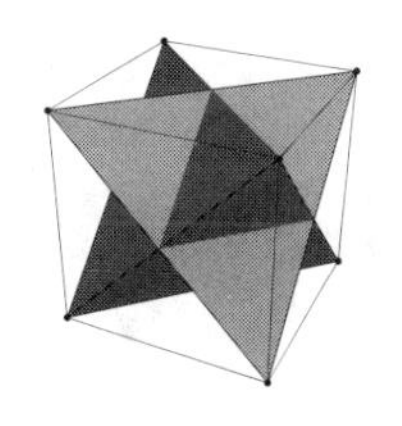

Chapter 11

Isometry Groups

11.1. More on Isometries of Euclidean Space

The goal of this section is to analyze the isometry group of Euclidean space. First, we will show that an isometry of Euclidean space which fixes the origin is actually a linear map. I'm going to choose a somewhat indirect but elegant way to do this, by using a uniqueness result: If an isometry fixes enough points, then it must fix all points, that is, it must be the identity map.

Definition 11.1.1. *A (linear) hyperplane* in $\mathbb{R}^n$ is the set of solutions to a non-trivial linear equation: i.e., $\{\boldsymbol{x} \in \mathbb{R}^n : \langle \boldsymbol{x}|\boldsymbol{\alpha}\rangle = 0\}$, for some non-zero $\boldsymbol{\alpha} \in \mathbb{R}^n$.

For example, in $\mathbb{R}^3$, a hyperplane is a plane through the origin.

Definition 11.1.2. *An affine hyperplane* in $\mathbb{R}^n$ is the translate of a linear hyperplane, $\boldsymbol{x}_0 + P$, where P is a linear hyperplane.

Write $P_{\boldsymbol{\alpha}}$ for $\{\boldsymbol{x} \in \mathbb{R}^n : \langle \boldsymbol{x}|\boldsymbol{\alpha}\rangle = 0\}$. The affine hyperplane $\boldsymbol{x}_0 + P_{\boldsymbol{\alpha}}$ is characterized as the set of points $\boldsymbol{x}$ such that $\langle \boldsymbol{x}|\boldsymbol{\alpha}\rangle = \langle \boldsymbol{x}_0|\boldsymbol{\alpha}\rangle$ (exercise). One can show that any n points in $\mathbb{R}^n$ lie on some affine hyperplane (exercise).

Proposition 11.1.3. *Let $\boldsymbol{a}$ and $\boldsymbol{b}$ be distinct points in $\mathbb{R}^n$. Then the set of points which are equidistant to $\boldsymbol{a}$ and $\boldsymbol{b}$ is the affine hyperplane $\boldsymbol{x}_0 + P_{\boldsymbol{\alpha}}$, where $\boldsymbol{x}_0 = (\boldsymbol{a}+\boldsymbol{b})/2$ and $\boldsymbol{\alpha} = \boldsymbol{b} - \boldsymbol{a}$.*

Proof. Exercise. □

Corollary 11.1.4. *Let $\boldsymbol{a}_i$ $(1 \le i \le n+1)$ be $n+1$ points in $\mathbb{R}^n$ which do not lie on any affine hyperplane. If $\boldsymbol{a}$ and $\boldsymbol{b}$ satisfy $d(\boldsymbol{a}, \boldsymbol{a}_i) = d(\boldsymbol{b}, \boldsymbol{a}_i)$ for $1 \le i \le n+1$, then $\boldsymbol{a} = \boldsymbol{b}$.*

Proof. This is just the contrapositive of the proposition. □

Corollary 11.1.5. *Let* $\boldsymbol{a}_i$ $(1 \leq i \leq n+1)$ *be* $n+1$ *points in* $\mathbb{R}^n$ *which do not lie on any affine hyperplane. If an isometry* τ *satisfies* $\tau(\boldsymbol{a}_i) = \boldsymbol{a}_i$ *for* $i \leq 1 \leq n+1$, *then* $\tau = \mathrm{id}$.

Proof. For any point $\boldsymbol{a} \in \mathbb{R}^n$, $d(\tau(\boldsymbol{a}), \boldsymbol{a}_i) = d(\tau(\boldsymbol{a}), \tau(\boldsymbol{a}_i)) = d(\boldsymbol{a}, \boldsymbol{a}_i)$ for $1 \leq i \leq n+1$. By the first corollary, $\tau(\boldsymbol{a}) = \boldsymbol{a}$. □

Theorem 11.1.6. *Let* τ *be an isometry of* $\mathbb{R}^n$ *such that* $\tau(\boldsymbol{0}) = \boldsymbol{0}$. *Then* τ *is linear.*

Proof. Let A be the matrix with columns $\{\tau(\hat{\boldsymbol{e}}_i)\}$, where the $\hat{\boldsymbol{e}}_i$ are the standard orthonormal basis of $\mathbb{R}^n$. By Lemma 4.4.2 and Exercise 4.4.3, A is an orthogonal matrix; hence, $\sigma : \boldsymbol{x} \mapsto A\boldsymbol{x}$ is a linear isometry such that $\sigma(\hat{\boldsymbol{e}}_i) = \tau(\hat{\boldsymbol{e}}_i)$ for $1 \leq i \leq n$. Of course, $\sigma(\boldsymbol{0}) = \tau(\boldsymbol{0}) = \boldsymbol{0}$ as well. Now the composed isometry $\sigma^{-1} \circ \tau$ fixes all the points $\boldsymbol{0}, \hat{\boldsymbol{e}}_1, \ldots, \hat{\boldsymbol{e}}_n$. This set of points does not lie on any affine hyperplane, by the linear independence of the $\hat{\boldsymbol{e}}_i$, so by Corollary 11.1.5, $\sigma^{-1} \circ \tau = \mathrm{id}$, or $\sigma = \tau$. □

Theorem 11.1.7.

(a) *Any linear isometry of* $\mathbb{R}^n$ *is a product of at most n orthogonal reflections.*

(b) *Any element of* $\mathrm{O}(n, \mathbb{R})$ *is a product of at most n reflection matrices.*

Proof. Let τ be a linear isometry of $\mathbb{R}^n$, let $\mathbb{E} = \{\hat{\boldsymbol{e}}_i\}$ denote the standard orthonormal basis of $\mathbb{R}^n$, and write $\hat{\boldsymbol{f}}_i = \tau(\hat{\boldsymbol{e}}_i)$ for $i \leq i \leq n$. I show by induction that, for $1 \leq p \leq n$. there is a product ρ_p of at most p orthogonal reflections satisfying $\tau_p(\hat{\boldsymbol{e}}_i) = \hat{\boldsymbol{f}}_i$ for $1 \leq i \leq p$. If $\hat{\boldsymbol{e}}_1 = \hat{\boldsymbol{f}}_1$, put $\rho_1 = \mathrm{id}$; otherwise, $\boldsymbol{0}$ lies on the hyperplane consisting of points equidistant to $\hat{\boldsymbol{e}}_1$ and $\hat{\boldsymbol{f}}_1$, and the reflection in this hyperplane maps $\hat{\boldsymbol{e}}_1$ to $\hat{\boldsymbol{f}}_1$. In this case, let ρ_1 be the orthogonal reflection in this hyperplane.

Now, suppose $\rho = \rho_{p-1}$ is a product of at most $p-1$ orthogonal reflections which maps $\hat{\boldsymbol{e}}_i$ to $\hat{\boldsymbol{f}}_i$ for $1 \leq i \leq p-1$. If also $\rho(\hat{\boldsymbol{e}}_p) = \hat{\boldsymbol{f}}_p$, then put $\rho_p = \rho$. Otherwise, observe that $\{\hat{\boldsymbol{f}}_1, \ldots, \hat{\boldsymbol{f}}_{p-1}, \hat{\boldsymbol{f}}_p\}$, and

$$\{\hat{\boldsymbol{f}}_1, \ldots, \hat{\boldsymbol{f}}_{p-1}, \rho(\hat{\boldsymbol{e}}_p)\} = \{\rho(\hat{\boldsymbol{e}}_1), \ldots, \rho(\hat{\boldsymbol{e}}_{p-1}), \rho(\hat{\boldsymbol{e}}_p)\}$$

are both orthonormal sets, hence, $\{\hat{\boldsymbol{f}}_1, \ldots, \hat{\boldsymbol{f}}_{p-1}\} \cup \{\boldsymbol{0}\}$ lies on the hyperplane of points equidistant to $\hat{\boldsymbol{f}}_p$ and $\rho(\hat{\boldsymbol{e}}_p)$. If σ is the orthogonal reflection in this hyperplane, then $\rho_p = \sigma \circ \rho$ has the desired properties. This completes the induction. Now $\tau = \rho_n$ by application of Corollary 11.1.5 to $\rho_n^{-1} \circ \tau$. □

Corollary 11.1.8.

(a) *An element of* $\mathrm{O(n, \mathbb{R})}$ *has determinant equal to* 1 *if and only if it is a product of an even number of reflection matrices.*

(b) *An element of* $\mathrm{O(n, \mathbb{R})}$ *has determinant equal to* -1 *if and only if it is a product of an odd number of reflection matrices.*

This should remind you of even and odd permutations; recall that a permutation is even if and only if it is a product of an even number of 2-cycles, and odd if and only if it is a product of an odd number of 2-cycles. The similarity is no coincidence; see exercise 11.1.7.

We now work out the structure of the group of all isometries of $\mathbb{R}^n$.

For $\boldsymbol{b} \in \mathbb{R}^n$, define the translation $\tau_{\boldsymbol{b}}$ by $\tau_{\boldsymbol{b}}(\boldsymbol{x}) = \boldsymbol{x} + \boldsymbol{b}$.

Lemma 11.1.9. *The set of translations of* $\mathbb{R}^n$ *is a normal subgroup of the group of isometries of* $\mathbb{R}^n$*, and is isomorphic to the additive group* $\mathbb{R}^n$*. For any isometry* σ*, and any* $\boldsymbol{b} \in \mathbb{R}^n$*, one has* $\sigma \tau_{\boldsymbol{b}} \sigma^{-1} = \tau_{\sigma(\boldsymbol{b})}$.

Proof. Exercise. □

Let τ be an isometry of $\mathbb{R}^n$, and let $\boldsymbol{b} = \tau(\boldsymbol{0})$. Then $\sigma = \tau_{-\boldsymbol{b}}\tau$ is an isometry satisfying $\sigma(\boldsymbol{0}) = \boldsymbol{0}$, so σ is linear by Theorem 11.1.6. Thus, $\tau = \tau_{\boldsymbol{b}}\sigma$ is a product of a linear isometry and a translation.

Theorem 11.1.10. *The group* $\mathrm{Isom(n)}$ *of isometries of* $\mathbb{R}^n$ *is the semi-direct product of the group of linear isometries and the translation group. Thus,* $\mathrm{Isom(n)} \cong \mathrm{O(n, \mathbb{R})} \ltimes \mathrm{R^n}$

Proof. We have just observed that the product of the group of translations and the group of linear isometries is the entire group of isometries. The intersection of these two subgroups is trivial, and the group of translations is normal. Therefore, the isometry group is a semi-direct product of the two subgroups. □

The remainder of this section contains some results on the geometric classification of isometries of $\mathbb{R}^2$ which we will use in Section 11.4 for classification of two-dimensional crystal groups.

An *affine reflection* in the hyperplane through a point $\boldsymbol{x}_0$ and perpendicular to $\hat{\boldsymbol{\alpha}}$ is given by $\boldsymbol{x} \mapsto \boldsymbol{x} - 2\langle \boldsymbol{x} - \boldsymbol{x}_0 | \hat{\boldsymbol{\alpha}} \rangle \hat{\boldsymbol{\alpha}}$.

A *glide-reflection* is the product $\tau_{\boldsymbol{a}}\sigma$, where σ is an affine reflection and $\boldsymbol{a}$ is parallel to the hyperplane of the reflection σ.

An *affine rotation* with center $\boldsymbol{x}_0$ is given by $\tau_{\boldsymbol{x}_0} R \tau_{-\boldsymbol{x}_0}$, where R is a rotation.

Proposition 11.1.11. *Every isometry of $\mathbb{R}^2$ is a translation, a glide-reflection, an affine reflection, or an affine rotation.*

Proof. Every isometry can be written uniquely as a product $\tau_{\boldsymbol{h}} B$, where B is a linear isometry. If $B = E$, then the isometry is a translation.

If $B = R_\theta$ is a rotation through an angle $0 < \theta < 2\pi$, then $E - R_\theta$ is invertible, so one can solve the equation $\boldsymbol{h} = \boldsymbol{x} - R_\theta \boldsymbol{x}$. Then $\tau_{\boldsymbol{h}} R_\theta = \tau_{\boldsymbol{x}} \tau_{-R_\theta(\boldsymbol{x})} R_\theta = \tau_{\boldsymbol{x}} R_\theta \tau_{-\boldsymbol{x}}$, an affine rotation.

If $B = j_{\hat{\boldsymbol{\alpha}}}$ is a reflection, write $\boldsymbol{h} = s\hat{\boldsymbol{\alpha}} + t\hat{\boldsymbol{\beta}}$, where $\hat{\boldsymbol{\beta}}$ is perpendicular to $\hat{\boldsymbol{\alpha}}$. Then note that $\tau_{s\hat{\boldsymbol{\alpha}}} j_{\hat{\boldsymbol{\alpha}}}(\boldsymbol{x}) = \boldsymbol{x} - 2\langle \boldsymbol{x} - (s/2)\hat{\boldsymbol{\alpha}} | \hat{\boldsymbol{\alpha}} \rangle \hat{\boldsymbol{\alpha}}$ is an affine reflection. Therefore, if $t = 0$, the result is an affine reflection, and otherwise the result is a glide-reflection. □

Exercises

Exercise 11.1.1. Suppose A, B, C, D are four non-coplanar points in $\mathbb{R}^3$. and $\alpha, \beta, \gamma, \delta$ are positive numbers. Show by elementary geometry that there is an most one point P such that $d(P, A) = \alpha$, $d(P, B) = \beta$, $d(P, C) = \gamma$, and $d(P, D) = \delta$. Now show that an isometry of $\mathbb{R}^3$ which fixes A, B, C, and D is the identity map.

Exercise 11.1.2. Show that $\boldsymbol{x}_0 + P_{\boldsymbol{\alpha}} = \{\boldsymbol{x} \in \mathbb{R}^n : \langle \boldsymbol{x} | \boldsymbol{\alpha} \rangle = \langle \boldsymbol{x}_0 | \boldsymbol{\alpha} \rangle\}$.

Exercise 11.1.3. Show that any n points in $\mathbb{R}^n$ lie on some affine hyperplane.

Exercise 11.1.4. Prove the Proposition 11.1.3 as follows: Show that $||\boldsymbol{x} - \boldsymbol{a}|| = ||\boldsymbol{x} - \boldsymbol{b}||$ is equivalent to $2\langle \boldsymbol{x} | \boldsymbol{b} - \boldsymbol{a} \rangle = ||b||^2 - ||a||^2 = 2\langle \boldsymbol{x}_0 | \boldsymbol{b} - \boldsymbol{a} \rangle$, where $\boldsymbol{x}_0 = (\boldsymbol{a} + \boldsymbol{b})/2$.

Exercise 11.1.5. Show that a set $\{\boldsymbol{a}_0, \boldsymbol{a}_1, \dots, \boldsymbol{a}_n\}$ does not lie on any affine hyperplane if and only if the set $\{\boldsymbol{a}_1 - \boldsymbol{a}_0, \boldsymbol{a}_2 - \boldsymbol{a}_0, \dots, \boldsymbol{a}_n - \boldsymbol{a}_0\}$ is linearly independent.

Exercise 11.1.6. Prove Theorem 11.1.6 more directly as follows. Using that τ preserves inner products, show that $||\tau(\boldsymbol{a}+\boldsymbol{b}) - \tau(\boldsymbol{a}) - \tau(\boldsymbol{b})|| = 0$, hence, τ is additive, $\tau(\boldsymbol{a}+\boldsymbol{b}) = \tau(\boldsymbol{a}) + \tau(\boldsymbol{b})$. It remains to show homogeneity, $\tau(s\boldsymbol{a}) = s\tau(\boldsymbol{a})$. Show that this follows for rational s from the additivity of τ, and use a continuity argument for irrational s.

Exercise 11.1.7.

(a) The homomorphism $T : S_n \longrightarrow \mathrm{GL}(n, \mathbb{R})$ described in Exercise 3.1.12 has range in $\mathrm{O(n, \mathbb{R})}$.
(b) For any 2-cycle (a, b), $T((a, b))$ is the orthogonal reflection in the hyperplane $\{\boldsymbol{x} : x_a = x_b\}$.
(c) $\pi \in S_n$ is even if and only if $\det(T(\pi)) = 1$.
(d) Any element of S_n is a product of at most n 2-cycles.

Exercise 11.1.8. Prove Lemma 11.1.9.

Exercise 11.1.9. $\mathrm{Isom(n)}$ is the subgroup of the affine group $\mathrm{Aff}(n)$ consisting of those affine transformations $T_{A,b}$ such that A is orthogonal. $\mathrm{Isom(n)}$ is isomorphic to the group of $(n+1)$-by-$(n+1)$ matrices $\begin{bmatrix} A & b \\ 0 & 1 \end{bmatrix}$ such that $A \in \mathrm{O(n, \mathbb{R})}$.

11.2. Euler's Theorem

In this section we discuss Euler's theorem on convex polyhedra, and show that there are only five regular polyhedra.

Theorem 11.2.1. *Let v, e, and f be respectively the number of vertices, edges, and faces of a convex polyhedron. Then $v - e + f = 2$.*

Let us assume this result for the moment, and see how it leads quickly to a classification of regular polyhedra. A regular polyhedron is one whose faces are mutually congruent regular polygons, and at each of whose vertices the same number of edges meet. Let p denote the number of edges on each face of a regular polyhedron, and q the valence of each vertex. For the known regular polyhedra one has the following data:

polyhedron	v	e	f	p	q
tetrahedron	4	6	4	3	3
cube	8	12	6	4	3
octahedron	6	12	8	3	4
dodecahedron	20	30	12	5	3
icosahedron	12	30	20	3	5

Since each edge is common to two faces and to two vertices, one has:

$$e = fp/2 = vq/2.$$

Lemma 11.2.2.

(a) *Solving the equations*

$$e = fp/2 = vq/2.$$

together with

$$2 = v - e + f$$

for v, e, f in terms of p and q gives

$$v = \frac{4p}{2p + 2q - qp}, \qquad e = \frac{2pq}{2p + 2q - qp}, \qquad f = \frac{4q}{2p + 2q - qp}.$$

(b) *It follows that that $2p + 2q > qp$.*

(c) *The only pairs (p, q) satisfying this inequality (as well as $p \geq 3$, $q \geq 3$) are $(3, 3)$, $(3, 4)$, $(4, 3)$, $(3, 5)$, and $(5, 3)$.*

Proof. Exercise. □

Corollary 11.2.3. *There are only five regular convex polyhedra.*

Next, we will take a brief side trip to prove Euler's Theorem, which we will interpret as a result of *graph theory.*

An *embedded graph* in $\mathbb{R}^3$ is a set of the form $V \cup E_1 \cup \cdots \cup E_n$, where V is a finite set of distinguished points (called *vertices*), and the E_i are smooth curves (called *edges*) such that

(a) Each curve E_i begins and ends at a vertex.
(b) The curves E_i have no self-intersections.
(c) Two curves E_i and E_j can intersect only at vertices, $E_i \cap E_j \subseteq V$.

The smoothness of the curves E_i and conditions (a) and (b) mean that for each i, there is a smooth injective function $\varphi_i : [0, 1] \to \mathbb{R}^3$ whose image is E_i such that $\varphi_i(0), \varphi_i(1) \in V$.

Figure 11.2.1 shows a graph.

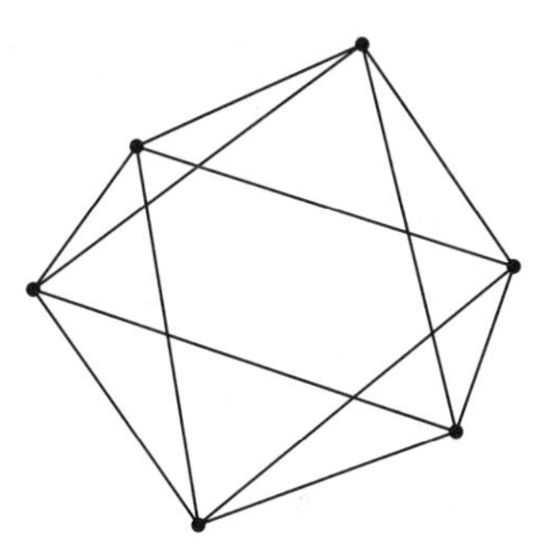

Figure 11.2.1. A Graph

Definition 11.2.4. The *valence* of a vertex on a graph is the number of edges containing that vertex as an endpoint.

I leave it to you to formulate precisely the notions of a *path* on a graph; a *cycle* is a path which begins and ends at the same vertex.

Definition 11.2.5. A graph which admits no cycles is called a *tree*.

Figure 11.2.2 shows a tree.

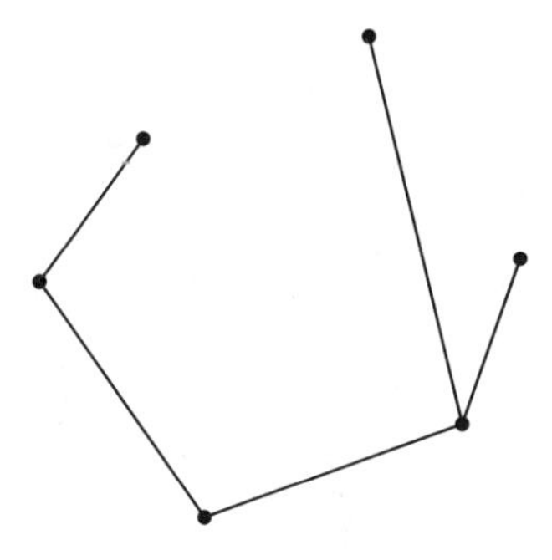

Figure 11.2.2. A Tree

The following is a graph theoretic version of Euler's Theorem:

Theorem 11.2.6. *Let $\mathcal{G}$ be a connected graph on the sphere $S = \{\boldsymbol{x} \in \mathbb{R}^3 : ||\boldsymbol{x}|| = 1\}$. Let e and v be the number of edges and vertices of $\mathcal{G}$, and let f be the number of connected components of $S \setminus \mathcal{G}$. Then $\epsilon(\mathcal{G}) = v - e + f = 2$.*

Proof. If $\mathcal{G}$ has exactly one edge, then $e = 1$, $v = 2$ and $f = 1$, so the formula is valid. So suppose that $\mathcal{G}$ has at least two edges, and that the result holds for all connected graphs on the sphere with fewer edges. If $\mathcal{G}$ admits a cycle, then choose some edge belonging to a cycle, and let $\mathcal{G}'$ be the graph resulting from deleting the chosen edge (but not the endpoints of the edge) from $\mathcal{G}$. Then $\mathcal{G}'$ remains connected. Furthermore, $v(\mathcal{G}') = v$, $e(\mathcal{G}') = e - 1$, and $f(\mathcal{G}') = f - 1$. Hence, $\epsilon(\mathcal{G}) = \epsilon(\mathcal{G}') = 2$. If $\mathcal{G}$ is a tree, let $\mathcal{G}'$ be the graph resulting from deleting a vertex with valence 1 together with the edge containing that vertex (but not the other endpoint of the edge). The $v(\mathcal{G}') = v - 1$, $e(\mathcal{G}') = e - 1$, and $f(\mathcal{G}') = f = 0$. Hence, $\epsilon(\mathcal{G}) = \epsilon(\mathcal{G}') = 2$. This completes the proof. □

Proof of Euler's Theorem: One can suppose without loss of generality that $\boldsymbol{0}$ is contained in the interior of the convex polyhedron, which is in turn contained in the interior of the unit sphere S. The projection $\boldsymbol{x} \mapsto \boldsymbol{x}/||\boldsymbol{x}||$ onto the sphere maps the polyhedron bijectively onto the sphere. The image of the edges and vertices is a connected graph $\mathcal{G}$ on the sphere with e edges and v vertices, and each face of the polyhedron projects onto a connected component in $S \setminus \mathcal{G}$. Hence, Euler's Theorem for polyhedra follows from Theorem 11.2.6. □

The projection of a dodecahedron onto the sphere is shown in Figure 11.2.3.

Figure 11.2.3. Dodecahedron Projected on the Sphere

Exercises

Exercise 11.2.1. Prove Lemma 11.2.2.

Exercise 11.2.2. Explain how the data $p \geq 3$, $q = 2$ can be sensibly interpreted, in the context of Lemma 11.2.2.

Exercise 11.2.3. Determine by case-by-case inspection that the symmetry groups of the five regular polyhedra satisfy

$$\#G = vq = fp.$$

Find an explanation for these formulas. *Hint:* Consider actions of G on vertices and on faces.

Exercise 11.2.4. A tree always has a vertex with valence 1. In a connected tree, there is exactly one path between any two vertices.

11.3. Finite Rotation Groups

In this section we will classify the finite subgroups of $SO(3, \mathbb{R})$ and $O(3, \mathbb{R})$, largely by means of exercises.

It's easy to obtain the corresponding result for 2 dimensions: A finite subgroup of $SO(2, \mathbb{R})$ is cyclic. A finite subgroup of $O(2, \mathbb{R})$ is either cyclic or a dihedral group D_n for $n \geq 1$ (exercise).

The classification for $SO(3, \mathbb{R})$ proceeds by analyzing the action of the finite group on the set of points left fixed by some element of the group. Let G be a finite subgroup of the rotation group $SO(3, \mathbb{R})$. Each non-identity element $g \in G$ is a rotation about some axis; thus, g has two fixed points on the unit sphere S, called the *poles* of g. The group G acts on the set $\mathcal{P}$ of poles, since if $\boldsymbol{x}$ is a pole of g and $h \in G$, then $h\boldsymbol{x}$ is a pole of hgh^{-1}. You are asked to show in the exercises that the stabilizer of a pole in G is a non-trivial cyclic group.

Let M denote the number of orbits of G acting on $\mathcal{P}$. Applying Burnside's Lemma 5.2.2 to this action gives:

$$M = \frac{1}{\#G}(\#\mathcal{P} + 2(\#G - 1)),$$

because the identity of G fixes all elements of $\mathcal{P}$, while each non-identity element fixes exactly two elements. We can rewrite this equation as

$$(M - 2)\#G = \#\mathcal{P} - 2, \tag{11.3.1}$$

which shows us that

$$M \geq 2.$$

We have the following data for several known finite subgroups of the rotation group:

group	#G	*orbits*	*orbit sizes*	*stabilizer sizes*
$\mathbb{Z}_n$	n	2	$1, 1$	n, n
D_n	$2n$	3	$2, n, n$	$n, 2, 2$
tetrahedron	12	3	$6, 4, 4$	$2, 3, 3$
cube/octahedron	24	3	$12, 8, 6$	$2, 3, 4$
dodec/icosahedron	60	3	$30, 20, 12$	$2, 3, 5$

Theorem 11.3.1. *The finite subgroups of* $\mathrm{SO}(3, \mathbb{R})$ *are the cyclic groups* $\mathbb{Z}_n$*, the dihedral groups* D_n*, and the rotation groups of the regular polyhedra.*

Proof. Let $\boldsymbol{x}_1, \ldots, \boldsymbol{x}_M$ be representatives of the M orbits. We rewrite

$$\#\mathcal{P}/\#G = \sum_{i=1}^{M} \frac{\#O(\boldsymbol{x}_i)}{\#G} = \sum_{i=1}^{M} \frac{1}{\#\mathrm{Stab}(\boldsymbol{x}_i)}.$$

Putting this into the orbit counting equation, writing M as $\sum_{i=1}^{M} 1$, and rearranging gives

$$\sum_{i=1}^{M} (1 - \frac{1}{\#\mathrm{Stab}(\boldsymbol{x}_i)}) = 2(1 - \frac{1}{\#G}). \tag{11.3.2}$$

Now, the right-hand side is strictly less than 2, and each term on the left is at least $1/2$, since the size of each stabilizer is at least 2, so we have

$$2 \leq M \leq 3.$$

If $M = 2$, we obtain from equation (11.3.1) or (11.3.2) that there are exactly two poles, and thus, two orbits of size 1. Hence, there is only one rotation axis for the elements of G, and G must be a finite cyclic group.

If $M = 3$, write $2 \leq a \leq b \leq c$ for the sizes of the stabilizers for the three orbits. The equation (11.3.2) rearranges to

$$\frac{1}{a} + \frac{1}{b} + \frac{1}{c} = 1 + \frac{2}{\#G}. \tag{11.3.3}$$

Since the right side is greater than 1 and no more than 2, the only solutions are $(2, 2, n)$ for $n \geq 2$, $(2, 3, 3)$, $(2, 3, 4)$, and $(2, 3, 5)$.

The rest of the proof consists of showing that the only groups which can realize these data are the dihedral group D_n acting as rotations of the regular n-gon, and the rotation groups of the tetrahedron, the octahedron, and the icosahedron. The idea is to construct the geometric figures from the data on the orbits of poles.

It's a little tedious to read the details of the four cases, but fun to work out the details for yourself. The exercises provide a guide. □

Exercises

Exercise 11.3.1. A finite subgroup of $\mathrm{SO}(2, \mathbb{R})$ is cyclic. Consequently, a finite subgroup of $\mathrm{SO}(3, \mathbb{R})$ which consists of rotations about a single axis is cyclic. A finite subgroup of $\mathrm{O}(2, \mathbb{R})$ is either cyclic or a dihedral group D_n for $n \geq 1$.

Exercise 11.3.2. The stabilizer of any pole is a cyclic group of order at least 2. The stabilizer of a pole $\boldsymbol{x}$ is the same as the stabilizer of the pole $-\boldsymbol{x}$, so the orbits of $\boldsymbol{x}$ and of $-\boldsymbol{x}$ have the same size. Consider the example of the rotation group G of the tetrahedron. What are the orbits of G acting on the set of poles of G? For which poles $\boldsymbol{x}$ is it true that $\boldsymbol{x}$ and $-\boldsymbol{x}$ belong to the same orbit?

Exercise 11.3.3. Let G be a finite subgroup of $\mathrm{SO}(3, \mathbb{R})$ with the data $M = 3$, $(a, b, c) = (2, 2, n)$ with $n \geq 2$. Show that $\#G = 2n$, and the sizes of the three orbits of G acting on $\mathcal{P}$ are n, n, and 2. The case $n = 2$ is a bit special, so consider first the case $n \geq 3$. There is one orbit of size 2, which must consist of a pair of poles $\{\boldsymbol{x}, -\boldsymbol{x}\}$; and the stabilizer of this pair is a cyclic group of rotations about the axis determined by $\{\boldsymbol{x}, -\boldsymbol{x}\}$. Show that there is an n-gon in the plane through the origin perpendicular to $\boldsymbol{x}$, whose n vertices consist of an orbit of poles of G, and show that G is the rotation group of this n-gon.

Exercise 11.3.4. Extend the analysis of the last exercise to the case $n = 2$. Show that G must be $D_2 \cong \mathbb{Z}_2 \times \mathbb{Z}_2$, acting as symmetries of a rectangular card.

Exercise 11.3.5. Let G be a finite subgroup of $\mathrm{SO}(3, \mathbb{R})$ with the data $M = 3$, $(a, b, c) = (2, 3, 3)$. Show that $\#G = 12$, and the size of the three orbits of G acting on $\mathcal{P}$ are 6, 4, and 4. Consider an orbit $\mathcal{O} \subseteq \mathcal{P}$ of size 4, and let $\boldsymbol{u} \in \mathcal{O}$. Choose a vector $\boldsymbol{v} \in \mathcal{O} \setminus \{\boldsymbol{u}, -\boldsymbol{u}\}$. Let $g \in G$ generate the stabilizer of $\boldsymbol{u}$ (so $o(g) = 3$).

Conclude that $\{\boldsymbol{u}, \boldsymbol{v}, g\boldsymbol{v}, g^2\boldsymbol{v}\} = \mathcal{O}$. Deduce that $\{\boldsymbol{v}, g\boldsymbol{v}, g^2\boldsymbol{v}\}$ are the three vertices of an equilateral triangle, which lies in a plane perpendicular to $\boldsymbol{u}$, and that $||\boldsymbol{u} - \boldsymbol{v}|| = ||\boldsymbol{u} - g\boldsymbol{v}|| = ||\boldsymbol{u} - g^2\boldsymbol{v}||$.

Now, let $\boldsymbol{v}$ play the role of $\boldsymbol{u}$, and conclude that the four points of $\mathcal{O}$ are equidistant and are, therefore, the four vertices of a regular tetrahedron

$\mathcal{T}$. Hence, G acts as symmetries of $\mathcal{T}$; since $\#G = 12$, conclude that G is the rotation group of $\mathcal{T}$.

Exercise 11.3.6. Let G be a finite subgroup of $\mathrm{SO}(3, \mathbb{R})$ with the data $M = 3$, $(a, b, c) = (2, 3, 4)$. Show that $\#G = 24$, and the size of the three orbits of G acting on $\mathcal{P}$ are 12, 8, and 6.

Consider the orbit $\mathcal{O} \subseteq \mathcal{P}$ of size 6. Since there is only one such orbit, $\mathcal{O}$ must contain together with any of its elements $\boldsymbol{x}$ the opposite vector $-\boldsymbol{x}$.

Let $\boldsymbol{u} \in \mathcal{O}$. Choose a vector $\boldsymbol{v} \in \mathcal{O} \setminus \{\boldsymbol{u}, -\boldsymbol{u}\}$. The stabilizer of $\{\boldsymbol{u}, -\boldsymbol{u}\}$ is cyclic of order 4; let g denote a generator of this cyclic group.

Show that $\{\boldsymbol{v}, g\boldsymbol{v}, g^2\boldsymbol{v}, g^3\boldsymbol{v}\}$ is the set of vertices of a square which lies in a plane perpendicular to $\boldsymbol{u}$. Show that $-\boldsymbol{v} = g^2\boldsymbol{v}$ and that the plane of the square bisects the segment $[\boldsymbol{u}, -\boldsymbol{u}]$.

Using rotations about the axis through $\boldsymbol{v}, -\boldsymbol{v}$, show that $\mathcal{O}$ consists of the 6 vertices of a regular octahedron. Show that G is the rotation group of this octahedron.

Exercise 11.3.7. Let G be a finite subgroup of $\mathrm{SO}(3, \mathbb{R})$ with the data $M = 3$, $(a, b, c) = (2, 3, 5)$.Show that $\#G = 60$, and the size of the three orbits of G acting on $\mathcal{P}$ are 30, 20, and 12.

Consider the orbit $\mathcal{O} \subseteq \mathcal{P}$ of size 12. Since there is only one such orbit, $\mathcal{O}$ must contain together with any of its elements $\boldsymbol{x}$ the opposite vector $-\boldsymbol{x}$.

Let $\boldsymbol{u} \in \mathcal{O}$ and let $\boldsymbol{v} \in \mathcal{O}$ satisfy $||\boldsymbol{u} - \boldsymbol{v}|| \le ||\boldsymbol{u} - \boldsymbol{y}||$ for all $\boldsymbol{y} \in \mathcal{O} \setminus \{\boldsymbol{u}\}$.

Let g be a generator of the stabilizer of $\{\boldsymbol{u}, -\boldsymbol{u}\}$, $o(g) = 5$. Show that the 5 points $\{g^i\boldsymbol{v} : 0 \le i \le 4\}$ are the vertices of a regular pentagon which lies on a plane perpendicular to $\boldsymbol{u}$.

Show that the plane of the pentagon *cannot* bisect the segment $[\boldsymbol{u}, -\boldsymbol{u}]$, that $\boldsymbol{u}$ and the vertices of the pentagon lie all to one side of the bisector of $[\boldsymbol{u}, -\boldsymbol{u}]$, and finally that the 12 points $\{\pm\boldsymbol{u}, \pm g^i\boldsymbol{v} : 0 \le i \le 4\}$ comprise the orbit $\mathcal{O}$. Show that these 12 points are the vertices of a regular icosahedron and that G is the rotation group of this icosahedron.

It is not very much work to extend our classification results to finite subgroups of $\mathrm{O}(3, \mathbb{R})$. If G is a finite subgroup of $\mathrm{O}(3, \mathbb{R})$, then $H = G \cap \mathrm{SO}(3, \mathbb{R})$ is a finite rotation group, so it is on the list of Theorem 11.3.1.

Exercise 11.3.8. If G is a finite subgroup of $\mathrm{O}(3, \mathbb{R})$, and the inversion $i : \boldsymbol{x} \mapsto -\boldsymbol{x}$ is an element of G, then $G = H \cup iH \cong H \times \mathbb{Z}_2$, where $H = G \cap \mathrm{SO}(3, \mathbb{R})$. Conversely, for any H on the list of Theorem 11.3.1, $G = H \cup iH \cong H \times \mathbb{Z}_2$ is a subgroup of $\mathrm{O}(3, \mathbb{R})$

Exercise 11.3.9.

(a) Suppose G is a finite subgroup of $\mathrm{O}(3, \mathbb{R})$, G is not contained in $\mathrm{SO}(3, \mathbb{R})$, and the inversion is not an element of G. Let $H = G \cap$

SO$(3,\mathbb{R})$. Show that $\psi : a \mapsto \det(a)a$ is an isomorphism of G onto a subgroup $\tilde{G}$ of SO$(3,\mathbb{R})$, and $H \subseteq \tilde{G}$ is an index 2 subgroup.

(b) Conversely, if $H \subseteq \tilde{G}$ is an index 2 pair of subgroups of SO$(3,\mathbb{R})$, let $R \in \tilde{G} \setminus H$, and define $G = H \cup (-R)H \subseteq \mathrm{O}(3,\mathbb{R})$. Show that G is a subgroup of O$(3,\mathbb{R})$, G is not contained in SO$(3,\mathbb{R})$, and the inversion is not an element of G. Furthermore, $\psi(G) = \tilde{G}$.

(c) Show that the complete list of index 2 pairs in SO$(3,\mathbb{R})$ is
   (i) $\mathbb{Z}_n \subseteq \mathbb{Z}_{2n}, n \geq 1$.
   (ii) $\mathbb{Z}_n \subseteq D_n, n \geq 2$.
   (iii) $D_n \subseteq D_{2n}, n \geq 2$.
   (iv) The rotation group of the tetrahedron contained in the rotation group of the cube.

This exercise completes the classification of finite subgroups of O$(3,\mathbb{R})$. Note that this is more than a classification up to group isomorphism; the groups are classified by their mode of action. For example, the abstract group Z_{2n} acts as a rotation group but also as a group of rotations and reflection-rotations, due to the pair $\mathbb{Z}_n \subseteq \mathbb{Z}_{2n}$ on the list of the previous exercise. More precisely, the classification is *up to conjugacy*: Recall that two subgroups A and B of O$(3,\mathbb{R})$ are conjugate if there is a $g \in \mathrm{O}(3,\mathbb{R})$ such that $A = gBg^{-1}$; conjugacy is an equivalence relation on subgroups. Our results classify the conjugacy classes of subgroups of O$(3,\mathbb{R})$.

11.4. Crystals

In this section, we shall investigate crystals in two and three dimensions, with the goal of analyzing their symmetry groups. We will encounter several new phenomena in group theory in the course of this discussion: Let's first say what we mean by a crystal.

Definition 11.4.1. A *lattice* L in a real vector space V (for us, $V = \mathbb{R}^2$ or $V = \mathbb{R}^3$) is the set of integer linear combinations of some basis of V.

For example, take the basis $\boldsymbol{a} = \begin{bmatrix} 1 \\ 0 \end{bmatrix}, \boldsymbol{b} = \begin{bmatrix} 1/2 \\ \sqrt{3}/2 \end{bmatrix}$ in $\mathbb{R}^2$. Figure 11.4.4 shows (part of) the lattice generated by $\{\boldsymbol{a}, \boldsymbol{b}\}$.

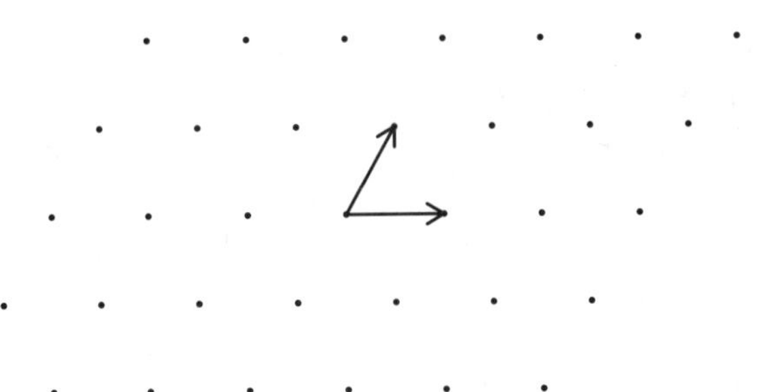

Figure 11.4.4. Hexagonal Lattice

Definition 11.4.2. A *fundamental domain* for a lattice L in V is a closed region D in V such that

1. $\bigcup_{x \in L} \boldsymbol{x} + D = V$.
2. For $\boldsymbol{x} \neq \boldsymbol{y}$ in L, $(\boldsymbol{x} + D) \cap (\boldsymbol{y} + D)$ is contained in the boundary of $(\boldsymbol{x} + D)$.

That is, the translates of D by elements of L cover V, and two different translates of D can intersect only in their boundary. (I am not interested in using complicated sets for D; convex polygons in $\mathbb{R}^2$ and convex polyhedra in $\mathbb{R}^3$ will be general enough.)

For the lattice given above, two different fundamental domains are:

1. The parallelepiped spanned by $\{\boldsymbol{a}, \boldsymbol{b}\}$, namely,

$$\{s\boldsymbol{a} + t\boldsymbol{b} : 0 \leq s, t \leq 1\}.$$

2. A hexagon centered at the origin, two of whose vertices are at $(0, \pm 1/\sqrt{3})$.

Definition 11.4.3. A *crystal* consists of some geometric figure in a fundamental domain of a lattice together with all translates of this figure by elements of the lattice.

For example, take the hexagonal fundamental domain for the lattice L described above, and the geometric figure in the fundamental domain which is displayed in Figure 11.4.5. The crystal generated by translations of this pattern is shown in Figure 11.4.6.

Crystals in two dimensions (in the form of fabrics, wallpapers, carpets, quilts, tilework, basket weaves, and lattice work) are a mainstay of decorative art. See Figure 11.4.7 You can find many examples by browsing in your library under "design" and related topics. For some remarkable examples created by "chaotic" dynamical systems, see M. Field and M. Golubitsky, *Symmetry in Chaos*, Oxford University Press, 1992.

Figure 11.4.5. Pattern in Fundamental Domain

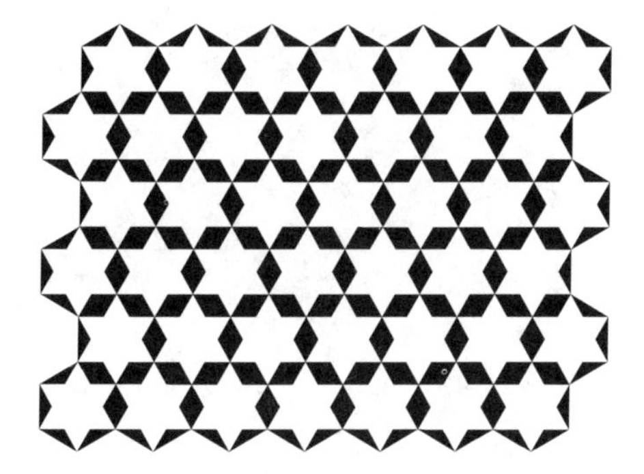

Figure 11.4.6. Hexagonal Crystal

What we have defined as a crystal in three dimensions is an idealization of physical crystals. Many solid substances are crystalline, consisting of arrangements of atoms which are repeated in a three-dimensional array. Solid table salt, for example, consists of huge arrays of sodium and chlorine atoms in a ratio of one to one. From x-ray diffraction investigation, one knows that the lattice of a salt crystal is cubic, that is, generated by three orthogonal vectors of equal length, which we can take to be one unit. A fundamental domain for this lattice is the unit cube. The sodium atoms occupy the eight vertices of the unit cube and the centers of the faces. The chlorine atoms occupy the centers of each edge as well as the center of the cube. (Each atom is surrounded by six atoms of the opposite sort, arrayed at the vertices of an octahedron.)

We turn to the main concern of this section: What is the group of isometries of a crystal? Of course, the symmetry group of a crystal built on a lattice L will include at least L, acting as translations. I will make the standing assumption that enough stuff was put into the pattern in the fundamental domain so that the crystal has no translational symmetries *other* than those in L.

Assumption: *L is the group of translations of the crystal.*

Figure 11.4.7. "Crystal" in Decorative Art

Now I want to define a "linear part" of the group of symmetries of a crystal, the so-called point group. One could be tempted by such examples as that in Figure 11.4.6 to define the point group of a crystal to be intersection of the symmetry group of the crystal with the orthogonal group. This is pretty close to being the right concept, but it is not quite right in general, as shown by the following example: Let L be the square lattice generated by $\{\hat{\boldsymbol{e}}_1, \hat{\boldsymbol{e}}_2\}$. Take as a fundamental domain the square with sides of length 1, centered at the origin. This square is divided into quarters by the coordinate axes. In the southeast quarter, put a small copy of the front page of today's New York Times. In the northwest corner, put a mirror image copy of the front page. The crystal generated by this data has no rotation or reflection symmetries. (See Figure 11.4.8.) Nevertheless, its symmetry group is larger than L, namely, $\tau_{\hat{\mathbf{e}}_2/2}\sigma$ is a symmetry, where σ is the reflection in the y-axis.

This example points us to the proper definition of the point group.

The symmetry group G of a crystal is a subgroup of Isom(V) (where $V = \mathbb{R}^2$ or $\mathbb{R}^3$). We know from Theorem 11.1.10 that Isom(V) is the semidirect product of the group V of translations and the orthogonal group $O(V)$. The lattice L is the intersection of G with the group of translations, and is a normal subgroup of G. The quotient group G/L is isomorphic to the image of G in $\text{Isom(V)}/\text{V} \cong \text{O(V)}$, by Proposition 3.4.18.

IIA ərlt ƨwəИ		IIA ərlt ƨwəИ		IIA ərlt ƨwəИ	
	All the News		All the News		All the News
IIA ərlt ƨwəИ		IIA ərlt ƨwəИ		IIA ərlt ƨwəИ	
	All the News		All the News		All the News

Figure 11.4.8. New York Times Crystal

Definition 11.4.4. The point group G^0 of a crystal is G/L.

Recall that the quotient map of Isom(V) onto $O(V)$ is given as follows: Each isometry of V can be written uniquely as a product $\tau\sigma$, where τ is a translation and σ is a linear isometry. The image of $\tau\sigma$ in $O(V)$ is σ. For the New York Times crystal described above, the symmetry group is generated by $\tau_{\hat{\mathbf{e}}_1}$, $\tau_{\hat{\mathbf{e}}_2}$, and $\tau_{\hat{\mathbf{e}}_2/2}\sigma$, where σ is the reflection in the y-axis. Therefore, the point group is the two element group generated by σ in $\mathrm{O}(2, \mathbb{R})$.

Of course, there are examples where $G^0 \cong G \cap O(V)$; this happens when G is the semi-direct product of L and $G \cap O(V)$ (exercise).

Now, consider $\boldsymbol{x} \in L$ and $A \in G^0$. By definition of G^0, there is an $\boldsymbol{h} \in V$ such that $\tau_{\boldsymbol{h}} A \in G$. Then $(\tau_{\boldsymbol{h}} A)\tau_{\boldsymbol{x}}(\tau_{\boldsymbol{h}} A)^{-1} = \tau_{\boldsymbol{h}}\tau_{A\boldsymbol{x}}\tau_{-\boldsymbol{h}} = \tau_{A\boldsymbol{x}} \in G$. It follows from our assumption that $A\boldsymbol{x} \in L$. Thus, we have proved:

Lemma 11.4.5. *L is invariant under G^0.*

Let $\mathbb{F}$ be a basis of L, that is, a basis of the ambient vector space V ($= \mathbb{R}^2$ or $\mathbb{R}^3$) such that L consists of *integer* linear combinations of $\mathbb{F}$. Since for $A \in G^0$ and $\boldsymbol{a} \in \mathbb{F}$, $A\boldsymbol{a} \in L$, it follows that the matrix of A with respect to the basis $\mathbb{F}$ is integer valued. This observation immediately yields a strong restriction on G^0:

Proposition 11.4.6. *Any rotation in G^0 is of order* 2, 3, 4, *or* 6.

Proof. On the one hand, a rotation $A \in G^0$ has an integer valued matrix with respect to $\mathbb{F}$ so, in particular, the trace of A is an integer. On the other hand, with respect to a suitable orthonormal basis of V, A has the matrix

$$\begin{bmatrix} \cos(\theta) & -\sin(\theta) & 0 \\ \sin(\theta) & \cos(\theta) & 0 \\ 0 & 0 & 1 \end{bmatrix}$$

in the three-dimensional case or

$$\begin{bmatrix} \cos(\theta) & -\sin(\theta) \\ \sin(\theta) & \cos(\theta) \end{bmatrix}$$

in the two-dimensional case. So it follows that $2\cos(\theta)$ is an integer, and, therefore, $\theta = \pm 2\pi/k$ for $k \in \{2, 3, 4, 6\}$. □

Corollary 11.4.7. *The point group of a two-dimensional crystal is one of the following ten groups (up to conjugacy in* $\mathrm{O}(2, \mathbb{R})$*):* $\mathbb{Z}_n$ *or* D_n *for* $n = 1, 2, 3, 4, 6$.

Proof. This follows from Exercise 11.3.1 and Proposition 11.4.6. □

We will call the classes of point groups, up to conjugacy in $\mathrm{O}(2, \mathbb{R})$. the *geometric point groups*.

Corollary 11.4.8.

(a) *The elements of infinite order in a two-dimensional crystal group are either translations or glide-reflections.*

(b) *An element of infinite order is a translation if and only if it commutes with the square of each element of infinite order.*

Proof. The first part follows from the classification of isometries of $\mathbb{R}^2$, together with the fact that rotations in a two-dimensional crystal group are of finite order. Since the square of a glide-reflection or of a translation is a translation, translations commute with the square of each element of infinite order. On the other hand, if τ is a glide-reflection, and σ is a translation in a direction not parallel to the line of reflection of τ, then τ does not commute with σ^2. □

Lemma 11.4.9. *Let L be a two-dimensional lattice. Let $\boldsymbol{a}$ be a vector of minimal length in L, and let $\boldsymbol{b}$ be a vector of minimal length in $L \setminus \mathbb{R}\boldsymbol{a}$. Then $\{\boldsymbol{a}, \boldsymbol{b}\}$ is a basis for L; that is, the integer linear span of $\{\boldsymbol{a}, \boldsymbol{b}\}$ is L.*

Proof. Exercise 11.4.4. □

Lemma 11.4.10. *Suppose the point group of a two-dimensional crystal contains a rotation R of order 3,4, or 6. Then the lattice L must be*

(a) *the hexagonal lattice spanned by two vectors of equal length at an angle of $\pi/3$, if R has order 3 or 6* or

(b) *the square lattice spanned by two orthogonal vectors of equal length, if R has order 4.*

Proof. Exercise 11.4.5. □

Lemma 11.4.11. *Let G_i be symmetry groups of two-dimensional crystals, with lattices L_i and point groups G_i^0, for $i = 1, 2$. Suppose $\varphi : G_1 \to G_2$ is a group isomorphism. Then:*

(a) $\varphi(L_1) = L_2$.

(b) *There is a $\Phi \in \mathrm{GL}(\mathbb{R}^2)$ such that $\varphi(\tau_{\boldsymbol{x}}) = \tau_{\Phi(\boldsymbol{x})}$ for $\boldsymbol{x} \in L_1$. The matrix of Φ with respect to bases of L_1 and L_2 is integer valued, and the inverse of this matrix is integer valued.*

(c) *φ induces an isomorphism $\tilde{\varphi} : G_1^0 \to G_2^0$. For $B \in G_1^0$, one has $\tilde{\varphi}(B) = \Phi B \Phi^{-1}$.*

(d) *φ maps affine rotations to affine rotations, affine reflections to affine reflections, translations to translations, and glide-reflections to glide-reflections.*

Proof. If τ is a translation in G_1, then τ commutes with the square of every element of infinite order in G_1. It follows that $\varphi(\tau)$ is an element of infinite order in G_2 with the same property, so $\varphi(\tau)$ is a translation, by Corollary 11.4.8. This proves part (a).

The isomorphism $\varphi : L_1 \to L_2$ extends to a linear isomorphism $\Phi : \mathbb{R}^2 \to \mathbb{R}^2$, whose matrix A with respect to bases of L_1 and L_2 is integer valued; applying the same reasoning to φ^{-1} shows that A^{-1} is also integer valued. This proves part (b).

If π_i denotes the quotient map from G_i to $G_i^0 = G_i/L_i$, for $i = 1, 2$, then $\pi_2 \circ \varphi : G_1 \to G_2^0$ is a surjective homomorphism with kernel L_1; therefore, there is an induced isomorphism $\tilde{\varphi} : G_1^0 = G_1/L_1 \to G_2^0$ such that $\tilde{\varphi} \circ \pi_1 = \pi_2 \circ \varphi$, by the Homomorphism Theorem 3.4.6.

On the other hand, G_i^0 can be identified with a finite subgroup of $\mathrm{O}(\mathbb{R}^2)$: G_i^0 is the set of $B \in \mathrm{O}(\mathbb{R}^2)$ such that there exists a $\boldsymbol{h} \in \mathbb{R}^2$ such that $\tau_{\boldsymbol{h}} B \in G_i$. So let $B \in G_1^0$ and let $\boldsymbol{h} \in \mathbb{R}^2$ satisfy $\tau_{\boldsymbol{h}} B \in G_1$. Then $\varphi(\tau_{\boldsymbol{h}} B) = \tau_{\boldsymbol{k}} \tilde{\varphi}(B)$ for some $\boldsymbol{k} \in \mathbb{R}^2$. Compute that $\tau_{B\boldsymbol{x}} = (\tau_{\boldsymbol{h}} B)\tau_{\boldsymbol{x}}(\tau_{\boldsymbol{h}} B)^{-1}$ for $\boldsymbol{x} \in L_1$. Applying φ to both sides gives $\tau_{\Phi(B\boldsymbol{x})} = \tau_{\boldsymbol{k}} \tilde{\varphi}(B) \tau_{\Phi(\boldsymbol{x})} \tilde{\varphi}(B)^{-1} \tau_{-\boldsymbol{k}} = \tau_{\tilde{\varphi}(B)\Phi(\boldsymbol{x})}$. Therefore, $\tilde{\varphi}(B) = \Phi B \Phi^{-1}$, which completes the proof of (c).

The isomorphism φ maps translations to translations, by part (a). The glide-reflections are the elements of infinite order which are not translations, so φ also maps glide-reflections to glide-reflections. The affine rotations in G_1 are elements of the form $g = \tau_h B$, where B is a rotation in G_1^0; for such an element, $\varphi(g) = \tau_k \Phi B \Phi^{-1}$ is also a rotation. Use a similar argument for affine reflections, or use the fact that affine reflections are the elements of finite order which are not affine rotations. □

Remark 11.4.12. One can show that any finite subgroup of $\mathrm{GL}(\mathbb{R}^n)$ is conjugate in $\mathrm{GL}(R^n)$ to a subgroup of $\mathrm{O}(\mathbb{R}^n)$, and subgroups of $\mathrm{O}(\mathbb{R}^n)$ which are conjugate in $\mathrm{GL}(\mathbb{R}^n)$ are also conjugate in $\mathrm{O}(\mathbb{R}^n)$. In particular, isomorphic two-dimensional crystal groups have point groups belonging to the same geometric class. These statements will be verified in the exercises.

Theorem 11.4.13. *There are exactly* 17 *isomorphism classes of two-dimensional crystal groups. They are distributed among the geometric point group classes as in Table* 11.4.1.

geometric class	*number of isomorphism classes*
$\mathbb{Z}_1$	1
D_1	3
$\mathbb{Z}_2$	1
D_2	4
$\mathbb{Z}_3$	1
D_3	2
$\mathbb{Z}_4$	1
D_4	2
$\mathbb{Z}_6$	1
D_6	1

Table 11.4.1. Distribution of Classes of Crystal Groups

Proof. The strategy of the proof is to go through the possible geometric point group classes and produce for each a list of possible crystal groups; these are then shown to be mutually non-isomorphic by using Lemma 11.4.11. In the proof, G will always denote the crystal group, G^0 the point group, and L the lattice. The main difficulties are already met in the case $G^0 = D_1$.

Point group $\mathbb{Z}_1$. The lattice L is the entire symmetry groups. Any two lattices in $\mathbb{R}^2$ are isomorphic as groups. A crystal with trivial point group is displayed in Figure 11.4.9.

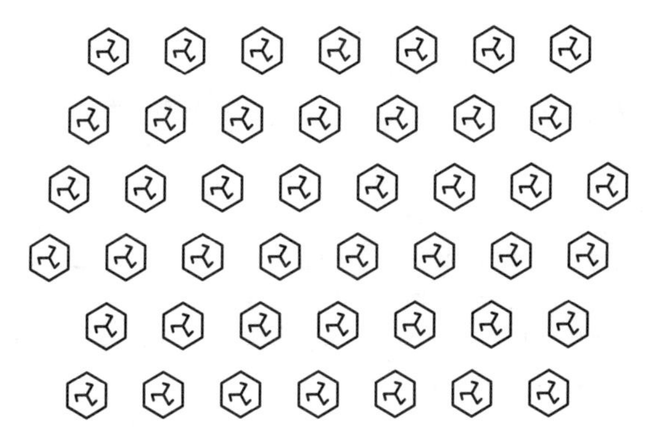

Figure 11.4.9. Crystal with Trivial Point Group

Point group D_1. The point group is generated by a single reflection σ in a line A through the origin. Let B be an orthogonal line through the origin. If $\boldsymbol{v}$ is any vector in L which is not in $A \cup B$, then $\boldsymbol{v} + \sigma(\boldsymbol{v})$ is in $L \cap A$ and $\boldsymbol{v} - \sigma(\boldsymbol{v})$ is in $L \cap B$. Let L_0 be the lattice generated by $L \cap (A \cup B)$.

Case: $L_0 = L$. In this case the lattice L is generated by orthogonal vectors $\boldsymbol{a} \in A$ and $\boldsymbol{b} \in B$.

If $\sigma \in G$, then G is the semi-direct product of L and D_1. This isomorphism class is denoted D_{1m}.

If $\sigma \notin G$, then G contains a glide-reflection $\tau_h\sigma$ in a line parallel to A; without loss of generality, one can suppose that the origin is on this line and that $g = \tau_{s\boldsymbol{a}}\sigma$, with $0 < |s| \leq 1/2$. Since $g^2 = \tau_{2s\boldsymbol{a}} \in G$, we have $|s| = 1/2$. Then G is generated by L and the glide-reflection $\tau_{\boldsymbol{a}/2}\sigma$. This isomorphism class is denoted by D_{1g}.

Case: $L_0 \neq L$. Let $\boldsymbol{u}$ be a vector of shortest length in $L \setminus L_0$; one can assume without loss of generality that $\boldsymbol{u}$ makes an acute angle with both the generators $\boldsymbol{a}$ and $\boldsymbol{b}$ of L_0. Put $\boldsymbol{v} = \sigma(\boldsymbol{u})$. One can now show that $\boldsymbol{u}$ and $\boldsymbol{v}$ generate L, $\boldsymbol{a} = \boldsymbol{u} + \boldsymbol{v}$, $\boldsymbol{b} = \boldsymbol{u} - \boldsymbol{v}$ (exercise).

As in the previous case, if G contains a glide-reflection, then we can assume without loss of generality that it contains the glide-reflection $\tau_{\boldsymbol{a}/2}\sigma = \tau_{(1/2)(\boldsymbol{u}+\boldsymbol{v})}\sigma$. But then G also contains the reflection

$$\tau_{-\boldsymbol{v}}\tau_{(1/2)(\boldsymbol{u}+\boldsymbol{v})}\sigma = \tau_{(1/2)(\boldsymbol{u}-\boldsymbol{v})}\sigma.$$

Thus, G is a semi-direct product of L and D_1. This isomorphism class is labeled D_{1c}.

The classes D_{1m}, D_{1g}, and D_{1c} are mutually non-isomorphic: D_{1m} and D_{1c} are both semi-direct products of L and D_1, while D_{1g} is not. In D_{1m}, the reflection σ is represented by the matrix

$$\begin{bmatrix} 1 & 0 \\ 0 & -1 \end{bmatrix}$$

with respect to a basis of L, and in D_{1g}, σ is represented by

$$\begin{bmatrix} 0 & 1 \\ 1 & 0 \end{bmatrix}.$$

These matrices are not conjugate in $\mathrm{GL}(2, \mathbb{Z})$, so by Lemma 11.4.11, the crystal groups are not isomorphic.

The New York Times crystal is of type D_{1g}. Figure 11.4.10 displays crystals of types D_{1m} and D_{1c}.

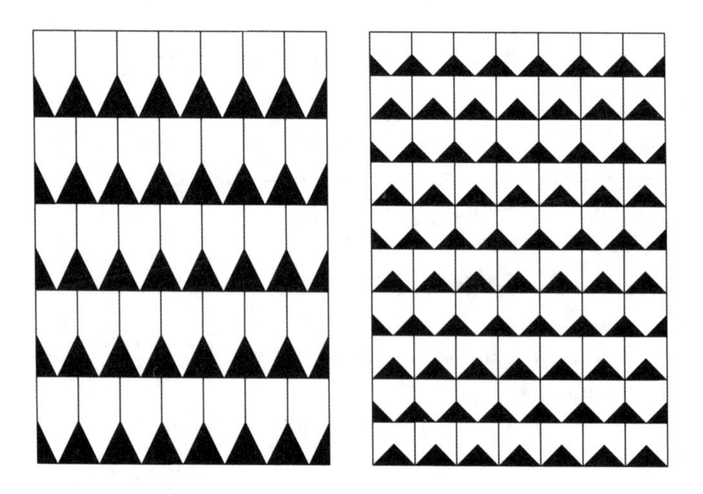

Figure 11.4.10. Crystals with Point Group D_1

Point group $\mathbb{Z}_2$. The point group is generated by the half-turn $-E$, so G contains an element $\tau_{\boldsymbol{h}}(-E)$, which is also a half-turn about some point. The group G is, thus, a semi-direct product of L and $\mathbb{Z}_2$; there is no restriction on the lattice. A crystal with symmetry type Z_2 is displayed in Figure 11.4.11.

Point group D_2. Here G^0 is generated by reflections in two orthogonal lines A and B. As for the point group D_1, one has to consider two cases. Let L_0 be the lattice generated by $L \cap (A \cup B)$.

Case: $L_0 = L$. The lattice L is generated by orthogonal vectors $\boldsymbol{a} \in A$ and $\boldsymbol{b} \in B$. Here there are three further possibilities: Let α and β denote the reflections in the lines A and B.

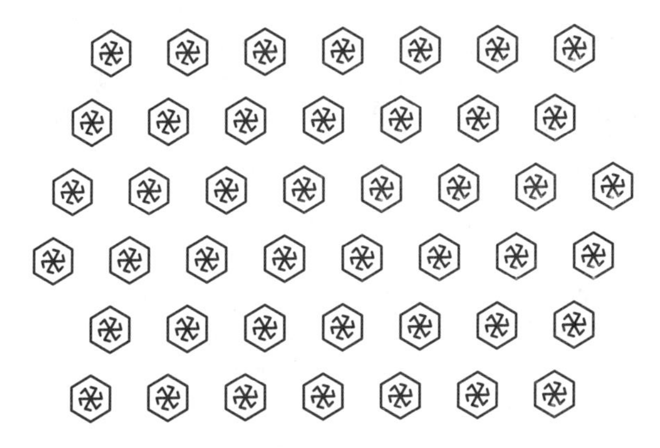

Figure 11.4.11. Crystal of Symmetry Type Z_2

If both α and β are contained in G, then G is the semi-direct product of L and D_2. The generators of D_2 are represented by matrices

$$\begin{bmatrix} 1 & 0 \\ 0 & -1 \end{bmatrix} \quad \text{and} \quad \begin{bmatrix} -1 & 0 \\ 0 & 1 \end{bmatrix}$$

with respect to a basis of L. This class is called D_{2mm}.

Suppose α is contained in G but not β. Then G is generated by L, α, and the glide-reflection $\tau_{\boldsymbol{b}/2}\beta$. This isomorphism class is called D_{2mg}.

If neither α nor β is contained in G, then G is generated by L and two glide-reflections $\tau_{\boldsymbol{a}/2}\alpha$ and $\tau_{\boldsymbol{b}/2}\beta$. This isomorphism class is called D_{2gg}.

Crystals of types D_{2mm}, D_{2mg}, and D_{2gg} are displayed in Figure 11.4.12.

Figure 11.4.12. Crystals with Point Group D_2

Case: $L_0 \neq L$. In this case, the lattice L is generated by two vectors of equal length $\boldsymbol{u}$ and $\boldsymbol{v}$ which are related to the generators $\boldsymbol{a}$ and $\boldsymbol{b}$ of L_0 by

$\boldsymbol{a} = \boldsymbol{u} + \boldsymbol{v}$ and $\boldsymbol{b} = \boldsymbol{u} - \boldsymbol{v}$. As in the analysis of case D_{1c}, one finds that G must contain reflections in orthogonal lines A and B. Thus, G is a semi-direct product of L and D_2, but the matrices of the generators of D_2 with respect to a basis of L are

$$\begin{bmatrix} 0 & 1 \\ 1 & 0 \end{bmatrix} \text{ and } \begin{bmatrix} 0 & 1 \\ -1 & 0 \end{bmatrix}.$$

This isomorphism class is called D_{2c}.

The four classes D_{2mm}, D_{2mg}, D_{2gg}, and D_{2c} are mutually non-isomorphic: The groups D_{2mm} and D_{2c} are semi-direct products of L and D_2, but the other two are not. The groups D_{2mm} and D_{2c} are non-isomorphic because they have matrix representations which are not conjugate in $\mathrm{GL}(2, \mathbb{Z})$. D_{2mg}, D_{2gg} are non-isomorphic because the former contains an element of order 2 while the latter does not.

A crystal of type D_{2c} is shown in Figure 11.4.13.

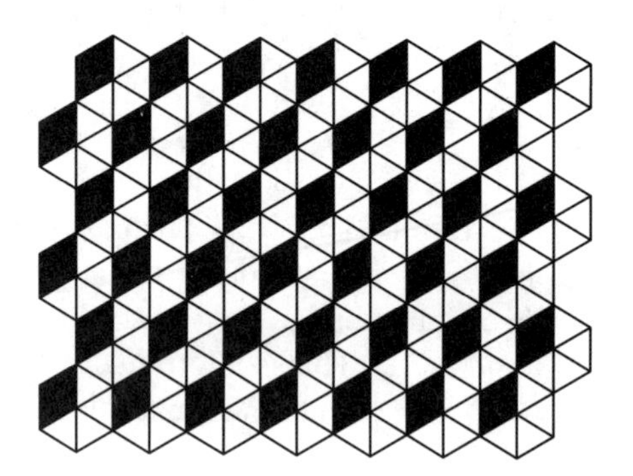

Figure 11.4.13. Crystal of Type D_{2c}

Point group $\mathbb{Z}_3$: By Exercise 11.4.5, the lattice is necessarily the hexagonal lattice generated by two vectors of equal length at an angle of $\pi/3$. G must contain an affine rotation of order 3 about some point, so G is a semi-direct product of L and $\mathbb{Z}_3$. A crystal with symmetry type Z_3 is displayed in Figure 11.4.14.

Point group D_3: As in the previous case, the lattice is necessarily the hexagonal lattice generated by two vectors of equal length at an angle of $\pi/3$.

Now, G^0 contains reflections in two lines forming an angle of $\pi/3$. One can argue as in case D_{1c} that G actually contains reflections in lines parallel to these lines. These two reflections generate a copy of D_3 in G, and hence G must be a semi-direct product of L and D_3.

There are still two possibilities for the action of D_3 on L: Let A and B be the lines fixed by two reflections in $D_3 \subseteq G$. As in the case D_1, the

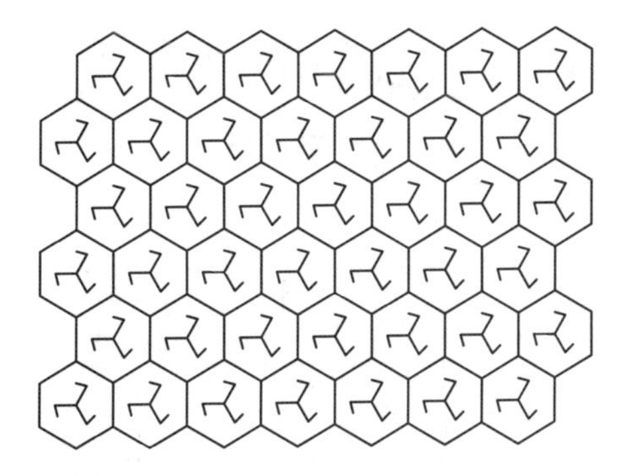

Figure 11.4.14. Crystal of Symmetry Type Z_3

lattice L must contain points on the lines A and B. Let L_0 be the lattice generated by $L \cap (A \cup B)$.

If $L_0 = L$, then D_3 is generated by reflections in the lines containing the six shortest vectors in the lattice. Generators for D_3 have matrices

$$\begin{bmatrix} 1 & 1 \\ 0 & -1 \end{bmatrix} \quad \text{and} \quad \begin{bmatrix} -1 & 0 \\ 1 & 1 \end{bmatrix}$$

with respect to a basis of L.

If $L_0 \neq L$, then, by the argument of case D_{1c}, L is generated by vectors which bisect the lines fixed by the reflections in D_3. In this case, generators of D_3 have matrices

$$\begin{bmatrix} 0 & 1 \\ 1 & 0 \end{bmatrix} \quad \text{and} \quad \begin{bmatrix} -1 & -1 \\ 0 & 1 \end{bmatrix}$$

with respect to a basis of L.

The two groups are non-isomorphic because the matrix representations are not conjugate in $\mathrm{GL}(2, \mathbb{Z})$.

Crystals of symmetry types D_{3m} and D_{3c} are displayed in Figure 11.4.15.

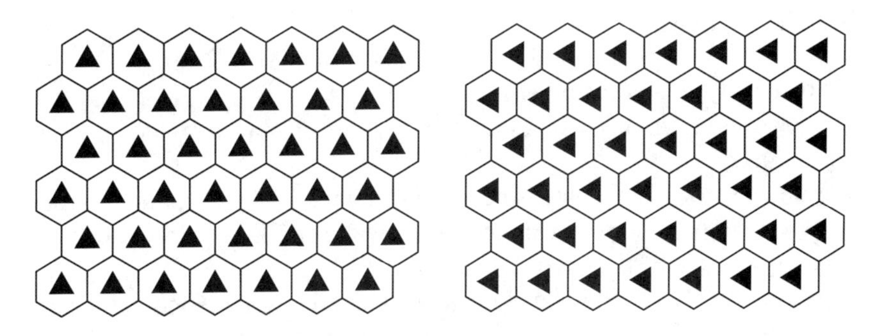

Figure 11.4.15. Crystals with Point Group D_3

Point group $\mathbb{Z}_4$: The group G must contain an affine rotation of order 4 about some point, so G is a semi-direct product of L and Z_4. The lattice is necessarily square, by Lemma 11.4.6. A crystal of symmetry type Z_4 is shown in Figure 11.4.16.

Figure 11.4.16. Crystal of Type Z_4

Point group D_4: The lattice is necessarily square, generated by orthogonal vectors of equal length $\boldsymbol{a}$ and $\boldsymbol{b}$. G contains a rotation of order 4. The point group G^0 contains reflections in the lines spanned by $\boldsymbol{a}, \boldsymbol{b}, \boldsymbol{a}+\boldsymbol{b}$, and $\boldsymbol{a}-\boldsymbol{b}$. By the argument for the case D_{1c}, G must actually contain reflections in the lines parallel to $\boldsymbol{a}+\boldsymbol{b}$ and $\boldsymbol{a}-\boldsymbol{b}$. Without loss of generality, one can suppose that the origin is the intersection of these two lines. There are still two possibilities:

Case: G contains reflections in the lines spanned by $\boldsymbol{a}$ and $\boldsymbol{b}$. Then G is a semi-direct product of L and D_4. This case is called D_{4m}.

Case: G does not contain a reflection in the line spanned by $\boldsymbol{a}$, but contains a glide-reflection $\tau_{\boldsymbol{a}/2}\sigma$, where σ is the reflection in the line spanned by $\boldsymbol{a}$. This case is called D_{4g}.

The two groups are non-isomorphic because one is a semi-direct product of L and D_4, and the other is not.

Crystals of symmetry types D_{4m} and D_{4g} are displayed in Figures 11.4.17 and 11.4.18.

Point group $\mathbb{Z}_6$: The lattice is hexagonal and the group G contains a rotation of order 6. G is a semi-direct product of L and $\mathbb{Z}_6$. A crystal of symmetry type Z_6 is displayed in Figure 11.4.19.

Point group D_6: The lattice is hexagonal and the group G contains a rotation of order 6. G also contains reflections in the lines spanned by the six shortest vectors in L, and in the lines spanned by sums and differences

Figure 11.4.17. Crystal with Point Group D_{4m}

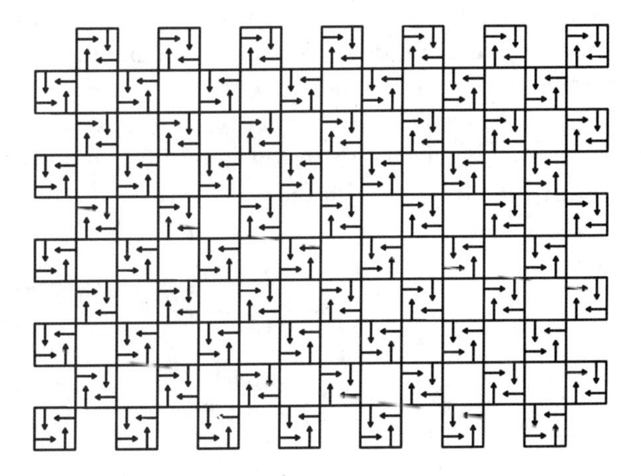

Figure 11.4.18. Crystal with Point Group D_{4g}

of these vectors. Observe that the example in Figure 11.4.6 has symmetry type D_6.

This completes the proof of Theorem 11.4.13. □

It is possible to classify three-dimensional crystals using similar principles. This was accomplished in 1890 by the crystallographer Fedorov (at a time when the atomic hypothesis was still definitely hypothetical). There are 32 geometric crystal classes (conjugacy classes of in $O(3, \mathbb{R})$ of finite groups which act on lattices); 73 arithmetic crystal classes (conjugacy classes in $GL(3, \mathbb{Z})$ of finite groups which act on lattices); and 230 isomorphism classes of crystal groups.

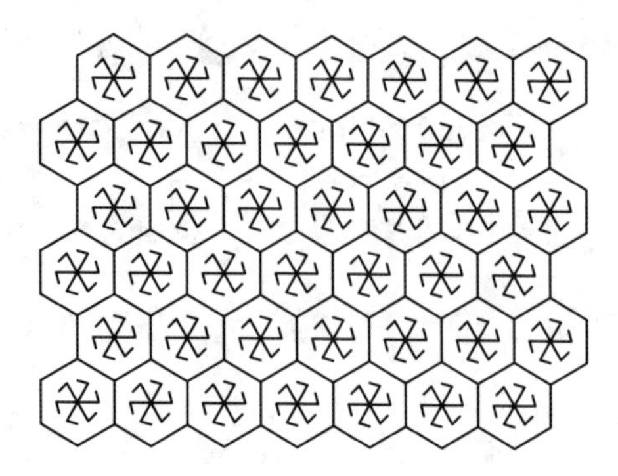

Figure 11.4.19. Crystal of Symmetry Type Z_6

About two decades after the theoretical classification of crystals by their symmetry type was achieved, developments in technology made experimental measurements of crystal structure possible. Namely, it was hypothesized by von Laue that the then newly discovered x-rays would have wave lengths comparable to the distance between atoms in crystalline substances and could be diffracted by crystal surfaces. Practical experimental implementation of this idea was developed by W. H. and W. L. Bragg (father and son) in 1912-1913.

Exercises

Exercise 11.4.1. Devise examples where the group of translation symmetries of a crystal built on a lattice L is strictly larger than L. (This can only happen if there is not enough "information" in a fundamental domain to prevent further translational symmetries.)

Exercise 11.4.2. If G is the semi-direct product of L and $G \cap O(V)$, then $G^0 \cong G \cap O(V)$.

Exercise 11.4.3. For several of the groups listed in Corollary 11.4.7, construct a two-dimensional crystal with that point group.

Exercise 11.4.4. Let L be a two-dimensional lattice. Let $\boldsymbol{a}$ be a vector of minimal length in L, and let $\boldsymbol{b}$ be a vector of minimal length in $L \setminus \mathbb{R}\boldsymbol{a}$.

(a) By using $||\boldsymbol{a} \pm \boldsymbol{b}|| \geq ||\boldsymbol{b}||$, show that $|\langle \boldsymbol{a}|\boldsymbol{b}\rangle| \leq ||\boldsymbol{a}||^2/2$.

(b) Let L_0 be the integer linear span of $\{\boldsymbol{a}, \boldsymbol{b}\}$. If $L_0 \neq L$, show that there is a non-zero vector $\boldsymbol{v} = s\boldsymbol{a} + t\boldsymbol{b} \in L \setminus L_0$ with $|s|, |t| \leq 1/2$.

(c) Using part (a), show that such a vector would satisfy

$$||\boldsymbol{v}||^2 \leq (3/4)||b||^2.$$

Show that this implies $\boldsymbol{v} = \boldsymbol{0}$.

(d) Conclude that $\{\boldsymbol{a}, \boldsymbol{b}\}$ is a basis for L; that is, the integer linear span of $\{\boldsymbol{a}, \boldsymbol{b}\}$ is L.

Exercise 11.4.5. Prove Lemma 11.4.10. *Hint:* Let $\boldsymbol{a}$ be a vector of minimal length in L; apply the previous exercise to $\{\boldsymbol{a}, R\boldsymbol{a}\}$.

Exercise 11.4.6. The purpose of this exercise and the next is to verify the assertions made in Remark 11.4.12. Let G be a finite subgroup of $\mathrm{GL}(\mathbb{R}^n)$. For $\boldsymbol{x}, \boldsymbol{y} \in \mathbb{R}^n$ define $\langle\langle \boldsymbol{x}|\boldsymbol{y}\rangle\rangle = \sum_{g \in G} \langle g\boldsymbol{x}|g\boldsymbol{y}\rangle$.

(a) Show that $(\boldsymbol{x}, \boldsymbol{y}) \mapsto \langle\langle \boldsymbol{x}|\boldsymbol{y}\rangle\rangle$ defines an inner product on $\mathbb{R}^n$. That is, this is a positive definite, bilinear function.

(b) If $g \in G$, then $\langle\langle g\boldsymbol{x}|g\boldsymbol{y}\rangle\rangle = \langle\langle \boldsymbol{x}|\boldsymbol{y}\rangle\rangle$.

(c) Conclude that the matrix of g with respect to an orthonormal basis for the inner product $\langle\langle \ | \ \rangle\rangle$ is orthogonal.

(d) Conclude that there is a matrix A (the change of basis matrix) such that AgA^{-1} is orthogonal, for all $g \in G$.

Exercise 11.4.7. Let T be an element of $\mathrm{GL}(n, \mathbb{R})$. It is known that T has a *polar decomposition* $T = U\sqrt{T^*T}$, where U is an orthogonal matrix and $\sqrt{T^*T}$ is a self-adjoint matrix which commutes with any matrix commuting with T^*T. Refer to a text on linear algebra for a discussion of these ideas.

(a) Suppose G_1 and G_2 are subgroups of $\mathrm{O}(\mathrm{n}, \mathbb{R})$ and that T is an element of $\mathrm{GL}(n, \mathbb{R})$ such that $TG_1T^{-1} = G_2$. Define $\varphi(g) = TgT^{-1}$ for $g \in G_1$, and verify that φ is an isomorphism of G_1 onto G_2.

(b) Use that fact that $G_i \subseteq \mathrm{O}(\mathrm{n}, \mathbb{R})$ to show that $\varphi(g^*) = \varphi(g)^*$ for $g \in G_1$. Here A^* is used to denote the transpose of the matrix A.

(c) Verify that $Tg = \varphi(g)T$ for $g \in G_1$. Apply transpose to both sides of this equation, and use that $g^* = g^{-1} \in G_1$ for all $g \in G_1$ to conclude that also $gT^* = T^*\varphi(g)$ for $g \in G_1$.

(d) Deduce that T^*T commutes with all elements of G_1 and, therefore, so does $\sqrt{T^*T}$. Conclude that $\varphi(g) = UgU^* = UgU^{-1}$. Thus, G_1 and G_2 are conjugate in $\mathrm{O}(\mathrm{n}, \mathbb{R})$.

Exercise 11.4.8. Consider the case of point group D_1 and $L_0 \neq L$. Let $\boldsymbol{u}$ be a vector of shortest length in $L \setminus L_0$; assume without loss of generality that $\boldsymbol{u}$ makes an acute angle with both the generators $\boldsymbol{a}$ and $\boldsymbol{b}$ of L_0. Put $\boldsymbol{v} = \sigma(\boldsymbol{u})$. Show that $\boldsymbol{u}$ and $\boldsymbol{v}$ generate L, $\boldsymbol{a} = \boldsymbol{u} + \boldsymbol{v}$, $\boldsymbol{b} = \boldsymbol{u} - \boldsymbol{v}$.

Exercise 11.4.9. Fill in the details of the analysis for D_3. Give examples of crystals of both symmetry types D_{3m} and D_{3c}.

Exercise 11.4.10. Fill in the details of the analysis of point group D_4, and give examples of crystals of both D_4 symmetry types.

Exercise 11.4.11. Fill in the details for the case D_6.

Exercise 11.4.12. Of what symmetry types are the following crystals?

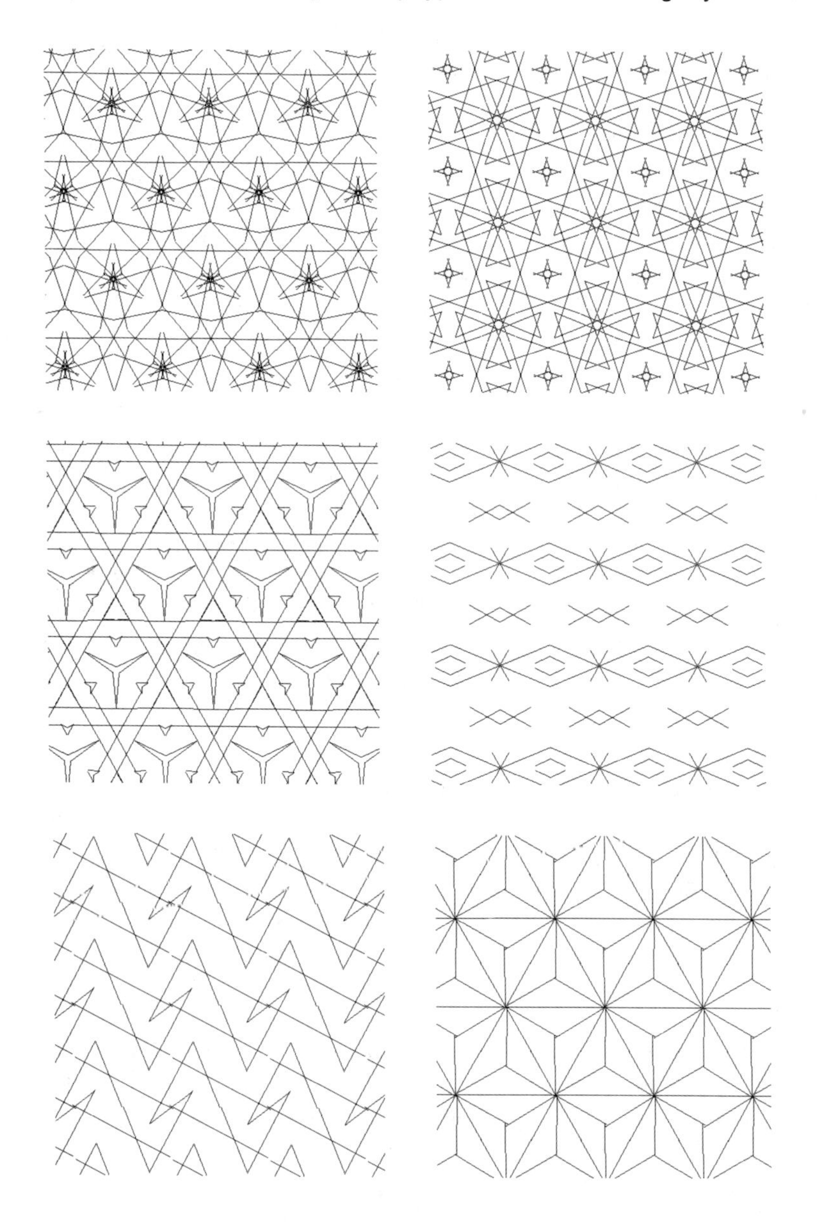

Appendix A

Almost Enough About Logic

The purpose of this course is to develop an appreciation for, and a facility in, the *practice* of mathematics. We will need to pay some attention to the language of mathematical discourse, so I include informal essays on logic and on the language of sets. These are intended to observe and describe the use of logic and set language in everyday mathematics, rather than to examine the logical or set theoretic foundations of mathematics.

To practice mathematics, one needs only very little set theory or logic; but that little one needs to use accurately. This requires developing a respect for precision in language.

You can find a more thorough, but still fairly brief and informal treatment of logic and sets, for example, in Keith Devlin, *Sets, Functions, and Logic, An Introduction to Abstract Mathematics*, 2nd ed., Chapman and Hall, London, 1992.

A.1. Statements

Logic, first of all, concerns itself with the logical structure of statements, with the construction of complex statements from simple components. A statement is a declarative sentence, which is supposed to be either true or false.

Whether a given sentence is sufficiently unambiguous to qualify as a statement, and whether it is true or false may well depend upon context. For example, the assertion *"He is my brother"* standing alone is neither true nor false, because we do not know from context to whom *"he"* refers, nor who is the speaker. The sentence becomes a statement only when both the speaker and the *"he"* have been identified. The same thing happens frequently in mathematical writing, with sentences which contain variables, for example: $x > 0$.

Such a sentence, which is capable of becoming a statement when the variables have been adequately identified, is called a *predicate* or, perhaps less pretentiously, a *statement-with-variables*.

I should remark that the validity of a statement (without variables) may also depend on the context. For example, *"The square of any number is positive"* is a true statement if it has been established by context that the numbers referred to are real numbers, but false if the numbers in question are complex numbers.

It is the job of every writer of mathematics (for example, you, when you write your homework papers!) to strive to abolish ambiguity. In particular, it is never permissible to introduce a symbol (except a symbol with a standard meaning such as $\mathbb{R}$) without declaring what it should stand for; otherwise, one is certain to end up with meaningless statements-with-variables.

A.2. Logical Connectives

Statements can be combined or modified by means of *logical connectives* to form new statements; the validity of such a composite statement depends only on the validity of its components. The basic logical connectives are *and*, *or*, *not*, and *if...then*. We consider these in turn.

A.2.1. The conjunction *and*. For statements A and B, the statement "A and B" is true exactly when both A and B are true. This is conventionally illustrated by a *truth table*:

A	B	A and B
t	t	t
t	f	f
f	t	f
f	f	f

The table contains one row for each of the four possible combinations of truth values of A and B; the last entry of the row is the corresponding truth value of "A and B." (The logical connective *and* thus defines a function from ordered pairs truth values to truth values, i.e., from $\{t, f\} \times \{t, f\}$ to $\{t, f\}$.)

A.2.2. The disjunction *or*. For statements A and B, the statement "A or B" is true when at least one of the component statements is true.

A	B	A or B
t	t	t
t	f	t
f	t	t
f	f	f

In everyday speech, "or" sometimes is taken to mean "one or the other, but not both," but in mathematics we insist that "or" always means "one or the other or both."

A.2.3. The negation *not*. The negation "not(A)" of a statement A is true when A is false and false when A is true.

A	not(A)
t	f
f	t

Of course, given an actual statement A, we do not generally negate it by writing "not(A)"; we employ instead one of various means afforded by our natural language. The negation of

- *I am satisfied with your explanation.*

is

- *I am not satisfied with your explanation.*

The statement

- *All the committee members supported the decision.*

has various negations:

- *Not all the committee members supported the decision.*
- *At least one of the committee members did not support the decision.*
- *It is not true that all the committee members supported the decision.*

The following is not a correct negation. Why not?

- *All of the committee members did not support the decision.*

At this point we might try to combine the negation "not" with the conjunction "and" or the disjunction "or." We compute the truth table of "not(A and B)," as follows:

A	B	A and B	not(A and B)
t	t	t	f
t	f	f	t
f	t	f	t
f	f	f	t

Next, we observe that "not(A) or not(B)" has the same truth table as "not(A and B)," i.e., defines the same function from $\{t, f\} \times \{t, f\}$ to $\{t, f\}$.

A	B	not(A)	not(B)	not(A) or not(B)
t	t	f	f	f
t	f	f	t	t
f	t	t	f	t
f	f	t	t	t

We say that two *statement formulas* such as "not(A and B)" and "not(A) or not(B)" are *logically equivalent* if they have the same truth table; when we substitute actual statements for A and B in the logically equivalent statement forms, we end up with two composite statements with exactly the same meaning.

Exercise A.1. Check similarly that "not(A or B)" is logically equivalent to "not(A) and not(B)" . Also verify that "not(not(A))" is equivalent to A.

A.2.4. The implication *if...then*. Next, we consider the implication

- *if A, then B*

or

- *A implies B.*

We define "if A, then B" to mean "not(A and not(B))," or, equivalently, "not(A) or B"; this is fair enough, since we want "if A, then B" to mean that one cannot have A without also having B. The negation of "A implies B" is thus "A and not(B)".

Exercise A.2. Write out the truth table for "A implies B" and for its negation.

Exercise A.3. Sometimes students jump to the conclusion that "A implies B" is equivalent to one or another of the following: "A and B," "B implies A,", or "not(A) implies not(B)." Check that in fact "A implies B" is not equivalent to any of these by writing out the truth tables and noticing the differences.

Exercise A.4. However "A implies B" *is* equivalent to its *contrapositive* "not(B) implies not(A)." Write out the truth tables to verify this.

Exercise A.5. Verify that "A implies (B implies C)" is logically equivalent to "(A and B) implies C."

Exercise A.6. Verify that "A or B" is equivalent to "not(A) implies B".

Often a statement of the form "A or B" is most conveniently proved by assuming A does not hold, and proving B.

The use of the connectives "and," "or," and "not" in logic and mathematics coincide with their use in everyday language, and their meaning is clear. Since "if ... then" has been defined in terms of these other connectives, its meaning ought to be just as clear. However, the use of "if ... then" in everyday language often differs (somewhat subtly) from that prescribed here, and we ought to clarify this by looking at an example. Sentences using "if ... then" in everyday speech frequently concern the uncertain future, for example:

$(*)$ *If it rains tomorrow, our picnic will be ruined.*

At first glance, "if ... then" seems to be used here with the prescribed meaning:

- *It is not true that it will rain tomorrow without our picnic being ruined.*

However, one notices something amiss if one forms the negation. (When one is trying to understand an assertion, it is often illuminating to consider the negation.) According to our prescription, the negation ought to be:

- *It will rain tomorrow, and our picnic will not be ruined.*

However, the actual negation of the sentence $(*)$ ought to comment on the consequences of the weather without predicting the weather:

$(**)$ *It is possible that it will rain tomorrow, and our picnic will not be ruined.*

What is going on here? Any sentence about the future must at least implicitly take account of uncertainty; the purpose of the original sentence $(*)$ is to deny uncertainty, by issuing an absolute prediction:

- *Under all circumstances, if it rains tomorrow, our picnic will be ruined.*

The negation $(**)$ readmits uncertainty.

The example above is distinctly non-mathematical, but actually something rather like this also occurs in mathematical usage of "if ... then". Very frequently "if ... then" sentences in mathematics also involve the *universal quantifier* "for every".

- *For every x, if $x \neq 0$, then $x^2 > 0$.*

Quite often the quantifier is only implicit; one writes instead

- *If $x \neq 0$, then $x^2 > 0$.*

The negation of this is not

- $x \neq 0$ *and* $x^2 \leq 0$,

as one would expect if one ignored the (implicit) quantifier. Because of the quantifier, the negation is actually

- *There exists an x such that $x \neq 0$ and $x^2 \leq 0$.*

Quantifiers and the negation of quantified statements are discussed more thoroughly below.

Here are a few logical locutions:

- "A if B" means "B implies A."
- "A only if B" means "A implies B."
- "A if, and only if, B" means "A implies B, and B implies A."
- "Unless" means "if not," but "if not" is equivalent to "or." (Check this!)
- Sometimes "but" is used instead of "and" for emphasis.

A.3. Quantifiers

We now turn to a more systematic discussion of quantifiers. Frequently, statements in mathematics assert that all objects of a certain type have a property, or that there exists at least one object with a certain property.

A.3.1. Universal Quantifier. A statement with a *universal quantifier* typically has the form:

- *For all x, $P(x)$,*

where $P(x)$ is a predicate containing the variable x. Here are some examples:

- *For every x, if $x \neq 0$, then $x^2 > 0$.*
- *For each f, if f is a differentiable real valued function on $\mathbb{R}$, then f is continuous.*

I have already mentioned that it is not unusual to omit the important introductory phrase "for all ... ." When the variable x has not already been specified, a sentence such as

- *If $x \neq 0$, then $x^2 > 0$.*

is conventionally interpreted as containing the universal quantifier. Another usual practice is to include part of the hypothesis in the introductory phrase, thus limiting the scope of the variable:

- *For every non-zero x, the quantity x^2 is positive.*
- *For every $f : \mathbb{R} \to \mathbb{R}$, if f is differentiable then f is continuous.*

It is actually preferable style not to use a variable at all, when this is possible:

- *The square of a non-zero real number is positive.*

A.3.2. Existential Quantifier. Statements containing the *existential quantifier* typically have the form:

- *There exists an x such that $P(x)$,*

where $P(x)$ is a predicate containing the variable x. Examples:

- *There exists an x such that x is positive and $x^2 = 2$.*
- *There exists a continuous function $f : (0, 1) \to \mathbb{R}$ which is not bounded.*

A.3.3. Negation of Quantified Statements. Let us consider how to form the negation of sentences containing quantifiers. The negation of the assertion that every x has a certain property is that *some* x does not have this property; thus the negation of

- *For every x, $P(x)$.*

is

- *There exists an x such that not $P(x)$.*

Consider the examples above of (true) statements containing universal quantifiers; their (false) negations are:

- *There exists a non-zero x such that $x^2 \leq 0$.*
- *There exists a function $f : \mathbb{R} \to \mathbb{R}$ such that f is differentiable and f is not continuous.*

Similarly the negation of a statement

- *There exists an x such that $P(x)$.*

is

- *For every x, not $P(x)$.*

Consider the examples above of (true) statements containing existential quantifiers; their (false) negations are:

- *For every x, $x \leq 0$ or $x^2 \neq 2$.*
- *For every function $f : (0, 1) \to \mathbb{R}$, if f is continuous, then f is bounded.*

These examples require careful thought; you should think about them until they become absolutely clear.

The various logical elements which we have discussed can be combined as necessary to express complex concepts. An example which is familiar to you from your calculus course is the definition of "the function f is continuous at the point y" :

- *For every $\epsilon > 0$, there exists a $\delta > 0$ such that (for all x) if $|x - y| < \delta$, then $|f(x) - f(y)| < \epsilon$.*

Exercise A.7. Form the negation of this statement.

A.3.4. Order of quantifiers. It is important to realize that the order of universal and existential quantifiers cannot be changed without utterly changing the meaning of the sentence. This is a true sentence:

- *For every integer n, there exists an integer m such that $m > n$.*

This is a false sentence, obtained by reversing the order of the quantifiers:

- *There exists an integer m such that for every integer n, $m > n$.*

A.4. Deductions

Logic concerns not only statements but also deductions. Basically there is only one rule of deduction:

- *If A, then B. A. Therefore B.*

For quantified statements this takes the form:

- *For all x, if $A(x)$, then $B(x)$. $A(\alpha)$. Therefore $B(\alpha)$.*

Example:

- *Every subgroup of an abelian group is normal. $\mathbb{Z}$ is an abelian group, and $n\mathbb{Z}$ is a subgroup. Therefore $n\mathbb{Z}$ is a normal subgroup of $\mathbb{Z}$.*

If you don't know what this means, it doesn't matter: You don't *have* to know what it means in order to appreciate its form. Here is another example of exactly the same form:

- *Every car will eventually end up as a pile of rust. My brand new blue Miata is a car. Therefore it will eventually end up as a pile of rust.*

As you begin to read proofs, you should look out for the verbal variations which this one form of deduction takes, and make note of them for your own use.

Most statements requiring proof are "if ... then" statements. To prove "if A, then B," one has to assume A, and prove B under this assumption. To prove "For all x, $A(x)$ implies $B(x)$," one assumes that $A(\alpha)$ holds for a particular (but arbitrary) α, and proves $B(\alpha)$ for this particular α. There are many examples of such proofs in the main text.

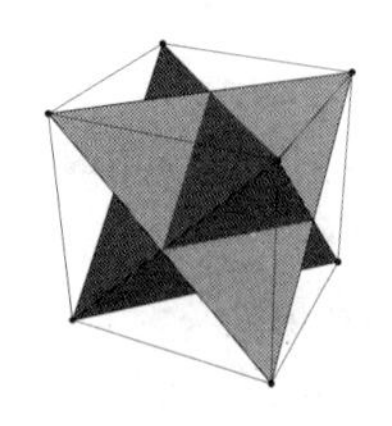

Appendix B

Almost Enough about Sets

This is an essay on the language of sets. As with logic, we treat the subject of sets in an intuitive fashion rather than as an axiomatic theory.

A *set* is a collection of (mathematical) objects. The objects contained in a set are called its *elements*. We write $x \in A$ if x is an element of the set A. Two sets are equal if they contain exactly the same elements. Very small sets can be specified by simply listing their elements, for example $A = \{1, 5, 7\}$. For sets A and B, we say that A is *contained* in B, and we write $A \subseteq B$ if each element of A is also an element of B. That is, if $x \in A$ then $x \in B$. (Because of the implicit universal quantifier, the negation of this is that there exists an element of A which is not an element of B.)

Two sets are *equal* if they contain exactly the same elements. This might seem like a quite stupid thing to mention, but in practice one often has two quite different descriptions of the same set, and one has to do a lot of work to show that the two sets contain the same elements. To do this, it is often convenient to show that each is contained in the other. That is, $A = B$ if, and only if, $A \subseteq B$ and $B \subseteq A$.

Subsets of a given set are frequently specified by a property or predicate; for example, $\{x \in \mathbb{R} : 1 \leq x \leq 4\}$ denotes the set of all real numbers between 1 and 4. Note that set containment is related to logical implication in the following fashion: If a property $P(x)$ implies a property $Q(x)$, then the set corresponding to $P(x)$ is contained the set corresponding to $Q(x)$. For example, $x < -2$ implies that $x^2 > 4$, so $\{x \in \mathbb{R} : x < -2\} \subseteq \{x \in \mathbb{R} : x^2 > 4\}$.

The *intersection* of two sets A and B, written $A \cap B$, is the set of elements contained in both sets. $A \cap B = \{x : x \in A \quad \text{and} \quad x \in B\}$. Note the relation between intersection and the logical conjunction. If $A = \{x \in C : P(x)\}$ and $B = \{x \in C : Q(x)\}$, then $A \cap B = \{x \in C : P(x) \quad \text{and} \quad Q(x)\}$.

The *union* of two sets A and B, written $A \cup B$, is the set of elements contained in at least one of the two sets. $A \cup B = \{x : x \in A \quad \text{or} \quad x \in B\}$. Set union and the logical disjunction are related as are set intersection and logical conjunction. If $A = \{x \in C : P(x)\}$ and $B = \{x \in C : Q(x)\}$, then $A \cup B = \{x \in C : P(x) \quad \text{or} \quad Q(x)\}$.

Given finitely many sets, for example, five sets A, B, C, D, E, one similarly defines their intersection $A \cap B \cap C \cap D \cap E$ to consist of those elements which are in all of the sets, and the union $A \cup B \cup C \cup D \cup E$ to consist of those elements which are in at least one of the sets.

There is a unique set with no elements at all which is called the *empty set*, or the *null set* and usually denoted $\emptyset$.

Proposition B.1. *The empty set is a subset of every set.*

Proof. Given an arbitrary set A, we have to show that $\emptyset \subseteq A$; that is, for every element $x \in \emptyset$, one has $x \in A$. The negation of this statement is that there exists an element $x \in \emptyset$ such that $x \notin A$. But this negation is false, because there are no elements at all in $\emptyset$! So the original statement is true. □

If the intersection of two sets is the empty set, we say that the sets are *disjoint*, or *non-intersecting*.

Here is a small theorem concerning the properties of set operations.

Proposition B.2. *For all sets A, B, C,*

(a) $A \cup A = A$, *and* $A \cap A = A$.

(b) $A \cup B = B \cup A$, *and* $A \cap B = B \cap A$.

(c) $(A \cup B) \cup C = A \cup B \cup C = A \cup (B \cup C)$, *and* $(A \cap B) \cap C = A \cap B \cap C = A \cap (B \cap C)$.

(d) $A \cap (B \cup C) = (A \cap B) \cup (A \cap C)$, *and* $A \cup (B \cap C) = (A \cup B) \cap (A \cup C)$.

The proofs are just a matter of checking definitions.

Given two sets A and B, we define the *relative complement* of B in A, denoted $A \setminus B$, to be the elements of A which are not contained in B. That is, $A \setminus B = \{x \in A : x \notin B\}$.

In general, all the sets appearing in some particular mathematical discussion are subsets of some "universal" set U; for example, we might be discussing only subsets of the real numbers $\mathbb{R}$. (However, there is no universal set once and for all, for all mathematical discussions; the assumption of a "set of all sets" leads to contradictions.) It is customary and convenient to use some special notation such as $\mathcal{C}(B)$ for the complement of B relative to U, and to refer to $\mathcal{C}(B) = \mathcal{U} \setminus B$ simply as *the complement of B*. (The notation $\mathcal{C}(B)$ is not standard.)

Exercise B.3. The sets $A \cap B$ and $A \setminus B$ are disjoint and have union equal to A.

Exercise B.4 (de Morgan's laws)**.** For any sets A and B, one has:

$$C(A \cup B) = C(A) \cap C(B),$$

and

$$C(A \cap B) = C(A) \cup C(B).$$

Exercise B.5. For any sets A and B, $A \setminus B = A \cap C(B)$.

Exercise B.6. For any sets A and B,

$$(A \cup B) \setminus (A \cap B) = (A \setminus B) \cup (B \setminus A).$$

Another important construction with sets is the Cartesian product. For any two sets A and B, an *ordered pair* of elements (a, b) is just a list with two items, the first from A and the second from B. The Cartesian product $A \times B$ is the set of all such pairs. Order counts, and $A \times B$ is not the same as $B \times A$, unless of course $A = B$. (The Cartesian product is named after Descarte, who realized the possibility of coordinatizing the plane as $\mathbb{R} \times \mathbb{R}$.)

We recall the notion of a *function from A to B* and some terminology regarding functions which is standard throughout mathematics. A function f from A to B is a rule which gives for each element of $a \in A$ an "outcome" in $f(a) \in B$. More formally, a function is a subset of $A \times B$ which contains for each element $a \in A$ exactly one pair (a, b); the subset contains (a, b) if, and only if, $b = f(a)$. A is called the *domain* of the function, B the *co-domain*, $f(a)$ is called the *value* of the function at a, and the set of all values, $\{f(a) : a \in A\}$, is called the *range* of the function. In general, the range is only a subset of B; a function is said to be *surjective*, or *onto*, if its range is all of B; that is, for each $b \in B$, there exists an $a \in A$, such that $f(a) = b$. Figure B.1 exhibits a surjective function.

A function f is said to be *injective*, or *one-to-one*, if for each two distinct elements a and a' in A, one has $f(a) \neq f(a')$. Equivalently, $f(a) = f(a')$ implies that $a = a'$. Figure B.2 displays an injective and a non-injective function.

Finally f is said to be *bijective* if it is both injective and surjective. A bijective function (or *bijection*) is also said to be a *one-to-one correspondence* between A and B, since it matches up the elements of the two sets one-to-one. When f is bijective, there is an *inverse function* f^{-1} defined by $f^{-1}(b) = a$ if, and only if, $f(a) = b$. Figure B.3 displays a bijective function.

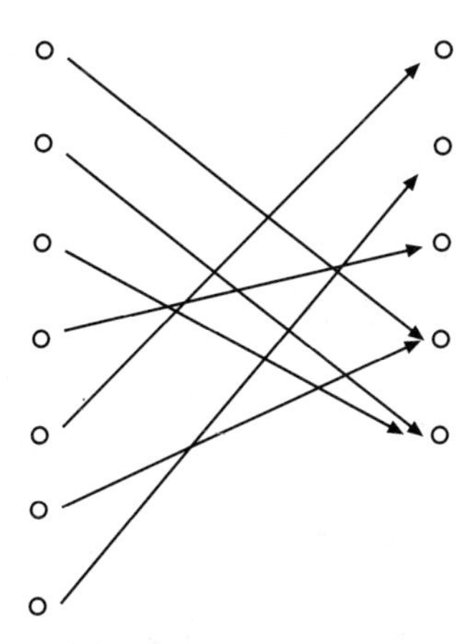

Figure B.1. A Surjection

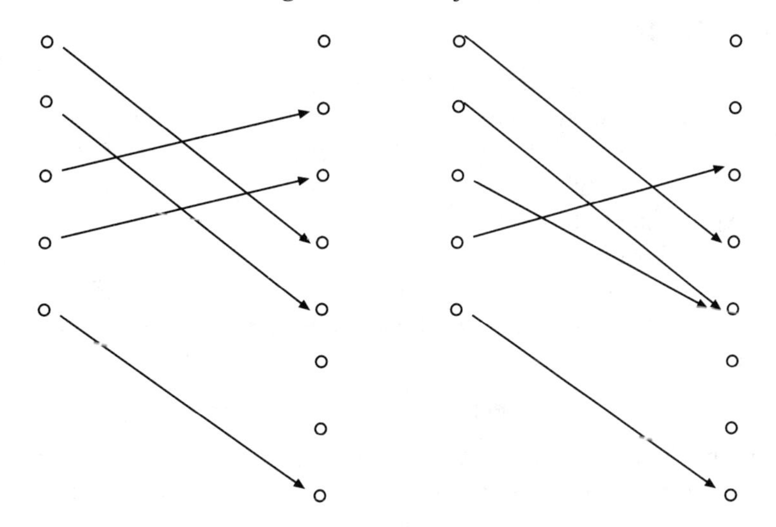

Figure B.2. Injective and Non-injective functions

If $f : X \to Y$ is a function and A is a subset of X, we write $f(A)$ for $\{f(a) : a \in A\} = \{y \in Y :$ there exists $a \in A$ such that $y = f(a)\}$. We refer to $f(A)$ as the *image of A under f*. If B is a subset of Y, we write $f^{-1}(B)$ for $\{x \in X : f(x) \in B\}$. We refer to $f^{-1}(B)$ as the *preimage of B under f*.

Exercise B.7. Let $f : X \to Y$ be a function, and let E and F be subsets of X. Then, $f(E) \cup f(F) = f(E \cup F)$. Also, $f(E) \cap f(F) \supseteq f(E \cap F)$; give an example to show that one can have strict containment.

Exercise B.8. Let $f : X \to Y$ be a function, and let E and F be subsets of Y. Then, $f^{-1}(E) \cup f^{-1}(F) = f^{-1}(E \cup F)$. Also. $f^{-1}(E) \cap f^{-1}(F) = f^{-1}(E \cap F)$. Finally, $f^{-1}(E \setminus F) = f^{-1}(E) \setminus f^{-1}(F)$.

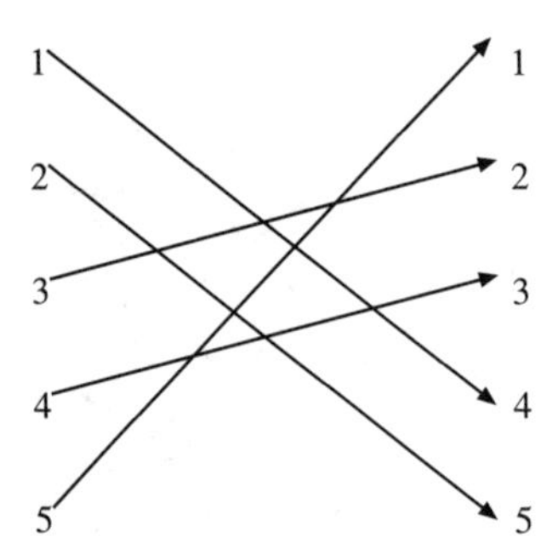

Figure B.3. A Bijection

B.1. Families of Sets; Unions and Intersections

The elements of a set can themselves be sets! This is not at all an unusual situation in mathematics. For example for each natural number n we could take the set $X_n = \{x : 0 \leq x \leq n\}$ and consider the collection of all of these sets X_n, which we call $\mathcal{F}$. Thus, $\mathcal{F}$ is a set whose members are sets. In order to at least try not to be confused, we often use words such as collection or family in referring to a set whose elements are sets.

Given a family $\mathcal{F}$ of sets, we define the union of $\mathcal{F}$, denoted $\cup\mathcal{F}$, to be the set of elements which are contained in at least one of the members of $\mathcal{F}$; that is,

$$\cup\mathcal{F} = \{x : \text{ there exists } A \in \mathcal{F} \text{ such that } x \in A\}.$$

Similarly, the intersection of the family $\mathcal{F}$, denoted $\cap\mathcal{F}$, is the set of elements which are contained in every member of $\mathcal{F}$; that is,

$$\cap\mathcal{F} = \{x : \text{ for every } A \in \mathcal{F}, \quad x \in A\}.$$

In the example we have chosen, $\mathcal{F}$ is an *indexed family of sets*, indexed by the natural numbers; that is, $\mathcal{F}$ consists of one set X_n for each natural number n. For indexed families we often denote the union and intersection as follows:

$$\cup\mathcal{F} = \bigcup_{n\in\mathbb{N}} X_n = \bigcup_{n=1}^{\infty} X_n$$

and

$$\cap\mathcal{F} = \bigcap_{n\in\mathbb{N}} X_n = \bigcap_{n=1}^{\infty} X_n.$$

It is a fact, perhaps obvious to you, that $\bigcup_{n\in\mathbb{N}} X_n$ is the set of all non-negative real numbers, whereas $\bigcap_{n\in\mathbb{N}} X_n = X_1$.

B.2. Finite and Infinite Sets

I include here a brief discussion of finite and infinite sets. I assume an intuitive familiarity with the natural numbers $\mathbb{N} = \{1, 2, 3, \ldots\}$. A set S is said to be *finite* if there is some natural number n and a bijection between S and the set $\{1, 2, \ldots, n\}$. A set is *infinite* otherwise. A set S is said to be *countably infinite* if there is a bijection between S and the set of natural numbers. A set is said to be *countable* if it is either finite or countably infinite. *Enumerable* or *denumerable* are synonyms for countable. (Some authors prefer to make a distinction between denumerable, which they take to mean countably infinite, and countable, which they use as we have.) It is a bit of a surprise to most people that infinite sets come in different sizes; in particular, there are infinite sets which are not countable. For example, using the completeness of the real numbers, one can show that the set of real numbers is not countable.

It is clear that every set is either finite or infinite, but it is not always possible to determine which! For example, it is pretty easy to show that there are infinitely many prime numbers, but it is *unknown* at present whether there are infinitely many twin primes, that is, successive odd numbers, such as 17 and 19, both of which are prime. Let us observe that although the even natural numbers $2\mathbb{N} = \{2, 4, 6, \ldots\}$, the natural numbers $\mathbb{N}$, and the integers $\mathbb{Z} = \{0, \pm 1, \pm 2, \ldots\}$ satisfy

$$2\mathbb{N} \subseteq \mathbb{N} \subseteq \mathbb{Z},$$

and no two of the sets are equal, they all are countable sets, so all have the same size or cardinality. Two sets are said to have the *same cardinality* if there is a bijection between them. One characterization of infinite sets is that a set is infinite if, and only if, it has a proper subset with the same cardinality.

Here are some theorems about finite and infinite sets which we will not prove here.

Theorem B.9. *A subset of a finite set is finite. A subset of a countable set is countable. A set S is countable if, and only if, there is an injective function with domain S and range contained in $\mathbb{N}$.*

Theorem B.10. *A union of finitely many finite sets is finite. A union of countably many countable sets is countable.*

Using this result, we see that the rational numbers $\mathbb{Q} = \{a/b : a, b \in \mathbb{Z}, b \neq 0\}$ is a countable set. Namely, $\mathbb{Q}$ is the union of the sets

$$\begin{aligned} A_1 &= \mathbb{Z} \\ A_2 &= (1/2)\mathbb{Z} = \{0, \pm 1/2, \pm 2/2, \pm 3/2 \ldots\} \\ &\ldots \\ A_n &= (1/n)\mathbb{Z} = \{0, \pm 1/n, \pm 2/n, \pm 3/n \ldots\} \\ &\ldots, \end{aligned}$$

each of which is countably infinite.

Appendix C

Induction

C.1. Proof by Induction

Suppose you need to climb a ladder. If you are able to reach the first rung of the ladder and you are also able to get from any one rung to the next, then there is nothing to stop you from climbing the whole ladder. This is called *the principle of mathematical induction.*

Mathematical induction is often used to prove statements about the natural numbers, or about families of objects indexed by the natural numbers. Suppose that you need to prove a statement of the form:

- *For all $n \in \mathbb{N}$, $P(n)$,*

where $P(n)$ is a predicate. Examples of such statements are:

- *For all $n \in \mathbb{N}$, $1 + 2 + \cdots + n = (n)(n+1)/2$.*
- *For all n and for all permutations $\pi \in S_n$, π has a unique decomposition as a product of disjoint cycles.* (See Section 2.2.)

To prove that $P(n)$ holds for all $n \in \mathbb{N}$, it suffices to show that $P(1)$ holds (you can reach the first rung), and that whenever $P(k)$ holds, then also $P(k+1)$ holds (you can get from any one rung to the next). Then, $P(n)$ holds for all n (you can climb the whole ladder).

Principle of Mathematical Induction *For the statement "For all $n \in \mathbb{N}$, $P(n)$" to be valid, it suffices that:*

1. *$P(1)$, and*
2. *For all $k \in \mathbb{N}$, $P(k)$ implies $P(k+1)$.*

To prove that "*For all $k \in \mathbb{N}$, $P(k)$ implies $P(k+1)$*," you have to assume $P(k)$ for a fixed but arbitrary value of k and prove $P(k+1)$ under this assumption. This sometimes seems like a big cheat to beginners, for we seem to be assuming what we want to prove, namely, that $P(n)$ holds. But it is not a cheat at all; we are just showing that it is possible to get from one rung to the next.

As an example we prove the identity

$$P(n) : 1 + 2 + \cdots + n = (n)(n+1)/2$$

by induction on n. The statement $P(1)$ reads

$$1 = (1)(2)/2,$$

which is evidently true. Now, we assume $P(k)$ holds for some k, that is,

$$1 + 2 + \cdots + k = (k)(k+1)/2,$$

and prove that $P(k+1)$ also holds. The assumption of $P(k)$ is called the *induction hypothesis*. Using the induction hypothesis, we have

$$1 + 2 + \cdots + k + (k+1) = (k)(k+1)/2 + (k+1) = \frac{(k+1)}{2}(k+2),$$

which is $P(k+1)$. This completes the proof of the identity.

The principle of mathematical induction is equivalent to the

Well ordering principle *Every non-empty subset of the natural numbers has a least element.*

Another form of the principle of mathematical induction is the following:

Principle of Mathematical Induction, 2nd form *For the statement "For all $n \in N$, $P(n)$" to be valid, it suffices that:*

1. *$P(1)$, and*
2. *For all $k \in \mathbb{N}$, if $P(r)$ for all $r \leq k$, then also $P(k+1)$.*

The two forms of the principle of mathematical induction and the well ordering principle are all equivalent statements about the natural numbers. That is, assuming any one of these principles, one can prove the other two. The proof of the equivalence is somewhat more abstract than the actual subject matter of this course, so I prefer to omit it. When you have more experience with doing proofs, you may wish to provide your own proof of the equivalence.

Recall that a natural number is *prime* if it is greater than 1 and not divisible by any natural number other than 1 and itself.

Proposition C.1. *Any natural number other than 1 can be written as a product of prime numbers.*

Proof. We prove this statement by using the second form of mathematical induction. Let $P(n)$ be the statement: "*n is a product of prime numbers.*" We have to show that $P(n)$ holds for all natural numbers $n \geq 2$. $P(2)$

is valid because 2 is prime. We make the inductive assumption that $P(r)$ holds for all $r \leq k$, and prove $P(k+1)$. $P(k+1)$ is the statement that $k+1$ is a product of primes. If $k+1$ is prime, there is nothing to show. Otherwise $k+1 = (a)(b)$, where $2 \leq a \leq k$ and $2 \leq b \leq k$. By the induction hypothesis, both a and b are products of prime numbers so $k+1 = ab$ is also a product of prime numbers. □

Remark C.2. It is a usual convention in mathematics to consider 0 to be the sum of an *empty* collection of numbers and 1 to be the product of an *empty* collection of numbers. This convention saves a lot of circumlocution and argument by cases. So we will consider 1 to have a prime factorization as well; it is the product of an empty collection of primes.

The following result is attributed to Euclid:

Theorem C.3. *There are infinitely many prime numbers.*

Proof. We prove for all natural numbers n the statement $P(n)$: *there are at least n prime numbers.* $P(1)$ is valid because 2 is prime. Assume $P(k)$ holds and let $2, 3, \dots, p_k$ be the first k prime numbers. Consider the natural number $M = (2)(3)\dots(p_k) + 1$. M is not divisible by any of the primes $2, 3, \dots, p_k$, so either M is prime, or M is a product of prime numbers each of which is greater than p_k. In either case, there must exist prime numbers which are greater than p_k, so there are at least $k+1$ prime numbers. □

C.2. Definitions by Induction

It is frequently necessary or convenient to define some sequence of objects (numbers, sets, functions, ...) *inductively* or *recursively*. That means the nth object is defined in terms of the first, second,$\dots$, $n-1$-st object, instead of there being a formula or procedure which tells you once and for all how to define the nth object. For example, the sequence of Fibonacci numbers is defined by the recursive rule:

$$f_1 = f_2 = 1 \quad f_n = f_{n-1} + f_{n-2} \quad \text{for } n \geq 3.$$

The well ordering principle, or the principle of mathematical induction, implies that such a rule suffices to define f_n for all natural numbers n. For f_1 and f_2 are defined by an explicit formula (we can get to the first rung), and if $f_1, \dots, f_k$ have been defined for some k, then the recursive rule $f_{k+1} = f_k + f_{k-1}$ also defines f_{k+1} (we can get from one rung to the next).

Principle of Inductive Definition: To define a sequence of objects $A_1, A_2, \ldots$ it suffices to have:

1. A definition of A_1.
2. For each $k \in \mathbb{N}$, a definition of A_{k+1} in terms of $\{A_1, \ldots, A_k\}$.

Here is an example relevant to this course: Suppose we are working in a system with an associative multiplication (perhaps a group, perhaps a ring, perhaps a field). Then, for an element a, we can define a^n for $n \in \mathbb{N}$ by the recursive rule: $a^1 = a$ and for all $k \in \mathbb{N}$, $a^{k+1} = a^k a$.

Here is another example where the objects being defined are intervals in the real numbers. The goal is to compute an accurate approximation to $\sqrt{7}$. We define a sequence of intervals $A_n = [a_n, b_n]$ with the properties:

1. $b_n - a_n = 6/2^n$,
2. $A_{n+1} \subseteq A_n$ for all $n \in \mathbb{N}$, and
3. $a_n^2 < 7$ and $b_n^2 > 7$ for all $n \in \mathbb{N}$.

Define $A_1 = [1, 7]$. If $A_1, \ldots, A_k$ have been defined, let $c_k = (a_k + b_k)/2$. If $c_k^2 < 7$, then define $A_{k+1} = [c_k, b_k]$. Otherwise, define $A_k = [a_k, c_k]$. (Remark that all the numbers a_k, b_k, c_k are rational, so it is never true that $c_k^2 = 7$.) You ought to do a proof (by induction) that the sets A_n defined by this procedure do satisfy the properties listed above. This example can easily be transformed into a computer program for calculating the square root of 7.

C.3. Multiple Induction

Let a be an element of an algebraic system with an associative multiplication (a group, a ring, a field). Consider the problem of showing that, for all natural numbers m and n, $a^m a^n = a^{m+n}$. It would seem that some sort of inductive procedure is appropriate, but two integer variables have to be involved in the induction. How is this to be done? Let $P(m, n)$ denote the predicate "$a^m a^n = a^{m+n}$." To establish that for all $m, n \in \mathbb{N}$, $P(m, n)$, I claim that it suffices to show that:

(1) $P(1, 1)$.
(2) For all $r, s \in \mathbb{N}$ if $P(r, s)$ holds, then $P(r + 1, s)$ holds.
(3) For all $r, s \in \mathbb{N}$ if $P(r, s)$ holds, then $P(r, s + 1)$ holds.

To justify this intuitively, we have to replace our image of a ladder with the grid of integer points in the quarter-plane, $\{(m, n) : m, n \in \mathbb{N}\}$. Showing $P(1, 1)$ gets us onto the grid. Showing (2) allows us to get from any point on the grid to the adjacent point to the right, and showing (3) allows us to get from any point on the grid to the adjacent point above. By taking steps to the right and up from $(1, 1)$, we can reach any point on the grid, so eventually $P(m, n)$ can be established for any (m, n).

As intuitive as this picture may be, it is not entirely satisfactory, because it seems to require a new principle of induction on two variables. And if we needed to do an induction on three variables, we would have to invent yet another principle. It is more satisfactory to justify the procedure by the principle of induction on one integer variable.

To do this, it is useful to consider the following very general situation. Suppose we want to prove a proposition of the form "for all $t \in T$, $P(t)$," where T is some set and P is some predicate. Suppose T can be written as a union of sets $T = \bigcup_1^\infty T_k$, where (T_k) is an increasing sequence of subsets $T_1 \subseteq T_2 \subseteq T_3 \dots$. According to the usual principle of induction it suffices to show

(a) For all $t \in T_1$, $P(t)$, and
(b) For all $k \in \mathbb{N}$, if $P(t)$ holds for all $t \in T_k$, then also $P(t)$ holds for all $t \in T_{k+1}$.

In fact, this suffices to show that for all $n \in N$, and for all $t \in T_n$, $P(t)$. But since each $t \in T$ belongs to some T_n, $P(t)$ holds for all $t \in T$.

Now, to apply this general principle to the situation of induction on two integer variable, we take T to be the set $\{(m, n) : m, n \in \mathbb{N}\}$. There are various ways in which we could choose the sequence of subsets T_k; for example we can take $T_k = \{(m, n) : m, n \in \mathbb{N} \text{ and } m \le k\}$. Now, suppose we have a predicate $P(m, n)$ for which we can prove:

(1) $P(1, 1)$
(2) For all $r, s \in \mathbb{N}$ if $P(r, s)$ holds then $P(r + 1, s)$ holds.
(3) For all $r, s \in \mathbb{N}$ if $P(r, s)$ holds then $P(r, s + 1)$ holds.

Using (1) and (3) and induction on one variable, we can conclude that $P(1, n)$ holds for all $n \in \mathbb{N}$; that is, $P(m, n)$ holds for all $(m, n) \in T_1$. Now, fix $k \in \mathbb{N}$, and suppose that $P(m, n)$ holds for all $(m, n) \in T_k$, that is, whenever $m \le k$. Then, in particular, $P(k, 1)$ holds. It then follows from (2) that $P(k + 1, 1)$ holds. Now, from this and (3) and induction on one variable, it follows that $P(k + 1, n)$ holds for all $n \in \mathbb{N}$. But then $P(m, n)$ holds for all (m, n) in $T_k \cup \{(k + 1, n) : n \in \mathbb{N}\} = T_{k+1}$. Thus, we can prove (a) and (b) for our sequence T_k.

Exercise C.4. Define $T_k = \{(m, n) : m, n \in \mathbb{N}, m \le k, \text{ and } n \le k\}$. Show how statements (1) through (3) above imply (a) and (b) for this choice of T_k.

Exercise C.5. Let a be a real number (or an element of a group, or an element of a ring). Show by induction on two variables that $a^m a^n = a^{m+n}$ for all $m, n \in \mathbb{N}$.

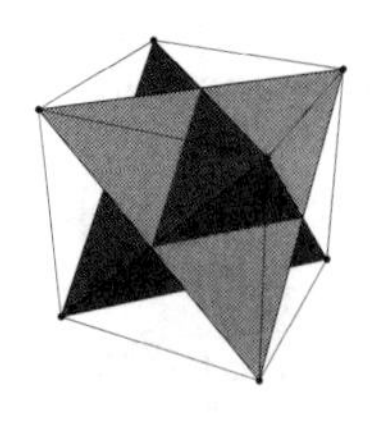

Appendix D

Complex Numbers

In this appendix, we review the construction of the complex numbers from the real numbers. As you know, the equation $x^2 + 1 = 0$ has no solution in the real numbers. However, it is possible to construct a field containing $\mathbb{R}$ by appending to $\mathbb{R}$ a solution of this equation.

To begin with, consider the set $\mathbb{C}$ of all formal sums $a + bi$, where a and b are in $\mathbb{R}$ and i is just a symbol. We give this set the structure of a two-dimensional real vector space in the obvious way: Addition is defined by $(a+bi) + (a' + b'i) = (a + a') + (b + b')i$ and multiplication with real scalars by $\alpha(a + bi) = \alpha a + \alpha bi$.

Next, we *try* to define a multiplication on $\mathbb{C}$ in such a way that the distributive law holds and also $i^2 = -1$, and $i\alpha = \alpha i$. These requirements force the definition: $(a+bi)(c+di) = (ac - bd) + (ad + bc)i$. Now it is completely straightforward to check that this multiplication is commutative and associative, and that the distributive law does indeed hold. Moreover, $(a + 0i)(c + di) = ac + adi$, so multiplication by $(a + 0i)$ coincides with scalar multiplication by $a \in \mathbb{R}$. In particular $1 = 1 + 0i$ is the multiplicative identity in $\mathbb{C}$, and we can identify $\mathbb{R}$ with the set of elements $a + 0i$ in $\mathbb{C}$.

To show that $\mathbb{C}$ is a field, it remains only to check that non-zero elements have multiplicative inverses. It is straightforward to compute that:

$$(a + bi)(a - bi) = a^2 + b^2 \in \mathbb{R}.$$

Hence, if not both a and b are zero, then

$$(a + bi)(\frac{a}{(a^2 + b^2)} - \frac{b}{(a^2 + b^2)}i) = 1.$$

It is a remarkable fact that every polynomial with complex coefficients has a complete set of roots in $\mathbb{C}$; that is, every polynomial with complex coefficients is a product of linear factors $x - \alpha$. This theorem is due to Gauss and is often know as the Fundamental Theorem of Algebra. All proofs of this theorem contain some analysis, and the most straightforward proofs involve some complex analysis; you can find a proof in any text on complex analysis. In general, a field K with the property that every polynomial with coefficients in K has a complete set of roots in K is called *algebraically closed.*

For any complex number $z = a + bi \in \mathbb{C}$, one defines the complex conjugate of z by $\bar{z} = a - bi$, and the modulus of z by $|z| = \sqrt{a^2 + b^2}$. Note that $z\bar{z} = |z|^2$, and $z^{-1} = \bar{z}/|z|^2$. The *real part* of z, denoted $\Re z$, is $a = (z + \bar{z})/2$, and the *imaginary part* of z, denoted $\Im z$ is $b = (z - \bar{z})/2i$. (Note that the imaginary part of z is a real number, and $z = \Re z + i\Im z$.)

If $z = a + bi$ is a complex number with modulus 1 (that is, $a^2 + b^2 = 1$), then there is a real number t such that $a = \cos t$ and $b = \sin t$; t is determined up to addition of an integer multiple of 2π.

Exercise D.0.1. Consider two complex numbers of modulus 1, namely, $\cos t + i \sin t$ and $\cos s + i \sin s$. Use trigonometric identities to verify that

$$(\cos t + i \sin t)(\cos s + i \sin s) = \cos(s + t) + i \sin(s + t).$$

We introduce the notation $e^{it} = \cos t + i \sin t$; then the result of the previous exercise is $e^{it}e^{is} = e^{i(t+s)}$.

For any complex number z, $z/|z|$ has modulus 1, so is of the form e^{it} for some t. Therefore, z itself can be written in the form $z = |z|(z/|z|) = |z|e^{it}$.

Exercise D.0.2. Write $5 - 3i$ in the form re^{it}, where $r > 0$ and $t \in \mathbb{R}$.

The form $z = re^{it}$ for a complex number is called the *polar form*. Multiplication of two complex numbers written in polar form is particularly easy to compute: $r_1 e^{it} r_2 e^{is} = r_1 r_2 e^{i(s+t)}$. It follows from this that for complex numbers z_1 and z_2, one has $|z_1 z_2| = |z_1||z_2|$.

For a complex number $z = re^{it}$, we have $z^n = r^n e^{int}$.

Exercise D.0.3. A complex number z satisfies $z^n = 1$ if, and only if, $z \in \{e^{i2\pi k/n} : 0 \le k \le n - 1\}$. Such a complex number is called an n-th root of unity.

Appendix E

Models of Regular Polyhedra

To make the models of the regular polyhedra, copy the patterns on the following pages onto heavy card stock (65 -70 pound stock), using as much magnification as possible. Then cut out the patterns (including the grey glue tabs), and score them along the internal lines with a utility knife. Glue them (using a slow-drying glue in order to give yourself time to adjust the position). The patterns for the icosahedron and dodecahedron come in two pieces, each of which makes one "hemisphere" of the polyhedron, which you have to glue together.

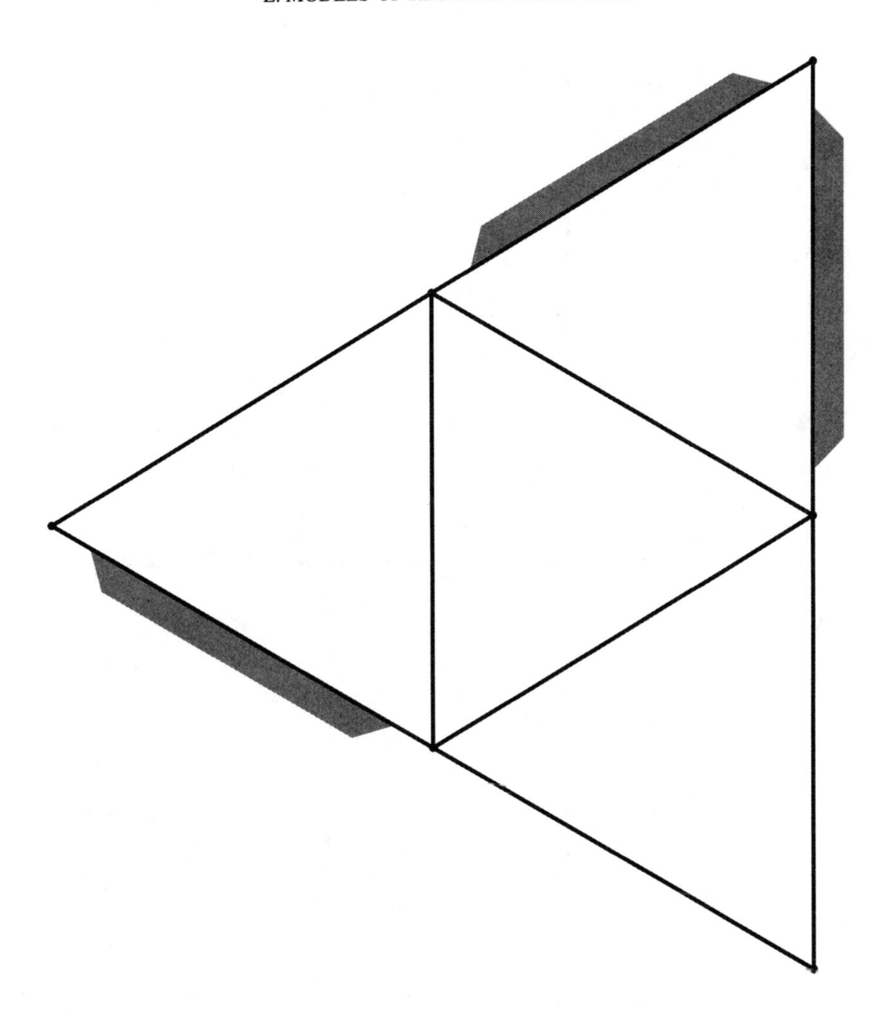

Figure E.1. Tetrahedron pattern

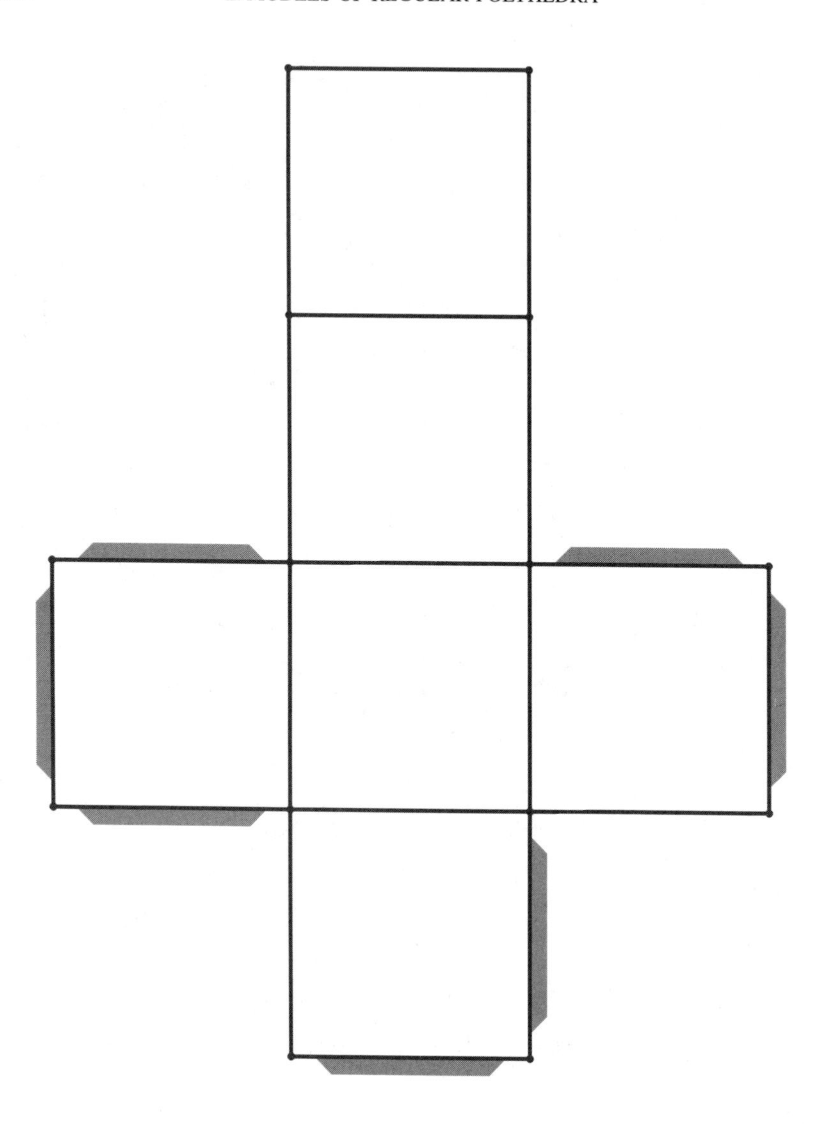

Figure E.2. Cube pattern

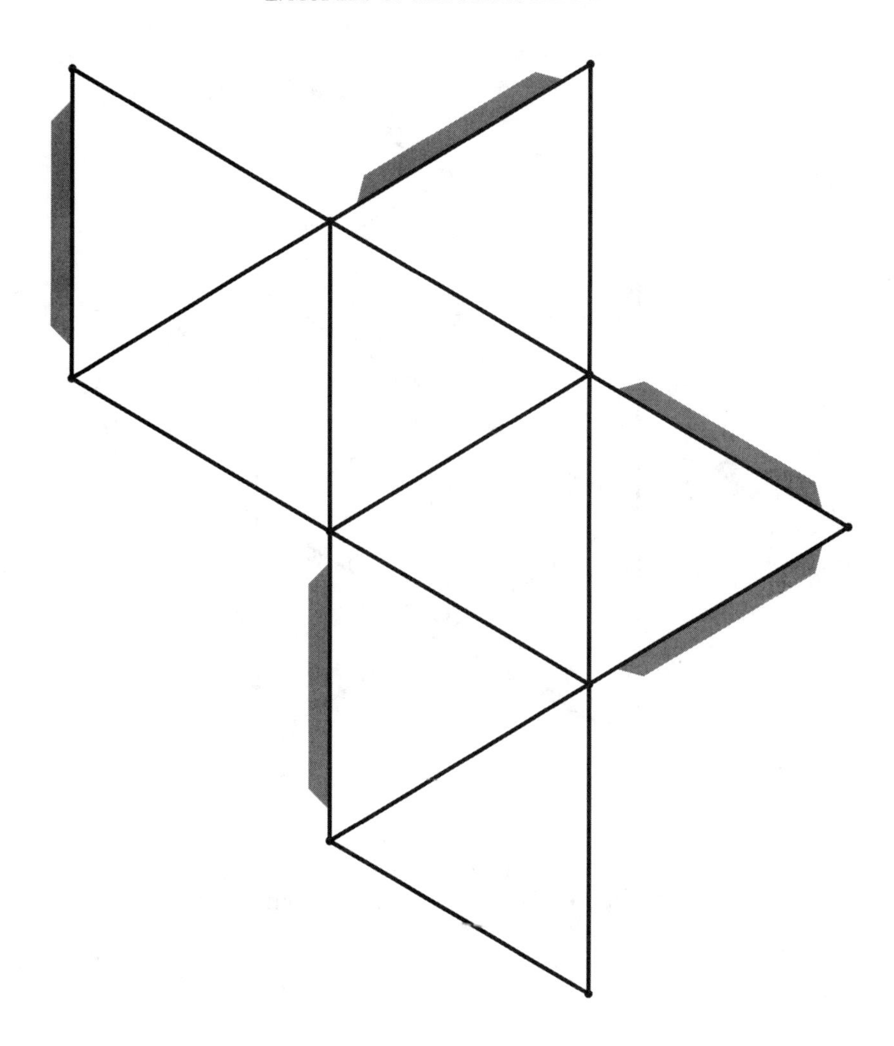

Figure E.3. Octahedron pattern

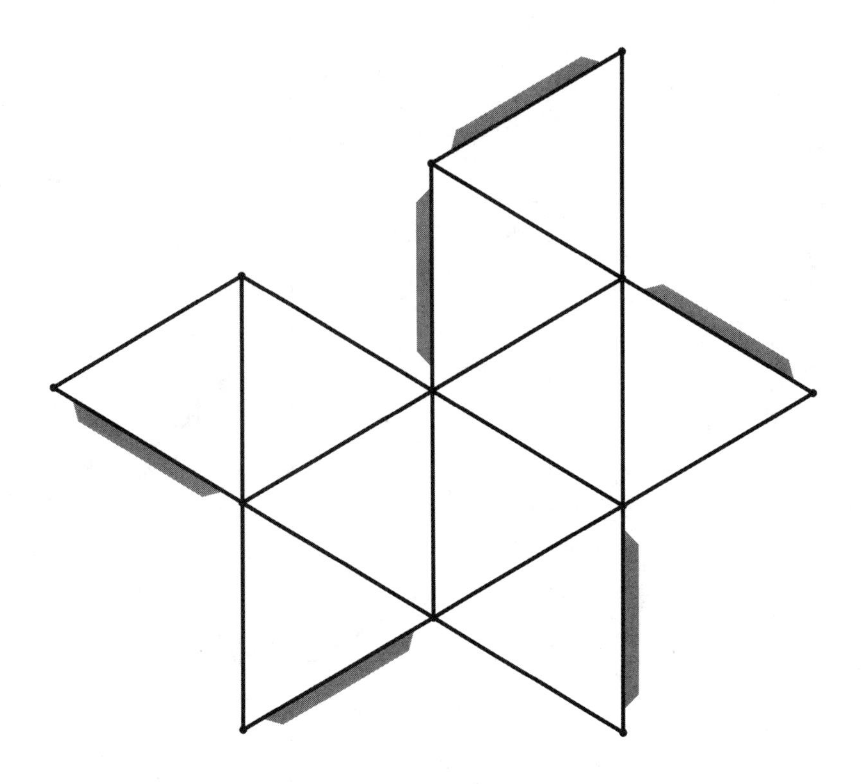

Figure E.4. Icosahedron top pattern

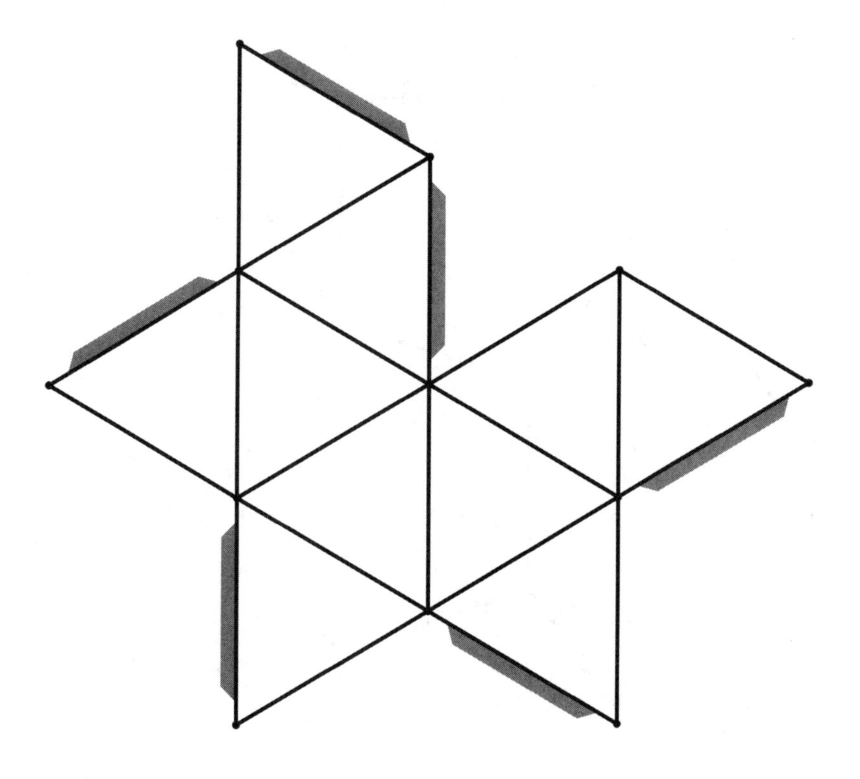

Figure E.5. Icosahedron bottom pattern

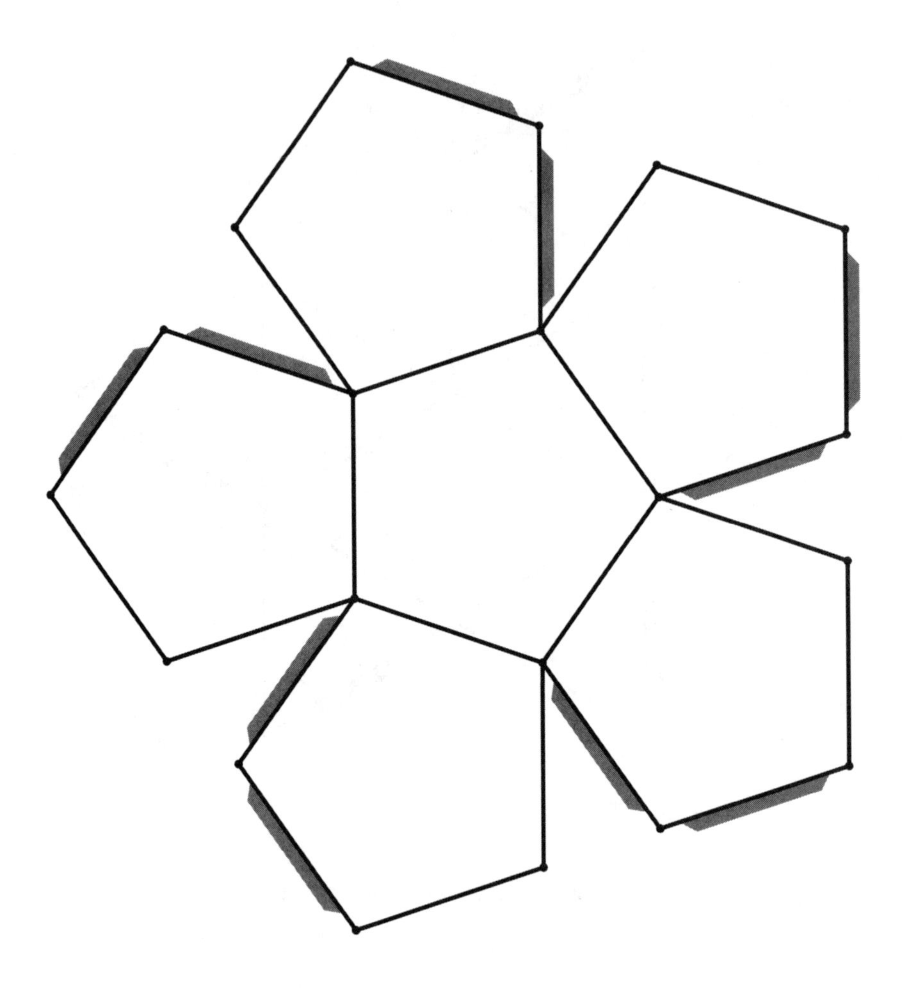

Figure E.6. Dodecahedron top pattern

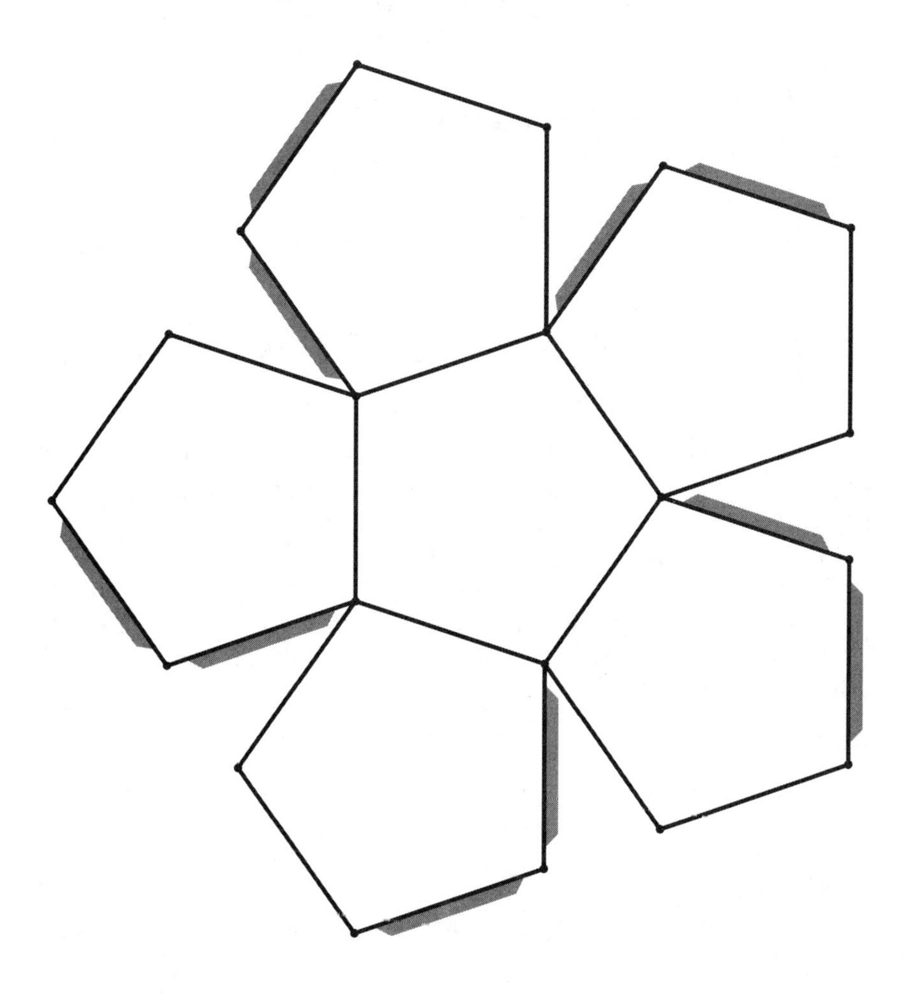

Figure E.7. Dodecahedron bottom pattern

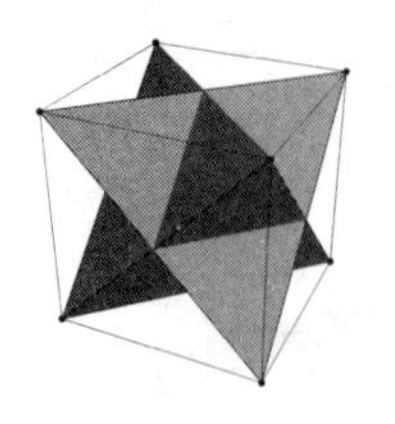

Appendix F

Suggestions for Further Study

The student who has worked his/her way through a substantial portion of this text stands on the threshold of modern mathematics and has access to many related topics.

Here are a few suggestions for further study.

First let me mention some algebra texts which you might use for collateral study, as a source of additional exercises, or for additional topics not discussed here:

Elementary:

- I. N. Herstein, *Abstract Algebra*, 3rd edition, Prentice-Hall, 1996.
- T. Shiffren, *Abstract Algebra* , Prentice Hall, 1995.

More advanced:

- D. S. Dummit and R. M. Foote, Abstract Algebra, Prentice Hall, 1991.
- I. N. Herstein, *Topics in Algebra*, 2nd edition, John-Wiley, 1975.
- M. Artin, *Algebra*, Prentice-Hall, 1995.

Linear (and multilinear) algebra is one of the most useful topics in algebra. For further study of mathematics and for applications of mathematics, eventually you need a better knowledge of linear algebra than you gained in your first course. For this you can look to:

- S. Axler, *Linear Algebra Done Right*, Springer Verlag, 1995.
- S. Lang, *Linear Algebra*, Springer Verlag, 1987.

Both group theory and field theory have substantial contact with number theory. For an introduction to number theory, see:

- G. Andrews, *Number Theory*, Dover Publications, 1994. (Original edition, Saunders, 1971.)

For a computational approach to polynomial rings in several variables and algebraic geometry, there is a beautiful and accessible text:

- D. Cox, J. Little, and D. O'Shea, *Ideals, Varieties and Algorithms*, Springer-Verlag, 1992.

For further study of group theory, my own preference is for the theory of representations and applications. I would recommend:

- W. Fulton and J. Harris, *Representation Theory, A First Course*, Springer-Verlag, 1991.
- B. Simon, *Representations of Finite and Compact Groups*, American Mathematical Society, 1996.
- S. Sternberg, *Group Theory and Physics*, Cambridge University Press, 1994.

These books are quite challenging, but they are accessible with a knowledge of this course, linear algebra, and undergraduate analysis.

Index

Symbols

A

B

C

D

N

O

P

Q

R